Springer Undergraduate Mathematics Series

For other titles published in this series, go to
www.springer.com/series/3423

Marek Capiński · Tomasz Zastawniak

Mathematics
for Finance

An Introduction
to Financial Engineering

Second Edition

 Springer

Marek Capiński
Faculty of Applied Mathematics
AGH University of Science and Technology
Al. Mickiewicza 30
30-059 Kraków
Poland

Tomasz Zastawniak
Department of Mathematics
University of York
Heslington, York
YO10 5DD
UK

Springer Undergraduate Mathematics Series ISSN 1615-2085
ISBN 978-0-85729-081-6
Springer London Dordrecht Heidelberg New York

British Library Cataloguing in Publication Data
A catalogue record for this book is available from the British Library

British Library Cataloguing in Publication Data
A catalogue record for this book is available from the British Library

Library of Congress Control Number: 2010938854

Mathematics Subject Classification (2010): 91G10, 91G20, 91G30, 97M30

Cover design: VTEX, Vilnius

Printed on acid-free paper

Springer is part of Springer Science+Business Media (www.springer.com)

Preface to the Second Edition

In the second edition the material has been thoroughly revised and rearranged. New features include:

- a new chapter on time-continuous models with intuitive outlines of the mathematical arguments and constructions;
- complete proofs of the two fundamental theorems of mathematical finance in a discrete setting;
- a case study to begin each chapter – a real-life situation motivating the development of theoretical tools;
- a detailed discussion of the case study at the end of each chapter.

In analysing a case study the first task is to pose suitable questions, which can then be cast within the framework of the current chapter. At this point, additional assumptions may be needed so the case can be formulated as a well-posed mathematical problem. Due to the flexibility involved there is no unique solution.

We would like to express our gratitude to the readers of the 1st edition for their invaluable feedback.

Marek Capiński and Tomasz Zastawniak
March 2010

www.springer.com

Preface to the First Edition

True to its title, this book itself is an excellent financial investment. For the price of one volume it teaches two Nobel Prize winning theories, with plenty more included for good measure. How many undergraduate mathematics textbooks can boast such a claim?

Building on mathematical models of bond and stock prices, these two theories lead in different directions: Black–Scholes arbitrage pricing of options and other derivative securities on the one hand, and Markowitz portfolio optimisation and the Capital Asset Pricing Model on the other hand. Models based on the Principle of No Arbitrage can also be developed to study interest rates and their term structure. These are three major areas of mathematical finance, all having an enormous impact on the way modern financial markets operate. This textbook presents them at a level aimed at second or third year undergraduate students, not only of mathematics but also, for example, business management, finance or economics.

The contents can be covered in a one-year course of about 100 class hours. Smaller courses on selected topics can readily be designed by choosing the appropriate chapters. The text is interspersed with a multitude of worked examples and exercises, complete with solutions, providing ample material for tutorials as well as making the book ideal for self-study.

Prerequisites include elementary calculus, probability and some linear algebra. In calculus we assume experience with derivatives and partial derivatives, finding maxima or minima of differentiable functions of one or more variables, Lagrange multipliers, the Taylor formula and integrals. Topics in probability include random variables and probability distributions, in particular the binomial and normal distributions, expectation, variance and covariance, conditional probability and independence. Familiarity with the Central Limit Theorem would be a bonus. In linear algebra the reader should be able to solve systems of linear equations, add, multiply, transpose and invert matrices, and compute determinants. In particular, as a reference in probability theory we recommend our book: M. Capiński and T. Zastawniak, *Probability Through Problems*, Springer-Verlag, New York, 2001.

In many numerical examples and exercises it may be helpful to use a computer with a spreadsheet application, though this is not absolutely essential. Microsoft Excel files with solutions to selected examples and exercises are available on the web page for this book, a link to which can be found at the address below.

We are indebted to Nigel Cutland for prompting us to steer clear of an inaccuracy frequently encountered in other texts, of which more will be said in Remark 4.1 (in the 1st edition). It is also a great pleasure to thank our students and colleagues for their feedback on preliminary versions of various chapters. We are also grateful to the readers of the first and second printings of this book, and particularly to Andrzej Palczewski, for indicating some mistakes, corrected in this printing, and suggesting various improvements.

Readers of this book are cordially invited to visit the accompanying web page to check for the latest downloads and corrections, or to contact the authors. Your comments will be greatly appreciated.

Marek Capiński and Tomasz Zastawniak
July 2004

www.springer.com

Contents

1
A Simple Market Model

Case 1

A UK company is preparing to purchase a piece of equipment in the US for
$100,000 in a year's time, the price guaranteed by the producer to remain
unchanged. Considering the current exchange rate of 1.62 dollars to a pound,
the manager of the company has reserved £64,000 in the budget to become
available at the time of purchase. Analyse this decision.

1.1 Basic Notions and Assumptions

Suppose that two assets are traded: one risk-free and one risky security. The
former can be thought of as a bank deposit or a bond issued by a government,
a financial institution, or a company. The risky security will typically be some
stock. It may also be a foreign currency, gold, a commodity or virtually any
asset whose future price is unknown today.

Throughout this chapter we restrict the time scale to two instants only:
today, $t = 0$, and some future time, $t = T$, say one year from now. More refined
and realistic situations will be studied in later chapters.

The position in risky securities can be specified as the number of shares
of stock held by an investor. The price of one share at time $t = 0, T$ will be
denoted by $S(t)$. The current stock price $S(0)$ is known to all investors, but

M. Capiński, T. Zastawniak, *Mathematics for Finance*,
Springer Undergraduate Mathematics Series,
© Springer-Verlag London Limited 2011

the future price $S(T)$ remains uncertain: it may go up as well as down. The difference $S(T) - S(0)$ as a fraction of the initial value represents the *return* on the stock,

$$K_S = \frac{S(T) - S(0)}{S(0)},$$

which is also uncertain.

The dynamics of stock prices will be discussed in Chapters 6, 7 and 8.

The risk-free position can be described as the amount held in a bank account. As an alternative to keeping money in a bank, investors may choose to invest in bonds. The price of one bond at time $t = 0, T$ will be denoted by $A(t)$. The current bond price $A(0)$ is known to all investors, just like the current stock price. However, in contrast to stock, the price $A(T)$ the bond will fetch at time T is also known with certainty, being guaranteed by the institution issuing bonds. The bond is said to mature at time T with face value $A(T)$. The return on bonds is defined in a similar way as that on stock,

$$K_A = \frac{A(T) - A(0)}{A(0)},$$

and it is called the risk-free return.

Chapters 2 and 9 give a detailed exposition of risk-free assets.

Our task is to build a mathematical model of a market of financial securities. A crucial first stage is concerned with the properties of the mathematical objects involved. This is done below by specifying a number of assumptions, the purpose of which is to find a compromise between the complexity of the real world and the limitations and simplifications of a mathematical model, imposed in order to make it tractable.

We have already observed that the future price $A(T)$ of the risk-free security is deterministic, that is, it is a number known in advance. Meanwhile, the future stock price $S(T)$, unknown at time 0, is a random variable with at least two different values. In reality the number of possible different values of $S(T)$ is finite because they are quoted to within a specified number of decimal places and because there is only a certain final amount of money in the whole world, supplying an upper bound for all prices. Therefore it is natural to require the future price $S(T)$ of a share of stock to be a random variable taking only finitely many values. However, in Chapter 8 this will not be the case.

All stock and bond prices must be strictly positive,

$$A(t) > 0 \quad \text{and} \quad S(t) > 0 \quad \text{for } t = 0, T.$$

The total wealth of an investor holding x stock shares and y bonds at a time instant $t = 0, T$ is

$$V(t) = xS(t) + yA(t). \tag{1.1}$$

The pair (x, y) is called a *portfolio*, $V(t)$ being the *value* of this portfolio or, in other words, the *wealth* of the investor at time t.

The jumps of asset prices between times 0 and T give rise to a change in the portfolio value

$$V(T) - V(0) = x(S(T) - S(0)) + y(A(T) - A(0)).$$

This difference as a fraction of the initial value represents the return on the portfolio,

$$K_V = \frac{V(T) - V(0)}{V(0)}.$$

The returns on bonds or stock are particular cases of the return on a portfolio (with $x = 0$ or $y = 0$, respectively). Note that because $S(T)$ is a random variable, so is $V(T)$ as well as the corresponding returns K_S and K_V. The return K_A on a risk-free investment is deterministic.

Example 1.1

Let $A(0) = 100$ and $A(T) = 110$ dollars. Then the return on an investment in bonds will be

$$K_A = 0.10 = 10\%.$$

Also, let $S(0) = 50$ dollars and suppose that the random variable $S(T)$ can take two values,

$$S(T) = \begin{cases} 52 & \text{with probability } p, \\ 48 & \text{with probability } 1 - p, \end{cases}$$

for some $0 < p < 1$. The return on stock will then be

$$K_S = \begin{cases} 4\% & \text{if stock goes up}, \\ -4\% & \text{if stock goes down}. \end{cases}$$

Example 1.2

Given the bond and stock prices in Example 1.1, the value at time 0 of a portfolio with $x = 20$ stock shares and $y = 10$ bonds is

$$V(0) = 2,000$$

dollars. The time T value of this portfolio will be

$$V(T) = \begin{cases} 2,140 & \text{if stock goes up}, \\ 2,060 & \text{if stock goes down}, \end{cases}$$

so the return on the portfolio will be

$$K_V = \begin{cases} 7\% & \text{if stock goes up,} \\ 3\% & \text{if stock goes down.} \end{cases}$$

Exercise 1.1

Let $A(0) = 90$, $A(T) = 100$, $S(0) = 25$ dollars and let

$$S(T) = \begin{cases} 30 & \text{with probability } p, \\ 20 & \text{with probability } 1 - p, \end{cases}$$

where $0 < p < 1$. For a portfolio with $x = 10$ shares and $y = 15$ bonds calculate $V(0)$, $V(T)$ and K_V.

Exercise 1.2

Given the same bond and stock prices as in Exercise 1.1, find a portfolio whose value at time T is

$$V(T) = \begin{cases} 1,160 & \text{if stock goes up,} \\ 1,040 & \text{if stock goes down.} \end{cases}$$

What is the value of this portfolio at time 0?

It is mathematically convenient and not too far from reality to allow arbitrary real numbers, including negative ones and fractions, to represent the risky and risk-free positions x and y in a portfolio, so in general no restrictions are imposed on the positions

$$x, y \in \mathbb{R}.$$

The fact that one can hold a fraction of a share or bond is referred to as *divisibility*. Almost perfect divisibility is achieved in real world dealings whenever the volume of transactions is large as compared to the unit prices.

The fact that no bounds are imposed on x or y is related to another market attribute known as *liquidity*. It means that any asset can be bought or sold on demand at the market price in arbitrary quantities. This is clearly a mathematical idealisation because in practice there may exist restrictions on the volume of trading, or large transactions may affect the prices. Modelling such effects requires sophisticated mathematics beyond the scope of this book.

If the number of securities of a particular kind held in a portfolio is positive, we say that the investor has a *long position*. Otherwise, we say that a *short position* is taken or that the asset is *shorted*. A short position in risk-free securities may involve issuing and selling bonds, but in practice the same financial effect is more easily achieved by borrowing cash. Repaying the loan with interest is referred to as *closing* the short position.

A short position in stock can be realised by *short selling*. This means that the investor borrows the stock, sells it, and uses the proceeds to make some other investment. The owner of the stock keeps all the rights to it. In particular, she is entitled to receive any dividends due and may wish to sell the stock at any time. Because of this, the investor must always have sufficient resources to fulfil the resulting obligations and, in particular, to *close* the short position in risky assets, that is, to repurchase the stock and return it to the owner. Similarly, the investor must always be able to close a short position in risk-free securities, by repaying the cash loan with interest. To reflect these practical restrictions sometimes it is assumed that the wealth of an investor must be non-negative at all times, $V(t) \geq 0$.

1.2 No-Arbitrage Principle

In this section we are going to formulate the most fundamental assumption about the market. In brief, we shall assume that the market does not allow risk-free profits with no initial investment. Profits of this kind may happen when some market participants make a mistake.

Example 1.3

Suppose that dealer A in New York offers to buy British pounds at a rate $d_A = 1.62$ dollars to a pound, while dealer B in London sells them at a rate $d_B = 1.60$ dollars to a pound. If this were the case, the dealers would, in effect, be handing out free money. An investor with no initial capital could realise a profit of $d_A - d_B = 0.02$ dollars per each pound traded by taking simultaneously a short position with dealer B and a long position with dealer A. The demand for their generous services would quickly compel the dealers to adjust the exchange rates so that this profitable opportunity would disappear.

Exercise 1.3

On 19 July 2002 dealer A in New York and dealer B in London used the following rates to change currency, namely euros (€), British pounds (£)

and US dollars (\$):

dealer A	buy	sell
€1.0000	\$1.0202	\$1.0284
£1.0000	\$1.5718	\$1.5844

dealer B	buy	sell
€1.0000	£0.6324	£0.6401
\$1.0000	£0.6299	£0.6375

Spot a chance of a risk-free profit without initial investment.

The next example illustrates a situation when a risk-free profit could be realised without initial investment in our simplified framework of a single time step.

Example 1.4

Suppose that dealer A in New York offers to buy British pounds a year from now at a rate $d_A = 1.58$ dollars to a pound, while dealer B in London would sell British pounds immediately at a rate $d_B = 1.60$ dollars to a pound. Suppose further that dollars can be borrowed at an annual rate of 4%, and British pounds can be invested in a bank account at 6%. This would also create an opportunity for a risk-free profit without initial investment, though perhaps not as obvious as before.

For instance, an investor could borrow 10, 000 dollars and convert them into 6, 250 pounds, which could then be deposited in a bank account. After one year interest of 375 pounds would be added to the deposit, and the whole amount could be converted back into 10, 467.50 dollars. (A suitable agreement would have to be signed with dealer A at the beginning of the year.) After paying back the dollar loan with interest of 400 dollars, the investor would be left with a profit of 67.50 dollars.

Apparently, one or both dealers have made a mistake in quoting their exchange rates, which can be exploited by investors. Once again, increased demand for their services will prompt the dealers to adjust the rates, reducing d_A and/or increasing d_B to a point when the profit opportunity disappears.

In the previous examples the profit was certain even though no initial investment was required. One should also consider the slightly more subtle possibility of making a profit that is no longer certain, but carries no risk of loss and requires no initial investment. Suppose a lottery has a hundred million tickets with just one award, a car. One would not expect these tickets to be offered for free even though in practice the chance of a profit is very slim indeed. A similar situation arises when a seller of some goods offers an additional free bonus of this kind (a scratch card where no proof of purchase is necessary). In reality, such a free bonus is never free, the premium being included in the price of the goods.

We shall make an assumption excluding situations similar to those mentioned above.

Assumption 1.5 (No-Arbitrage Principle)

There is no portfolio (x, y) with initial value $V(0) = 0$ such that $V(T) \geq 0$ with probability 1 and $V(T) > 0$ with non-zero probability, where $V(0), V(T)$ are given by (1.1).

In other words, if the initial value of a portfolio is zero, $V(0) = 0$ (no initial investment), and $V(T) \geq 0$ (no risk of a loss), then $V(T) = 0$ (no profit) with probability 1. This means that no investor can lock in a profit without risk and with no initial endowment. If a portfolio violating this principle did exist, we would say that an *arbitrage* opportunity were available.

Arbitrage opportunities rarely exist in practice. If and when they do, the gains are typically small as compared to the volume of transactions, making them beyond the reach of small investors. In addition, they can be more subtle than the examples above. Situations when the No-Arbitrage Principle is violated are typically short-lived and difficult to spot. The activities of investors (called arbitrageurs) pursuing arbitrage profits effectively make the market free of arbitrage opportunities.

The exclusion of arbitrage in the mathematical model is close enough to reality and turns out to be a critically important and fruitful assumption. Arguments based on the No-Arbitrage Principle are the main tools of financial mathematics.

1.3 One-Step Binomial Model

In this section we restrict ourselves to a very simple example, in which the stock price $S(T)$ takes only two different values. Despite its simplicity, this situation is sufficiently interesting to convey the flavour of the theory to be developed later on.

Example 1.6

Suppose that $S(0) = 100$ dollars and $S(T)$ can take two values,

$$S(T) = \begin{cases} 125 & \text{with probability } p, \\ 105 & \text{with probability } 1 - p, \end{cases}$$

where $0 < p < 1$, while the bond prices are $A(0) = 100$ and $A(T) = 110$ dollars. Thus, the return K_S on stock will be 25% if stock goes up, or 5% if stock goes down. The risk-free return will be $K_A = 10\%$. The stock prices are represented as a tree in Figure 1.1.

Figure 1.1   One-step binomial tree of stock prices

Observe that both stock prices at time T happen to be higher than that at time 0. Going 'up' or 'down' is relative to the other price at time T rather than to the price at time 0. In fact we could just as well allow the value in the 'up' scenario to be lower than the 'down' value. For instance, if the asset is a currency, some investors may regard a reduction in the exchange rate as favourable. In fact an increase in an exchange rate means a decrease in the converse rate. It is best to treat the references to 'up' and 'down' price movements as an abstract indication of two possible different outcomes.

In general, the choice of stock and bond prices in a binomial model is constrained by the No-Arbitrage Principle. Suppose that the possible up and down stock prices at time T are

$$S(T) = \begin{cases} S^u(T) & \text{with probability } p, \\ S^d(T) & \text{with probability } 1 - p, \end{cases}$$

where $S^d(T) < S^u(T)$ and $0 < p < 1$.

Proposition 1.7

The restriction

$$\frac{S^d(T)}{S(0)} < \frac{A(T)}{A(0)} < \frac{S^u(T)}{S(0)},$$

has to be imposed on the model or else an arbitrage opportunity would arise.

Proof

Suppose that $\frac{A(T)}{A(0)} \leq \frac{S^d(T)}{S(0)}$. In this case, at time 0:

- borrow the amount $S(0)$ risk free;
- buy one share of stock for $S(0)$.

This way, you will be holding a portfolio (x, y) with $x = 1$ shares of stock and $y = -\frac{S(0)}{A(0)}$ bonds. The time 0 value of this portfolio is

$$V(0) = 0.$$

At time T the value will become

$$V(T) = \begin{cases} S^u(T) - \frac{S(0)}{A(0)}A(T) & \text{if stock goes up,} \\ S^d(T) - \frac{S(0)}{A(0)}A(T) & \text{if stock goes down.} \end{cases}$$

The first of these two possible values is strictly positive, while the other one is non-negative, that is, $V(T)$ is a non-negative random variable such that $V(T) > 0$ with probability $p > 0$. The portfolio provides an arbitrage opportunity, violating the No-Arbitrage Principle (Assumption 1.5).

Now suppose that $\frac{A(T)}{A(0)} \geq \frac{S^u(T)}{S(0)}$. If this is the case, then at time 0:

- sell short one share for $S(0)$;
- invest the amount $S(0)$ risk free.

As a result, you will be holding a portfolio (x, y) with $x = -1$ and $y = \frac{S(0)}{A(0)}$, again of zero initial value,

$$V(0) = 0.$$

The final value of this portfolio will be

$$V(T) = \begin{cases} -S^u(T) + \frac{S(0)}{A(0)}A(T) & \text{if stock goes up,} \\ -S^d(T) + \frac{S(0)}{A(0)}A(T) & \text{if stock goes down,} \end{cases}$$

which is non-negative, with the second value being strictly positive. Thus, $V(T)$ is a non-negative random variable such that $V(T) > 0$ with probability $1 - p > 0$. Once again, this indicates an arbitrage opportunity, contrary to the No-Arbitrage Principle (Assumption 1.5). □

The common sense reasoning behind the above argument is straightforward: buy cheap assets and sell (or sell short) expensive ones, pocketing the difference.

1.4 Risk and Return

Let $A(0) = 100$ and $A(T) = 110$ dollars, as before, but $S(0) = 80$ dollars and

$$S(T) = \begin{cases} 100 & \text{with probability } 0.8, \\ 60 & \text{with probability } 0.2. \end{cases}$$

Suppose that you have $10,000 to invest in a portfolio. You decide to buy $x = 50$ shares, which fixes the risk-free investment at $y = 60$. Then

$$V(T) = \begin{cases} 11,600 & \text{if stock goes up,} \\ 9,600 & \text{if stock goes down,} \end{cases}$$

$$K_V = \begin{cases} 16\% & \text{if stock goes up,} \\ -4\% & \text{if stock goes down.} \end{cases}$$

The *expected return*, that is, the mathematical expectation of the return on the portfolio is

$$E(K_V) = 16\% \times 0.8 - 4\% \times 0.2 = 12\%.$$

The *risk* of this investment is defined to be the standard deviation of the random variable K_V:

$$\sigma_V = \sqrt{(16\% - 12\%)^2 \times 0.8 + (-4\% - 12\%)^2 \times 0.2} = 8\%.$$

Let us compare this with investments in just one type of security. If $x = 0$, then $y = 100$, that is, the whole amount is invested risk free, then the return is known with certainty to be $K_A = 10\%$ and the risk as measured by the standard deviation is zero, $\sigma_A = 0$.

On the other hand, if $x = 125$ and $y = 0$, the entire amount being invested in stock, then

$$V(T) = \begin{cases} 12,500 & \text{if stock goes up,} \\ 7,500 & \text{if stock goes down,} \end{cases}$$

and $E(K_S) = 15\%$ with $\sigma_S = 20\%$.

Given the choice between two portfolios with the same expected return, any investor would obviously prefer that involving lower risk. Similarly, if the risk levels were the same, any investor would opt for higher expected return. However, in the case in hand, higher expected return is associated with higher risk. In such circumstances the choice depends on individual preferences. These issues will be discussed in Chapter 3, where we shall also consider portfolios consisting of several risky securities. The emerging picture will show the power of portfolio selection and portfolio diversification as tools for managing risk versus expected return.

Exercise 1.4

For the above stock and bond prices, design a portfolio with initial wealth of $10,000$ split fifty-fifty between stock and bonds. Compute the expected return and risk as measured by standard deviation.

1.5 Forward Contracts

A *forward contract* is an agreement to buy or sell a risky asset at a specified future time, known as the *delivery date*, for a price F fixed at the present moment, called the *forward price*. An investor who agrees to buy the asset is said to *enter into a long forward contract* or to *take a long forward position*. If an investor agrees to sell the asset, we speak of a *short forward contract* or a *short forward position*. No money is paid at the time when a forward contract is exchanged.

Example 1.8

Suppose that the forward price is $80. If the market price of the asset (the *spot* price) turns out to be $84 on the delivery date, then the long forward contract holder will gain $4 by buying the asset for $80, while the holder of a short forward contract will suffer a loss of $4. However, if the spot price turns out to be $75, then the holder of a long forward contract will have to pay $80 for the asset, suffering a loss of $5, whereas the short contract holder will gain $5. In either case the loss of one party will be the gain of the other.

In general, the party holding a long forward contract with delivery date T will benefit if the future asset price $S(T)$ rises above the forward price F. If the asset price $S(T)$ falls below the forward price F, then the holder of a long forward contract will suffer a loss. The payoff for a long forward position is $S(T) - F$ (which can be positive, negative or zero). For a short forward position the payoff is $F - S(T)$.

Apart from stock and bonds, a portfolio held by an investor may contain forward contracts, in which case it will be described by a triple (x, y, z). Here x is the number of shares of stock and y the number of bonds, as before, and z is the number of forward contracts (positive for a long forward position and negative for a short position). Because no payment is due when a forward contract is exchanged, the initial value of such a portfolio is simply

$$V(0) = xS(0) + yA(0). \tag{1.2}$$

On the delivery date the value of the portfolio will become

$$V(T) = xS(T) + yA(T) + z(S(T) - F). \tag{1.3}$$

Having included forward contracts in the portfolio, we need to extend the No-Arbitrage Principle accordingly.

Assumption 1.9 (No-Arbitrage Principle)

There is no portfolio (x, y, z) that includes a position z in forward contracts and has initial value $V(0) = 0$ such that $V(T) \geq 0$ with probability 1 and $V(T) > 0$ with non-zero probability, where $V(0), V(T)$ are given by (1.2) and (1.3).

The No-Arbitrage Principle determines the forward price F. For simplicity we shall consider the simplest case when the risky security involves no cost of carry. A typical example is a stock paying no dividends. By contrast, a commodity will usually involve storage costs, whereas a foreign currency will earn interest, which can be regarded as negative cost of carry.

A long forward position guarantees that the asset can be bought for the forward price F at delivery. Alternatively, the asset can be bought now and held until delivery. However, if the initial cash outlay is to be zero, the purchase must be financed by a loan. The loan with interest, which will need to be repaid at the delivery date, is a candidate for the forward price. The following proposition shows that this is indeed the case.

Proposition 1.10

If the risky security involves no cost of carry, then the forward price must be

$$F = S(0)\frac{A(T)}{A(0)} = S(0)(1 + K_A) \tag{1.4}$$

or an arbitrage opportunity would exist otherwise.

Proof

Suppose that $F > S(0)(1 + K_A)$. Then, at time 0:
- borrow the amount $S(0)$;
- buy the asset for $S(0)$;
- enter into a short forward contract with forward price F and delivery date T.

The resulting portfolio $(1, -\frac{S(0)}{A(0)}, -1)$ consisting of stock, a risk-free position, and a short forward contract has initial value $V(0) = 0$. Then, at time T:

- close the short forward position by selling the asset for F;
- close the risk-free position by paying $\frac{S(0)}{A(0)}A(T) = S(0)(1 + K_A)$.

The final value of the portfolio,

$$V(T) = F - S(0)(1 + K_A) > 0$$

will be your profit, violating the No-Arbitrage Principle (Assumption 1.9). On the other hand, if $F < S(0)(1 + K_A)$, then at time 0:

- sell short the asset for $S(0)$;
- invest this amount risk free;
- take a long forward position in stock with forward price F and delivery date T.

The initial value of this portfolio $(-1, \frac{S(0)}{A(0)}, 1)$ is $V(0) = 0$. Subsequently, at time T:

- receive the amount $\frac{S(0)}{A(0)}A(T) = S(0)(1 + K_A)$ from the risk-free investment;
- buy the asset for F, closing the long forward position, and return the asset to the owner.

Your arbitrage profit will be

$$V(T) = S(0)(1 + K_A) - F > 0,$$

which once again violates the No-Arbitrage Principle (Assumption 1.9). It follows that the forward price must be $F = S(0)(1 + K_A)$. ☐

Exercise 1.5

Let $A(0) = 100$, $A(1) = 112$ and $S(0) = 34$ dollars. Is it possible to find an arbitrage opportunity if the forward price of stock is $F = 38.60$ dollars with delivery date 1?

Remark 1.11

Forward contracts make it possible to place Example 1.3 on a rigorous footing within the No-Arbitrage Principle framework. The price offered by either of the dealers A, B is a promise to exchange at rate d_A or d_B, respectively, these quotations being forward rates for instantaneous delivery. An arbitrage portfolio would be of the form $(x, y, z_1, z_2) = (0, 0, -1, 1)$ with the last two entries representing the positions z_1 taken with dealer A and z_2 with dealer B.

1.6 Call and Put Options

Let $A(0) = 100$, $A(T) = 110$, $S(0) = 100$ dollars and

$$S(T) = \begin{cases} 120 & \text{with probability } p, \\ 80 & \text{with probability } 1 - p, \end{cases}$$

where $0 < p < 1$.

A *call option* with *strike price* \$100 and *exercise time* T is a contract giving the holder the right (but no obligation) to purchase a share of stock for \$100 at time T.

If the stock price falls below the strike price, the option will be worthless. There would be little point in buying a share for \$100 if its market price is \$80, and no one would want to exercise the right. Otherwise, if the share price rises to \$120, which is above the strike price, the option will bring a profit of \$20 to the holder, who is entitled to buy a share for \$100 at time T and may sell it immediately at the market price of \$120. This is known as *exercising* the option. The option may just as well be exercised simply by collecting the difference of \$20 between the market price of stock (the *spot* price) and the strike price. In practice, the latter is often the preferred method because no stock needs to change hands.

As a result, the payoff of the call option, that is, its value at time T is a random variable

$$C(T) = \begin{cases} 20 & \text{if stock goes up,} \\ 0 & \text{if stock goes down.} \end{cases}$$

In general

$$C(T) = \max(S(T) - X, 0),$$

where X is the strike price. Meanwhile, $C(0)$ will denote the value of the option at time 0, that is, the price for which the option can be bought or sold today.

Remark 1.12

At first sight a call option might resemble a long forward position. Both involve buying an asset at a future date for a price fixed in advance. An essential difference is that the holder of a long forward contract is committed to buying the asset for the fixed price, whereas the owner of a call option has the right but no obligation to do so. Another difference is that an investor will need to pay a premium to purchase a call option, whereas no payment is due when exchanging a forward contract.

In a market in which options are available, it is possible to invest in a portfolio (x, y, z) consisting of x shares of stock, y bonds and z options. The time 0 value of such a portfolio is

$$V(0) = xS(0) + yA(0) + zC(0). \tag{1.5}$$

At time T it will be worth

$$V(T) = xS(T) + yA(T) + zC(T). \tag{1.6}$$

The No-Arbitrage Principle needs, once again, to be extended to cover portfolios of this kind.

Assumption 1.13 (No-Arbitrage Principle)

There is no portfolio (x, y, z) that includes a position z in call options and has initial value $V(0) = 0$ such that $V(T) \geq 0$ with probability 1 and $V(T) > 0$ with non-zero probability, where $V(0), V(T)$ are given by (1.5) and (1.6).

We need to understand how to compute the price $C(0)$ of the call option at time 0 consistent with the absence of arbitrage opportunities. Because the holder of a call option has a certain right, but never an obligation, it is reasonable to expect that $C(0)$ will be positive: one needs to pay a premium to acquire this right. We shall see that the option price $C(0)$ can be found in two steps:

Step 1
Construct an investment in x stocks and y bonds such that the value of the investment at time T is the same as that of the option,

$$xS(T) + yA(T) = C(T),$$

no matter whether the stock price $S(T)$ goes up to \$120 or down to \$80. This is known as *replicating* the option.

Step 2
Compute the time 0 value of the investment in stock and bonds. It will be shown in Proposition 1.14 that it must be equal to the option price,

$$xS(0) + yA(0) = C(0),$$

because an arbitrage opportunity would exist otherwise. This step will be referred to as *pricing* or *valuing* the option.

Step 1 (Replicating the Option)

The time T value of the investment in stock and bonds will be

$$xS(T) + yA(T) = \begin{cases} x120 + y110 & \text{if stock goes up,} \\ x80 + y110 & \text{if stock goes down.} \end{cases}$$

The equality $xS(T) + yA(T) = C(T)$ between the two random variables can be written as

$$\begin{cases} x120 + y110 = 20, \\ x80 + y110 = 0. \end{cases}$$

The first of these equations covers the case when the stock price goes up to $120. The other one applies when stock drops down to $80. Because we want the value of the investment in stock and bonds at time T to match exactly that of the option *no matter whether the stock price goes up or down*, these two equations are to be satisfied simultaneously. Solving for x and y, we find that

$$x = \frac{1}{2}, \quad y = -\frac{4}{11}.$$

To replicate the option we need to buy $\frac{1}{2}$ a share of stock and take a short position of $-\frac{4}{11}$ in bonds (or borrow $\frac{4}{11} \times 100 = \frac{400}{11}$ dollars in cash).

Step 2 (Pricing the Option)

We can compute the value of the investment in stock and bonds at time 0:

$$xS(0) + yA(0) = \frac{1}{2} \times 100 - \frac{4}{11} \times 100 \cong 13.6364$$

dollars. The following proposition shows that this must be equal to the price of the option.

Proposition 1.14

If the option can be replicated by investing in a portfolio (x, y) of stock and bonds, then $C(0) = xS(0) + yA(0)$, or else an arbitrage opportunity would exist.

Proof

Suppose that $C(0) > xS(0) + yA(0)$. If this is the case, then at time 0:

- issue and sell one option for $C(0)$ dollars;
- take a long position in the portfolio (x, y), which costs $xS(0) + yA(0)$ (for a call option this involves buying shares and borrowing cash).

The balance of these transactions is positive, $C(0) - xS(0) - yA(0) > 0$. Invest this amount risk free. The resulting portfolio has initial value $V(0) = 0$. Subsequently, at time T:

- if stock goes up, then settle the option by paying the difference of $S^u(T) - X$ between the spot price and the strike price; you will pay nothing if stock goes down; the cost to you will be $C(T)$, which covers both possibilities;
- close the position in stock and bonds, receiving the amount $xS(T) + yA(T)$.

The cash balance of these transactions will be zero, $-C(T) + xS(T) + yA(T) = 0$, regardless of whether stock goes up or down (the portfolio replicates the option). But you will be left with the initial risk-free investment of $C(0) - xS(0) - yA(0)$ plus interest, thus realising an arbitrage opportunity, contrary to the No-Arbitrage Principle (Assumption 1.13).

On the other hand, if $C(0) < xS(0) + yA(0)$, then, at time 0:

- buy one option for $C(0)$ dollars;
- take a short position in the portfolio (x, y) (which involves buying bonds and short selling stock).

The cash balance of these transactions is positive, $-C(0) + xS(0) + yA(0) > 0$, and can be invested risk free. In this way you will have constructed a portfolio with initial value $V(0) = 0$. Subsequently, at time T:

- if stock goes up, then exercise the option, receiving the difference of $S^u(T) - X$ between the spot price and the strike price; you will receive nothing if stock goes down; your income will be $C(T)$, which covers both possibilities;
- close the position in stock and bonds, paying $xS(T) + yA(T)$.

The cash balance of these transactions will be zero, $C(T) - xS(T) - yA(T) = 0$, regardless of whether stock goes up or down. But you will be left with an arbitrage profit resulting from the risk-free investment of $-C(0) + xS(0) + yA(0)$ plus interest, contradicting the No-Arbitrage Principle (Assumption 1.13). $\square$

Here we can see once again that the arbitrage strategy follows a common-sense pattern: sell (or sell short if necessary) expensive securities and buy inexpensive ones, as long as all your financial obligations arising in the process can be discharged regardless of what happens in the future.

Proposition 1.14 implies that today's price of the option must be

$$C(0) = \frac{1}{2}S(0) - \frac{4}{11}A(0) \cong 13.6364$$

dollars. Anyone who would offer to sell the option for less or to buy it for more than this price would be creating an arbitrage opportunity, which amounts to handing out free money.

Remark 1.15

Note that the probabilities p and $1-p$ of stock going up or down are irrelevant in pricing or replicating the option. This is a remarkable feature, and by no means a coincidence.

Remark 1.16

Options may appear to be superfluous in a market in which they can be replicated by stock and bonds. In the simplified one-step model this is indeed a valid objection. However, in a situation involving multiple time steps (or continuous time) replication becomes a much more onerous task. It requires rebalancing the positions in stock and bonds at every time instant at which there is a change in their prices, resulting in considerable management and transaction costs. In some cases it may not even be possible to replicate an option precisely. This is why the majority of investors prefer to buy or sell options, replication being normally undertaken by specialised dealers and institutions.

Exercise 1.6

Let the bond and stock prices $A(0)$, $A(T)$, $S(0)$, $S(T)$ be as above. Compute the price $C(0)$ of a call option with exercise time T and a) strike price \$90, b) strike price \$110.

Exercise 1.7

Let the prices $A(0)$, $S(0)$, $S(T)$ be as above. Compute the price $C(0)$ of a call option with strike price \$100 and exercise time T if a) $A(T) = 105$ dollars, b) $A(T) = 115$ dollars.

A *put* option with strike price \$100 and exercise time T gives the right (but no obligation) to *sell* one share of stock for \$100 at time T. This kind of option is worthless if the stock goes up, but it brings a profit otherwise, the payoff being

$$P(T) = \begin{cases} 0 & \text{if stock goes up,} \\ 20 & \text{if stock goes down,} \end{cases}$$

given that the possible values of $S(T)$ are the same as above. In general,

$$P(T) = \max(X - S(T), 0),$$

where X is the strike price. The notion of a portfolio can be extended to allow positions in put options, denoted by z, as before.

The replicating and pricing procedure for puts follows the same pattern as for call options. In particular, the price $P(0)$ of the put option is equal to the time 0 value of a replicating investment in stock and bonds.

Remark 1.17

There is some similarity between a put option and a short forward position: both involve selling an asset for a fixed price at a certain time in the future. However, an essential difference is that the holder of a short forward contract is committed to selling the asset for the fixed price, whereas the owner of a put option has the right but no obligation to sell. Moreover, an investor who wants to buy a put option will have to pay for it, whereas no payment is involved when a forward contract is exchanged.

Exercise 1.8

Let the bond and stock prices $A(0)$, $A(T)$, $S(0)$, $S(T)$ be as above. Formulate a version of the No-Arbitrage Principle for portfolios containing puts and compute the price $P(0)$ of a put option with strike price $100.

The general properties of options and forward contracts will be discussed in Chapters 4 and 5. In Chapters 6 and 7 the pricing and replicating schemes will be extended to more complicated though still discrete market models, as well as to other financial instruments.

1.7 Foreign Exchange

Foreign currency is an important example of a risky security. The basic novelty as compared to stock is that each unit of foreign currency held in a portfolio generates additional risk-free income. Here we assume that the currency is kept in a bank account or invested in risk-free bonds. The ability to generate additional income, also present in the stock market if dividends are paid to the shareholders, needs to be accounted for when considering derivative securities written on such assets.

First we begin with an adjustment to formula (1.4) for forward prices. Note that to have one unit of foreign currency at time T it is sufficient to buy a fraction of the unit, namely $\frac{A_f(0)}{A_f(T)} = \frac{1}{1+K_f}$, where the subscript f designates foreign currency bond prices and the return on them. Therefore, a candidate for the forward price (more precisely, *forward rate of exchange*) is

$$F = S(0)\frac{1 + K_h}{1 + K_f}, \tag{1.7}$$

where $1 + K_h = \frac{A_h(T)}{A_h(0)}$, the subscript h indicating the home currency.

Exercise 1.9

Give an arbitrage proof of formula (1.7).

Exercise 1.10

Suppose that the return on dollar bonds is $K_\$ = 5\%$, the present exchange rate is $S(0) = 1.6$ dollars to a pound, and the forward rate for delivery date $T = 1$ is $F = 1.50$ dollars to a pound. How much should a sterling bond cost today if it promises to pay £100 at time 1?

Our next task is to find prices of options written on foreign currency. This is best illustrated by an example.

Example 1.18

Consider pound sterling as a security on the US dollar market. Suppose that the return on US dollar bonds is $K_\$ = 5\%$ and on British pound bonds it is $K_£ = 3\%$. Consider a call option on pound sterling with strike price $X = 1.64$ dollars to a pound within a binomial model with $S(0) = 1.62$, $S^u(T) = 1.84$ and $S^d(T) = 1.46$ dollars to a pound. The replication step requires finding x, y such that

$$x(1 + K_£)S^u(T) + y(1 + K_\$) = S^u(T) - X,$$
$$x(1 + K_£)S^d(T) + y(1 + K_\$) = 0.$$

Note the difference as compared to the replication of stock options, where the factor $1 + K_£$ representing the income generated by the underlying security (here pound sterling) was absent. The solution of this system of equations is $x = 0.5110$ pounds and $y = -0.7318$ dollars, and consequently

$$C(0) = xS(0) + y = 0.0960$$

dollars.

Exercise 1.11

Find the price of a put written on pound sterling with strike $X = 1.71$ dollars to a pound using the data of Example 1.18.

1.8 Managing Risk with Options

The availability of options and other derivative securities extends the possible investment scenarios. Suppose that your initial wealth is $1,000$ and compare the following two investments in the setup of Section 1.6:

- buy 10 shares; at time T they will be worth

$$10 \times S(T) = \begin{cases} 1,200 & \text{if stock goes up,} \\ 800 & \text{if stock goes down;} \end{cases}$$

or

- buy $1,000/13.6364 \cong 73.3333$ call options; in this case your final wealth will be

$$73.3333 \times C(T) \cong \begin{cases} 1,466.67 & \text{if stock goes up,} \\ 0.00 & \text{if stock goes down.} \end{cases}$$

If stock goes up, the investment in options will produce a much higher return than shares, namely about 46.67%. However, it will be disastrous otherwise: you will lose all your money. Meanwhile, when investing in shares, you would just gain or lose 20%. Without specifying the probabilities we cannot compute the expected returns or standard deviations. Nevertheless, one would readily agree that investing in options is more risky than in stock. This can be exploited by adventurous investors.

Exercise 1.12

In the above setting, find the final wealth of an investor whose initial capital of $1,000$ is split fifty-fifty between stock and options.

Options can also be employed to reduce risk. Consider an investor planning to purchase stock in the future. The share price today is $S(0) = 100$ dollars, but the investor will only have funds available at a future time T, when the share price will become

$$S(T) = \begin{cases} 160 & \text{with probability } p, \\ 40 & \text{with probability } 1-p, \end{cases}$$

for some $0 < p < 1$. Assume, as before, that $A(0) = 100$ and $A(T) = 110$ dollars, and compare the following two strategies:

- wait until time T, when the funds become available, and purchase the stock for $S(T)$;

or

- at time 0 borrow money to buy a call option with strike price $100; then, at time T repay the loan with interest and purchase the stock, exercising the option if the stock price goes up.

The investor will be open to considerable risk if she chooses to follow the first strategy. On the other hand, following the second strategy, she will need to borrow $C(0) \cong 31.8182$ dollars to pay for the option. At time T she will have to repay $35 to clear the loan and may use the option to purchase the stock, hence the cost of purchasing one share will be

$$S(T) - C(T) + 35 = \begin{cases} 135 & \text{if stock goes up,} \\ 75 & \text{if stock goes down.} \end{cases}$$

Clearly, the risk is reduced, the spread between these two figures being narrower than before.

Exercise 1.13

Compute the risk as measured by the standard deviation of the cost of buying one share with and without the option if a) $p = 0.25$, b) $p = 0.5$, c) $p = 0.75$.

Exercise 1.14

Show that the risk (as measured by the standard deviation of the cost of buying one share) of the above strategy involving an option is half of that when no option is purchased, no matter what the probability $0 < p < 1$ is.

If two options are bought, then the risk will be reduced to nil:

$$S(T) - 2 \times C(T) + 70 = 110 \text{ with probability 1.}$$

This strategy turns out to be equivalent to a long forward contract, since the forward price of the stock is exactly $110 (see Section 1.5). It is also equivalent to borrowing money to purchase a share for $100 today and repaying $110 to clear the loan at time T.

Case 1: Discussion

The first task is to recognise the nature of the problem. Since the UK company will have £64,000 at their disposal in a year's time to purchase the piece of equipment for $100,000, the dollar price being guaranteed by the producer, the source of concern is the unknown exchange rate at the time of purchase. The amount budgeted for may turn out insufficient to purchase the equipment if the exchange rate falls below 1.5625 dollars to a pound.

To ensure that the project is successful a financial package needs to be designed to hedge the interest rate risk. Having followed this chapter, we are familiar with some tools to achieve this goal. It will be necessary to collect some additional market data to apply these tools.

The first approach is to employ forward contracts to convert pounds into dollars at the end of the year. The forward price F of pound sterling can be found using the methods discussed in Section 1.7. In order to do so, we need to look up the pound sterling and US dollar bond prices, or, equivalently, the corresponding risk-free returns. Suppose that these turn out to be $K_\$ = 5\%$ and $K_£ = 3\%$. This gives $F = 1.6515$ dollars to a pound. Converting £64,000 at this rate will give $105,693.20. The company will be able to purchase the piece of equipment, leaving it with a surplus of $5,693.20, irrespective of what happens to the exchange rate.

As an alternative to forward contracts, since the company will need to convert their currency into dollars, in effect selling pound sterling, we could employ put options on pound sterling.

This requires building a model of future random exchange rates. At this point all we can do is to use the single step binomial model. We know the spot rate $S(0) = 1.62$ dollars to a pound and the risk-free returns $K_\$ = 5\%$ and $K_£ = 3\%$. The two future rates $S^u(1), S^d(1)$ need to be calibrated to match market data. To this end we can use the prices of some traded options on pound sterling. Suppose that calls with strikes 1.6 and 1.7 dollars to a pound, selling at 1.1178 and 0.0624 dollars to a pound, are the most liquid (and therefore considered the most representative) options traded. Since these call prices can be expressed in terms of $S^u(1), S^d(1)$ just like in Example 1.18, we obtain a system of two equations, which can be solved to get $S^u(1) \cong 1.8126$ and $S^d(1) \cong 1.4273$ dollars to a pound.

Having calibrated the model, we need to decide the strike rate X for the put option to be used to hedge the interest rate risk. The strike X affects the option price P, which we shall denote by $P(X)$ for clarity. We need to purchase 64,000 puts to convert the available sum of £64,000 into dollars. This will cost $P(X) \times 64,000$ dollars, which we need to borrow. Suppose that the company's bank quotes a 10% interest rate for a US dollar loan. At the end of the year we shall need to repay $P(X) \times 64,000 \times 1.1$ dollars in addition to the price of the piece of equipment. This gives an equation for the strike price:

$$100,000 + P(X) \times 64,000 \times 1.1 = X \times 64,000.$$

The solution is $X \cong 1.6680$ pounds to a dollar. The price of a put is $P(X) \cong 0.0959$ dollars. We need to borrow $P(X) \times 64,000 \cong 6,135.32$ dollars to purchase the options, and will have to repay \$6,748.85 including interest at the end of the year. If the exchange rate goes up, we shall sell pound sterling at the market price with an additional profit of $64,000 \times 1.8126 - 106,748.85 \cong 9,259.81$ dollars. If the rate goes down, we shall exercise the options and break even.

The strategy involving options does not look as attractive as forward contracts. Hedging with puts will also enable the company to purchase the piece of equipment irrespective of any changes in the exchange rate, but the end financial result is unpredictable, depending on the fluctuations of the rate. A practical problem could be that put options with the ideal strike rate $X \cong 1.6680$ pounds to a dollar may not be available via exchange trading, necessitating an over-the-counter (OTC) transaction. Another obvious weakness of this approach to hedging with options is connected with the simplicity of the model of exchange rates. The case is certainly worth revisiting when we acquire better models. On the other hand, when using forwards, we did not need to adopt any particular model of interest rates, so the result is much more robust.

2
Risk-Free Assets

Case 2

Consider a do-it-yourself pension fund based on regular savings invested in a bank account attracting interest at 5% per annum. When you retire after 40 years, you want to receive a pension equal to 50% of your final salary and payable for 20 years. Your earnings are assumed to grow at 2% annually, and you want the pension payments to grow at the same rate.

2.1 Time Value of Money

It is a fact of life that $100 to be received after one year is worth less than the same amount today. The main reason is that money due in the future or locked in a fixed-term account cannot be spent right away. One would therefore expect to be compensated for postponed consumption. In addition, prices may rise in the meantime, and the amount will not have the same purchasing power as it would have at present. Moreover, there is always some risk, even if very low, that the money will never be received. Whenever a future payment is uncertain due to the possibility of default, its value today will be reduced to compensate for this. (However, here we shall consider situations free from default or credit risk.) As generic examples of risk-free assets we shall consider a bank deposit or a bond.

M. Capiński, T. Zastawniak, *Mathematics for Finance*,
Springer Undergraduate Mathematics Series,
© Springer-Verlag London Limited 2011

The way in which money changes its value in time is a complex issue of fundamental importance in finance. We shall be concerned mainly with two questions:

What is the future value of an amount invested or borrowed today?

What is the present value of an amount to be paid or received at a certain time in the future?

The answers depend on various factors, which will be discussed in the present chapter. This topic is often referred to as the *time value of money*.

2.1.1 Simple Interest

Suppose that an amount is paid into a bank account, where it is to earn *interest*. The *future value* of this investment consists of the initial deposit, called the *principal* and denoted by P, plus all the interest earned since the money was deposited in the account.

To begin with, we shall consider the case when interest is attracted only by the principal, which remains unchanged during the period of investment. For example, the interest earned may be paid out in cash, credited to another account attracting no interest, or credited to the original account after some longer period.

After one year the interest earned will be rP, where $r > 0$ is the *interest rate*. The value of the investment will thus become $V(1) = P + rP = (1 + r)P$. After two years the investment will grow to $V(2) = (1 + 2r)P$. Consider a fraction of a year. Interest is typically calculated on a daily basis: the interest earned in one day will be $\frac{1}{365}rP$. After n days the interest will be $\frac{n}{365}rP$, and the total value of the investment will become $V(\frac{n}{365}) = (1 + \frac{n}{365}r)P$. This motivates the following rule of *simple interest*: the value of the investment at time t, denoted by $V(t)$, is given by

$$V(t) = (1 + tr)P, \tag{2.1}$$

where time t, expressed in years, can be an arbitrary non-negative real number; see Figure 2.1. In particular, we have the obvious equality $V(0) = P$. The number $1 + rt$ is called the *growth factor*. Here we assume that the interest rate r is constant. If the principal P is invested at time s, rather than at time 0, then the value at time $t \geq s$ will be

$$V(t) = (1 + (t - s)r)P. \tag{2.2}$$

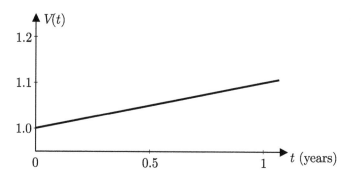

Figure 2.1 Principal attracting simple interest at 10% ($r = 0.1$, $P = 1$)

Throughout this book the unit of time will be one year. We shall transform any period of time expressed in other units (days, weeks, months) into a fraction of a year.

Example 2.1

Consider a deposit of $150 held for 20 days and attracting simple interest at 8%. This gives $t = \frac{20}{365}$ and $r = 0.08$. After 20 days the deposit will grow to $V(\frac{20}{365}) = (1 + \frac{20}{365} \times 0.08) \times 150 \cong 150.66$.

Remark 2.2

An important example of employing simple interest is provided by the LIBOR rate (London Interbank Offered Rate). This is the rate of interest valid for transactions between the largest London banks. LIBOR rates are quoted for various short periods of time up to one year, and have become reference values for a variety of transactions. For instance, the rate for a typical commercial loan may be formulated as a particular LIBOR rate plus some additional margin.

The *return* on an investment commencing at time s and terminating at time t will be denoted by $K(s,t)$. It is given by

$$K(s,t) = \frac{V(t) - V(s)}{V(s)}. \qquad (2.3)$$

In the case of simple interest

$$K(s,t) = (t - s)r,$$

which clearly follows from (2.2). In particular, the interest rate is equal to the return over one year,

$$K(0,1) = r.$$

As a general rule, interest rates will always refer to a period of one year, facilitating the comparison between different investments, independently of their actual maturity time. By contrast, the return reflects both the interest rate *and* the length of time for which the investment is held.

Exercise 2.1

A sum of $9,000 paid into a bank account for two months (61 days) to attract simple interest will produce $9,020 at the end of the term. Find the interest rate r and the return on this investment.

Exercise 2.2

How much would you pay today to receive $1,000 at a certain future date if you require a 2% return?

Exercise 2.3

How long will it take for $800 attracting simple interest to become $830 if the rate is 9%? Compute the return on this investment.

Exercise 2.4

Find the principal to be deposited in an account attracting simple interest at 8% if $1,000 is needed after three months (91 days).

The last exercise is concerned with an important general problem: find the initial amount whose value at time t is given. In the case of simple interest the answer is easily found by solving (2.1) for the principal, obtaining

$$V(0) = V(t)(1 + rt)^{-1}. \qquad (2.4)$$

This number is called the *present* or *discounted value* of $V(t)$, and $(1 + rt)^{-1}$ is called the *discount factor*.

Example 2.3

A *perpetuity* is a sequence of payments of a fixed amount to be made at equal time intervals and continuing indefinitely into the future. For example, suppose that payments of an amount C are to be made once a year, the first payment due a year hence. This can be achieved by depositing

$$P = \frac{C}{r}$$

in a bank account to earn simple interest at a constant rate r. Such a deposit will indeed produce a sequence of interest payments amounting to $C = rP$ payable every year.

In practice simple interest is used only for short-term investments and for certain types of loans and deposits. It is not a realistic description of the value of money in the longer term. In the majority of cases the interest already earned can be reinvested to attract even more interest, producing higher return than implied by (2.1). This will be analysed in detail in what follows.

2.1.2 Periodic Compounding

Once again, suppose that an amount P is deposited in a bank account, attracting interest at a constant rate $r > 0$. However, in contrast to the case of simple interest, we assume that the interest earned will now be added to the principal periodically, for example, annually, semi-annually, quarterly, monthly, or perhaps even on a daily basis. Subsequently, interest will be attracted not just by the original deposit, but also by all the interest added so far. In these circumstances we shall talk of *discrete* or *periodic compounding*.

Example 2.4

In the case of monthly compounding the first interest payment of $\frac{r}{12}P$ will be due after one month, increasing the principal to $(1 + \frac{r}{12})P$, all of which will attract interest in the future. The next interest payment, due after two months, will thus be $\frac{r}{12}(1 + \frac{r}{12})P$, and the capital will become $(1 + \frac{r}{12})^2 P$. After one year it will become $(1 + \frac{r}{12})^{12}P$, after n months it will be $(1 + \frac{r}{12})^n P$, and after t years $(1 + \frac{r}{12})^{12t}P$. The last formula accepts t equal to a whole number of months, that is, a multiple of $\frac{1}{12}$.

In general, if m interest payments are made per annum, the time between two consecutive payments measured in years will be $\frac{1}{m}$, the first interest payment being due at time $\frac{1}{m}$. Each interest payment will increase the principal by a factor $1 + \frac{r}{m}$. Given that the interest rate r remains unchanged, after t years the *future value* of an initial principal P will become

$$V(t) = \left(1 + \frac{r}{m}\right)^{tm} P \qquad (2.5)$$

because there will be tm interest payments during this period. In this formula t must be a whole multiple of the period $\frac{1}{m}$. The number $\left(1 + \frac{r}{m}\right)^{tm}$ is the *growth factor*.

Note that in each period we apply simple interest with the starting amount changing from period to period. Formula (2.5) can be equivalently written in a recursive way:

$$V\left(t+\frac{1}{m}\right) = V(t)\left(1+\frac{r}{m}\right)$$

with $V(0) = P$.

The exact value of the investment may sometimes need to be known at time instants between interest payments. In particular, this may be so if the account is closed on a day when no interest payment is due. For example, what is the value after 10 days of a deposit of \$100 subject to monthly compounding at 12%? One possible answer is \$100, since the first interest payment would be due only after one whole month. This suggests that (2.5) should be extended to arbitrary values of t by means of a step function with steps of length $\frac{1}{m}$, as shown in Figure 2.2. Later on, in Remark 2.21 we shall see that the extension consistent with the No-Arbitrage Principle should use the right-hand side of (2.5) for all $t \geq 0$.

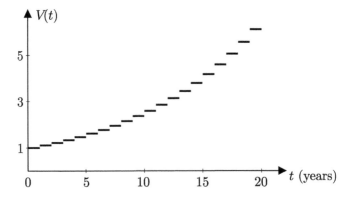

Figure 2.2 Annual compounding at 10% ($m = 1$, $r = 0.1$, $P = 1$)

Exercise 2.5

How long will it take to double a deposit attracting interest at 6% compounded daily?

Exercise 2.6

What is the interest rate if a deposit subject to annual compounding is doubled after 10 years?

Exercise 2.7

Find and compare the future value after two years of a deposit of $100 attracting interest at 10% compounded a) annually and b) semi-annually.

Proposition 2.5

The future value $V(t)$ increases if any one of the parameters m, t, r or P increases, the others remaining unchanged.

Proof

It is immediately obvious from (2.5) that $V(t)$ increases if t, r or P increases. To show that $V(t)$ increases as the compounding frequency m increases, we need to verify that if $m < k$, then

$$\left(1 + \frac{r}{m}\right)^{tm} < \left(1 + \frac{r}{k}\right)^{tk}.$$

The latter reduces to

$$\left(1 + \frac{r}{m}\right)^{m} < \left(1 + \frac{r}{k}\right)^{k},$$

which can be verified directly:

$$\left(1 + \frac{r}{m}\right)^{m} = 1 + r + \frac{1 - \frac{1}{m}}{2!}r^2 + \cdots + \frac{\left(1 - \frac{1}{m}\right) \times \cdots \times \left(1 - \frac{m-1}{m}\right)}{m!}r^m$$

$$\leq 1 + r + \frac{1 - \frac{1}{k}}{2!}r^2 + \cdots + \frac{\left(1 - \frac{1}{k}\right) \times \cdots \times \left(1 - \frac{m-1}{k}\right)}{m!}r^m$$

$$< 1 + r + \frac{1 - \frac{1}{k}}{2!}r^2 + \cdots + \frac{\left(1 - \frac{1}{k}\right) \times \cdots \times \left(1 - \frac{k-1}{k}\right)}{k!}r^k$$

$$= \left(1 + \frac{r}{k}\right)^{k}.$$

The first inequality holds because each term of the sum on the left-hand side is less than or equal to the corresponding term on the right-hand side. The second inequality is true because the sum on the right-hand side contains $m - k$ additional positive terms as compared to the sum on the left-hand side. In both equalities we use the binomial formula

$$(a + b)^m = \sum_{i=0}^{m} \frac{m!}{i!(m - i)!}a^i b^{m-i}.$$

This completes the proof.   □

Exercise 2.8

Which will deliver higher future value after one year, a deposit of $1,000 attracting interest at 15% compounded daily, or at 15.5% compounded semi-annually?

Exercise 2.9

What initial investment subject to annual compounding at 12% is needed to produce $1,000 after two years?

The last exercise touches upon the problem of finding the present value of an amount payable at some future time instant in the case when periodic compounding applies. Here the formula for the *present* or *discounted value* of $V(t)$ is

$$V(0) = V(t)\left(1 + \frac{r}{m}\right)^{-tm},$$

the number $(1 + \frac{r}{m})^{-tm}$ being the *discount factor*.

Remark 2.6

Fix the terminal value $V(t)$ of an investment. It is an immediate consequence of Proposition 2.5 that the present value increases if any one of the factors r, t, m decreases, the other ones remaining unchanged.

Exercise 2.10

Find the present value of $100,000 to be received after 100 years if the interest rate is assumed to be 5% throughout the whole period and a) daily or b) annual compounding applies.

One often requires the value $V(t)$ of an investment at an intermediate time $0 < t < T$, given the value $V(T)$ at some fixed future time T. This can be achieved by computing the present value of $V(T)$, taking it as the principal, and running the investment forward up to time t. Under periodic compounding with frequency m and interest rate r, this gives

$$V(t) = \left(1 + \frac{r}{m}\right)^{-(T-t)m} V(T). \qquad (2.6)$$

To find the return on a deposit attracting interest compounded periodically we use the general formula (2.3) and readily arrive at

$$K(s,t) = \frac{V(t) - V(s)}{V(s)} = \left(1 + \frac{r}{m}\right)^{(t-s)m} - 1.$$

In particular,

$$K\left(0, \frac{1}{m}\right) = \frac{r}{m},$$

which provides a simple way of computing the interest rate given the return.

Exercise 2.11

Find the return over one year under monthly compounding with $r = 10\%$.

Exercise 2.12

Which is greater, the interest rate r or the return $K(0,1)$ if the compounding frequency m is greater than 1?

Remark 2.7

The return on a deposit subject to periodic compounding is *not* additive. Take, for simplicity, $m = 1$. Then

$$K(0,1) = K(1,2) = r,$$
$$K(0,2) = (1+r)^2 - 1 = 2r + r^2,$$

and clearly $K(0,1) + K(1,2) \neq K(0,2)$.

2.1.3 Streams of Payments

An *annuity* is a sequence of finitely many payments of a fixed amount due at equal time intervals. Suppose that payments of an amount C are to be made once a year for n years, the first one due a year hence. Assuming that annual compounding applies, we shall find the present value of such a stream of payments. We compute the present values of all payments and add them up to get

$$\frac{C}{1+r} + \frac{C}{(1+r)^2} + \frac{C}{(1+r)^3} + \cdots + \frac{C}{(1+r)^n}.$$

It is sometimes convenient to introduce the following seemingly cumbersome piece of notation:

$$\text{PA}(r,n) = \frac{1}{1+r} + \frac{1}{(1+r)^2} + \cdots + \frac{1}{(1+r)^n}. \tag{2.7}$$

This number is called the *present value factor for an annuity*. It allows us to express the present value of an annuity in a concise form:

$$\text{PA}(r,n) \times C.$$

The expression for $\text{PA}(r,n)$ can be simplified by using the formula

$$a + qa + q^2 a + \cdots + q^{n-1} a = a \frac{1-q^n}{1-q}. \tag{2.8}$$

In our case $a = \frac{1}{1+r}$ and $q = \frac{1}{1+r}$, hence

$$\text{PA}(r,n) = \frac{1 - (1+r)^{-n}}{r}. \tag{2.9}$$

Remark 2.8

Note that an initial bank deposit of

$$P = \text{PA}(r,n) \times C = \frac{C}{1+r} + \frac{C}{(1+r)^2} + \cdots + \frac{C}{(1+r)^n}$$

attracting interest at a rate r compounded annually would produce a stream of n annual payments of C each. A deposit of $C(1+r)^{-1}$ would grow to C after one year, which is just what is needed to cover the first annuity payment. A deposit of $C(1+r)^{-2}$ would become C after two years to cover the second payment, and so on. Finally, a deposit of $C(1+r)^{-n}$ would deliver the last payment of C due after n years.

Example 2.9

Consider a loan of $\$1,000$ to be paid back in 5 equal instalments due at yearly intervals. The instalments include both the interest payable each year calculated at 15% of the current outstanding balance and the repayment of a fraction of the loan. A loan of this type is called an *amortised loan*. The amount of each instalment can be computed as

$$\frac{1,000}{\text{PA}(15\%,5)} \cong 298.32.$$

This is because the loan is equivalent to an annuity from the point of view of the lender.

Exercise 2.13

What is the amount of interest included in each instalment? How much of the loan is repaid as part of each instalment? What is the outstanding balance of the loan after each instalment is paid?

Exercise 2.14

How much can you borrow if the interest rate is 18%, you can afford to pay $10,000 at the end of each year, and you want to clear the loan in 10 years?

Exercise 2.15

Suppose that you deposit $1,200 at the end of each year for 40 years, subject to annual compounding at a constant rate of 5%. Find the balance after 40 years.

Exercise 2.16

Suppose that you took a mortgage of $100,000 on a house to be paid back in full by 10 equal annual instalments, each consisting of the interest due on the outstanding balance plus a repayment of a part of the amount borrowed. If you decided to clear the mortgage after eight years, how much money would you need to pay on top of the eighth instalment, assuming that a constant annual compounding rate of 6% applies throughout the period of the mortgage?

Recall that a *perpetuity* is an infinite sequence of payments of a fixed amount C occurring at the end of each year. The formula for the present value of a perpetuity can be obtained from (2.7) by letting $n \to \infty$:

$$\lim_{n \to \infty} \mathrm{PA}(r, n) \times C = \frac{C}{1+r} + \frac{C}{(1+r)^2} + \frac{C}{(1+r)^3} + \cdots = \frac{C}{r}. \qquad (2.10)$$

The limit amounts to taking the sum of a geometric series.

Remark 2.10

The present value of a perpetuity is given by the same formula as in Example 2.3, even though periodic compounding has been used in place of simple interest. In both cases the annual payment C is exactly equal to the interest earned throughout the year, and the amount remaining to earn interest in the following year is always $\frac{C}{r}$. Nevertheless, periodic compounding allows us to view the same sequence of payments in a different way: the present value $\frac{C}{r}$ of the perpetuity is decomposed into infinitely many parts, as in (2.10), each responsible for producing one future payment of C.

Remark 2.11

Formula (2.9) for the annuity factor is easier to memorise in the following way, using the formula for a perpetuity: the sequence of n payments of $C = 1$ can be represented as the difference between two perpetuities, one starting now and the other after n years. (Cutting off the tail of a perpetuity, we obtain an annuity.) In doing so we need to compute the present value of the latter perpetuity. This can be achieved by means of the discount factor $(1 + r)^{-n}$. Hence,

$$\mathrm{PA}(r, n) = \frac{1}{r} - \frac{1}{r} \times \frac{1}{(1 + r)^n} = \frac{1 - (1 + r)^{-n}}{r}.$$

Exercise 2.17

Find a formula for the present value of an infinite stream of payments of the form C, $C(1 + g)$, $C(1 + g)^2, \ldots$, growing at a constant rate g. By the tail-cutting procedure find a formula for the present value of n such payments.

2.1.4 Continuous Compounding

Formula (2.5) for the future value at time t of a principal P attracting interest at a rate $r > 0$ compounded m times a year can be written as

$$V(t) = \left[\left(1 + \frac{r}{m} \right)^{\frac{m}{r}} \right]^{tr} P.$$

In the limit as $m \to \infty$, we obtain

$$V(t) = \mathrm{e}^{tr} P, \qquad (2.11)$$

where

$$\mathrm{e} = \lim_{x \to \infty} \left(1 + \frac{1}{x} \right)^x$$

is the natural logarithm base. This is known as *continuous compounding*. The corresponding *growth factor* is e^{tr}. A typical graph of $V(t)$ is shown in Figure 2.3.

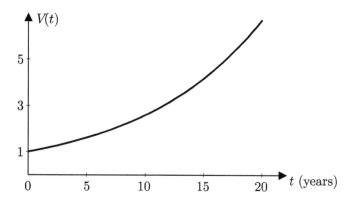

Figure 2.3 Continuous compounding at 10% ($r = 0.1$, $P = 1$)

The derivative of $V(t) = e^{tr}P$ is

$$V'(t) = re^{tr}P = rV(t).$$

In the case of continuous compounding the rate of the growth is proportional to the current wealth.

Formula (2.11) is a good approximation of the case of periodic compounding when the frequency m is large. It is simpler and lends itself more readily to transformations than the formula for periodic compounding.

Exercise 2.18

How long will it take to earn $1 in interest if $1,000,000 is deposited at 10% compounded continuously?

Exercise 2.19

In 1626 Peter Minuit, governor of the colony of New Netherland, bought the island of Manhattan from Indians paying with beads, cloth, and trinkets worth $24. Find the value of this sum in the year 2000 at 5% compounded a) continuously and b) annually.

Proposition 2.12

Continuous compounding produces higher future value than periodic compounding with any frequency m, given the same principal P and interest rate r.

Proof

It suffices to verify that

$$e^{tr} > \left(1 + \frac{r}{m}\right)^{tm} = \left[\left(1 + \frac{r}{m}\right)^{\frac{m}{r}}\right]^{rt}.$$

The inequality holds because the sequence $\left(1 + \frac{r}{m}\right)^{\frac{m}{r}}$ is increasing and converges to e as $m \nearrow \infty$.   □

Exercise 2.20

What will be the difference between the value after one year of $100 deposited at 10% compounded monthly and compounded continuously? How frequent should periodic compounding be for the difference to be less than $0.01?

The present value under continuous compounding is given by

$$V(0) = V(t)e^{-tr}.$$

In this case the *discount factor* is e^{-tr}. Given the terminal value $V(T)$, we clearly have

$$V(t) = e^{-r(T-t)}V(T). \tag{2.12}$$

Exercise 2.21

Find the present value of $1,000,000 to be received after 20 years assuming continuous compounding at 6%.

Exercise 2.22

Given that the future value of $950 subject to continuous compounding will be $1,000 after half a year, find the interest rate.

The return $K(s,t)$ defined by (2.3) on an investment subject to continuous compounding fails to be additive, just like in the case of periodic compounding. It proves convenient to introduce the *logarithmic return*

$$k(s,t) = \ln \frac{V(t)}{V(s)}. \tag{2.13}$$

Proposition 2.13

The logarithmic return is additive,

$$k(s,t) + k(t,u) = k(s,u).$$

Proof

This is an easy consequence of (2.13):

$$\begin{aligned}
k(s,t) + k(t,u) &= \ln \frac{V(t)}{V(s)} + \ln \frac{V(u)}{V(t)} \\
&= \ln \frac{V(t)}{V(s)} \frac{V(u)}{V(t)} = \ln \frac{V(u)}{V(s)} = k(s,u).
\end{aligned}$$

$\square$

If $V(t)$ is given by (2.11), then $k(s,t) = r(t-s)$, which enables us to recover the interest rate

$$r = \frac{k(s,t)}{t-s}.$$

Exercise 2.23

Suppose that the logarithmic return over 2 months on an investment subject to continuous compounding is 3%. Find the interest rate.

2.1.5 How to Compare Compounding Methods

As we have already noticed, frequent compounding will produce higher future value than less frequent compounding if the interest rates and the principal are the same. We shall consider the general circumstances in which one compounding method will produce either the same or higher future value than another method, given the same principal.

Example 2.14

Suppose that certificates promising to pay $120 after one year can be purchased or sold now, or at any time during this year, for $100. This is consistent with a constant interest rate of 20% under annual compounding. If an investor decided to sell such a certificate half a year after the purchase, what price would it fetch? Suppose it is $110, a frequent first guess based on halving the annual profit of

$20. However, this turns out to be too high a price, leading to the following arbitrage strategy:

- borrow $1,000 to buy 10 certificates for $100 each;
- after six months sell the 10 certificates for $110 each and buy 11 new certificates for $100 each; the balance of these transactions is nil;
- after another six months sell the 11 certificates for $110 each, cashing $1,210 in total, and pay $1,200 to clear the loan with interest; the balance of $10 would be the arbitrage profit.

A similar argument shows that the certificate price after six months cannot be too low, say, $109.

The price of a certificate after six months is related to the interest rate under semi-annual compounding: if this rate is r, then the price will be $100\left(1 + \frac{r}{2}\right)$ dollars, and vice versa. Arbitrage will disappear if the corresponding growth factor $\left(1 + \frac{r}{2}\right)^2$ over one year is equal to the growth factor 1.2 under annual compounding,

$$\left(1 + \frac{r}{2}\right)^2 = 1.2,$$

which gives $r \cong 0.1909$, or 19.09%. If so, then the certificate price after six months should be $100\left(1 + \frac{0.1909}{2}\right) \cong 109.54$ dollars.

Growth factors over a fixed period, typically one year, can be used to compare any two compounding methods.

Definition 2.15

We say that two compounding methods are *equivalent* if the corresponding growth factors over a period of one year are the same. If one of the growth factors exceeds the other, then the corresponding compounding method is said to be *preferable*.

Example 2.16

Semi-annual compounding at 10% is equivalent to annual compounding at 10.25%. Indeed, in the former case the growth factor over a period of one year is

$$\left(1 + \frac{0.1}{2}\right)^2 = 1.1025,$$

which is the same as the growth factor in the latter case. Both are preferable to monthly compounding at 9%, for which the growth factor over one year is only

$$\left(1 + \frac{0.09}{12}\right)^{12} \cong 1.0938.$$

We can freely switch from one compounding method to another equivalent method by recalculating the interest rate. In the chapters to follow we shall normally use either annual or continuous compounding.

Exercise 2.24

Find the rate for continuous compounding equivalent to monthly compounding at 12%.

Exercise 2.25

Find the frequency of periodic compounding at 20% to be equivalent to annual compounding at 21%.

Instead of comparing the growth factors, it is often convenient to compare the so-called effective rates as defined below.

Definition 2.17

For a given compounding method with interest rate r the *effective rate* r_e is one that gives the same growth factor over a one year period under annual compounding.

In particular, in the case of periodic compounding with frequency m and rate r the effective rate r_e satisfies

$$\left(1 + \frac{r}{m}\right)^m = 1 + r_e.$$

In the case of continuous compounding with rate r

$$e^r = 1 + r_e.$$

Example 2.18

In the case of semi-annual compounding at 10% the effective rate is 10.25%, see Example 2.16.

Proposition 2.19

Two compounding methods are equivalent if and only if the corresponding effective rates r_e and r'_e are equal, $r_e = r'_e$. The compounding method with effective rate r_e is preferable to the other method if and only if $r_e > r'_e$.

Proof

This is because the growth factors over one year are $1 + r_e$ and $1 + r'_e$, respectively. $\qquad\qquad\qquad\qquad\qquad\qquad\qquad\qquad\qquad\qquad\qquad\qquad\qquad\qquad$ $\square$

Example 2.20

In Exercise 2.8 we have seen that daily compounding at 15% is preferable to semi-annual compounding at 15.5%. The corresponding effective rates r_e and r'_e can be found from

$$1 + r_e = \left(1 + \frac{0.15}{365}\right)^{365} \cong 1.1618,$$

$$1 + r'_e = \left(1 + \frac{0.155}{2}\right)^{2} \cong 1.1610.$$

This means that r_e is about 16.18% and r'_e about 16.10%.

Remark 2.21

Recall that formula (2.5) for periodic compounding, that is,

$$V(t) = \left(1 + \frac{r}{m}\right)^{tm} P,$$

admits only time instants t being whole multiples of the compounding period $\frac{1}{m}$. An argument similar to that in Example 2.14 shows that the appropriate no-arbitrage value of an initial sum P at any time $t \geq 0$ should be $\left(1 + \frac{r}{m}\right)^{tm} P$. A reasonable extension of (2.5) is therefore to use the right-hand side for all $t \geq 0$ rather than just for whole multiples of $\frac{1}{m}$. From now on we shall always use this extension.

In terms of the effective rate r_e the future value can be written as

$$V(t) = (1 + r_e)^t P.$$

for all $t \geq 0$. This applies both to continuous compounding and to periodic compounding extended to arbitrary times as in Remark 2.21. Proposition 2.19 implies that, given the same initial principal, equivalent compounding methods will produce the same future value for all times $t \geq 0$. Similarly, a compounding method preferable to another one will produce higher future values for all $t > 0$.

Remark 2.22

Simple interest does not fit into the scheme for comparing compounding methods. In this case the future value $V(t)$ is a linear function of time t, whereas it is an exponential function if either continuous or periodic compounding applies.

Exercise 2.26

What is the present value of an annuity consisting of monthly payments of an amount C continuing for n years? Express the answer in terms of the effective rate r_e.

Exercise 2.27

What is the present value of a perpetuity consisting of bimonthly payments of an amount C? Express the answer in terms of the effective rate r_e.

2.2 Money Market

The money market consists of risk-free (more precisely, default-free) securities. An example is a *bond*, which is a financial security promising the holder a sequence of guaranteed future payments. Risk-free means here that these payments will be delivered with certainty. (Nevertheless, even in this case risk cannot be completely avoided, since the market prices of such securities may fluctuate unpredictably; see Chapter 9.) There are many kinds of bonds like treasury bills and notes, treasury, mortgage and debenture bonds, commercial papers, and others with various particular arrangements concerning the issuing institution, maturity, number of payments, embedded rights and guarantees.

2.2.1 Zero-Coupon Bonds

The simplest case of a bond is a *zero-coupon bond*, which involves just a single payment. The issuing institution (for example, a government, a bank or a company) promises to exchange the bond for a certain amount of money F, called the *face value*, on a given day T, called the *maturity date*. Typically, the life span of a zero-coupon bond is up to one year, the face value being some round figure, for example 100. In effect, the person or institution who buys the bond is lending money to the bond writer.

Given the interest rate, the present value of such a bond can be computed. Suppose that a bond with face value $F = 100$ dollars is maturing in one year, and the annual compounding rate r is 12%. Then the present value of the bond should be

$$V(0) = F(1 + r)^{-1} \cong 89.29$$

dollars.

In reality, the opposite happens: bonds are freely traded and their prices are determined by market forces, whereas the interest rate is implied by the bond prices,

$$r = \frac{F}{V(0)} - 1. \tag{2.14}$$

This formula gives the implied annual compounding rate. For instance, if a one-year bond with face value \$100 is being traded at \$91, then the implied rate is 9.89%.

For simplicity, we shall consider *unit bonds* with face value equal to one unit of the home currency, $F = 1$.

Typically, a bond can be sold at any time prior to maturity at the market price. This price at time t is denoted by $B(t, T)$. In particular, $B(0, T)$ is the current, time 0 price of the bond, and $B(T, T) = 1$ is equal to the face value. Again, these prices determine the interest rates by (2.6) and (2.12) with $V(t) = B(t, T)$ and $V(T) = 1$. For example, the implied annual compounding rate satisfies the equation

$$B(t, T) = (1 + r)^{-(T-t)}.$$

The last formula has to be suitably modified if a different compounding method is used. Using periodic compounding with frequency m, we need to solve the equation

$$B(t, T) = \left(1 + \frac{r}{m}\right)^{-m(T-t)}.$$

In the case of continuous compounding the equation for the implied rate satisfies

$$B(t, T) = e^{-r(T-t)}.$$

Of course all these different implied rates are equivalent to one another, since the bond price does not depend on the compounding method used.

Remark 2.23

In general, the implied interest rate may depend on the trading time t as well as on the maturity time T. This is an important issue, which will be discussed in Chapter 9. For the time being, we adopt the simplifying assumption that the interest rate remains constant throughout the period up to maturity.

Exercise 2.28

An investor paid $95 for a bond with face value $100 maturing in six months. When will the bond value reach $99 if the interest rate remains constant?

Exercise 2.29

Find the interest rates for annual, semi-annual and continuous compounding implied by a unit bond with $B(0.5, 1) = 0.9455$.

Note that $B(0, T)$ is the discount factor and $B(0, T)^{-1}$ is the growth factor for each compounding method. These universal factors are all that is needed to compute the time value of money, without resorting to the corresponding interest rates. However, interest rates are useful because they are more intuitive. For the average bank customer the information that a one-year $100 bond is selling for $92.59 may not be as clear as the equivalent statement that a deposit will earn 8% interest if kept for one year.

2.2.2 Coupon Bonds

Bonds promising a sequence of payments are called *coupon bonds*. These payments consist of the face value due at maturity, and *coupons* paid regularly, typically annually, semi-annually, or quarterly, the last coupon due at maturity. The assumption of constant interest rates allows us to compute the price of a coupon bond by discounting the future payments.

Example 2.24

Consider a bond with face value $F = 100$ dollars maturing in five years, $T = 5$, with coupons of $C = 10$ dollars paid annually, the last one at maturity. This means a stream of payments of $10, 10, 10, 10, 110$ dollars at the end of each consecutive year. Given the continuous compounding rate r, say 12%, we can find the price of the bond:

$$V(0) = 10e^{-r} + 10e^{-2r} + 10e^{-3r} + 10e^{-4r} + 110e^{-5r} \cong 90.27$$

dollars.

Exercise 2.30

Find the price of a bond with face value $100 and $5 annual coupons that matures in four years, given that the continuous compounding rate is a) 8% or b) 5%.

Exercise 2.31

Sketch the graph of the price of the bond in Exercise 2.30 as a function of the continuous compounding rate r. What is the value of this function for $r = 0$? What is the limit as $r \to \infty$?

Example 2.25

We continue Example 2.24. After one year, once the first coupon is cashed, the bond becomes a four-year bond worth

$$V(1) = 10e^{-r} + 10e^{-2r} + 10e^{-3r} + 110e^{-4r} \cong 91.78$$

dollars. Observe that the total wealth at time 1 is

$$V(1) + C = V(0)e^r.$$

Six months later the bond will be worth

$$V(1.5) = 10e^{-0.5r} + 10e^{-1.5r} + 10e^{-2.5r} + 110e^{-3.5r} \cong 97.45$$

dollars. After four years the bond will become a zero-coupon bond with face value $110 and price

$$V(4) = 110e^{-r} \cong 97.56$$

dollars.

An investor may choose to sell the bond at any time prior to maturity. The price at that time can once again be found by discounting all the payments due at later times.

Exercise 2.32

Sketch the graph of the price of the coupon bond in Examples 2.24 and 2.25 as a function of time.

Exercise 2.33

How long will it take for the price of the coupon bond in Examples 2.24 and 2.25 to reach $95 for the first time?

Remark 2.26

The practice with respect to quoting bond prices between coupon payments is somewhat complicated. The present value of future payments, called the *dirty price*, is the price the bond would fetch when sold between coupon payments. *Accrued interest* accumulated since the last coupon payment is evaluated by applying the simple interest rule. The *clean price* is then quoted by subtracting accrued interest from the dirty price.

The coupon can be expressed as a fraction of the face value. Assuming that coupons are paid annually, we shall write $C = iF$, where i is called the *coupon rate*.

Proposition 2.27

Whenever coupons are paid annually, the coupon rate is equal to the interest rate for annual compounding if and only if the price of the bond is equal to its face value. In this case we say that the bond is trading *at par*.

Proof

To avoid cumbersome notation we restrict ourselves to an example. Suppose that annual compounding with $r = i$ applies, and consider a bond with face value $F = 100$ maturing in three years, $T = 3$. Then the price of the bond is

$$\frac{C}{1+r} + \frac{C}{(1+r)^2} + \frac{F+C}{(1+r)^3} = \frac{rF}{1+r} + \frac{rF}{(1+r)^2} + \frac{F(1+r)}{(1+r)^3}$$

$$= \frac{rF}{1+r} + \frac{rF}{(1+r)^2} + \frac{F}{(1+r)^2} = \frac{rF}{1+r} + \frac{F(1+r)}{(1+r)^2} = F.$$

Conversely, note that

$$\frac{C}{1+r} + \frac{C}{(1+r)^2} + \frac{F+C}{(1+r)^3}$$

is one-to-one as a function of r (in fact, a strictly decreasing function), so it assumes the value F exactly once, and we know this happens for $r = i$. □

Remark 2.28

If a bond sells below the face value, it means that the implied interest rate is higher than the coupon rate (since the bond price decreases when the interest rate goes up). If the bond price is higher than the face value, it means that the interest rate is lower than the coupon rate. This is important information in circumstances when the bond price is determined by the market and gives an indication of the level of interest rates.

Exercise 2.34

A bond with face value $F = 100$ and annual coupons $C = 8$ maturing after three years, at $T = 3$, is trading at par. Find the implied continuous compounding rate.

2.2.3 Money Market Account

An investment in the money market can be realised by means of a financial intermediary, typically an investment bank, who buys and sells bonds on behalf of its customers (thus reducing transaction costs). The risk-free position of an investor is given by the level of his or her account with the bank. It is convenient to think of this account as a tradable asset, since the bonds themselves are tradable. A long position in the money market involves buying the asset, that is, investing money. A short position amounts to borrowing money.

First, consider an investment in a zero-coupon bond closed prior to maturity. An initial amount $A(0)$ invested in the money market makes it possible to purchase $A(0)/B(0,T)$ bonds. The value of each bond will fetch

$$B(t,T) = e^{-(T-t)r} = e^{rt}e^{-rT} = e^{rt}B(0,T)$$

at time t. As a result, the investment will reach

$$A(t) = \frac{A(0)}{B(0,T)}B(t,T) = A(0)e^{rt}$$

at time $t \leq T$.

Exercise 2.35

Find the return on a 75-day investment in zero-coupon bonds if $B(0,1) = 0.89$.

Exercise 2.36

The return on a bond over six months is 7%. Find the implied continuous compounding rate.

Exercise 2.37

After how many days will a bond purchased for $B(0,1) = 0.92$ produce a 5% return?

The investment in a bond has a finite time horizon. It will be terminated with $A(T) = A(0)e^{rT}$ at the maturity time T of the bond. To extend the position in the money market beyond T one can reinvest the amount $A(T)$ into a new bond issued at time T maturing at $T' > T$. Taking $A(T)$ as the initial investment with T playing the role of the starting time, we have

$$A(t') = A(T)e^{r(t'-T)} = A(0)e^{rt'}$$

for $T \le t' \le T'$. By repeating this argument, we readily arrive at the conclusion that an investment in the money market can be prolonged for as long as required, the formula

$$A(t) = A(0)e^{rt} \qquad (2.15)$$

being valid for all $t \ge 0$.

Exercise 2.38

Suppose that one dollar is invested in zero-coupon bonds maturing after one year. At the end of each year the proceeds are reinvested in new bonds of the same kind. How many bonds will be purchased at the end of year nine? Express the answer in terms of the implied continuous compounding rate.

We can also consider coupon bonds as a tool to manufacture an investment in the money market. Suppose that the first coupon C is due at time t. At time 0 we buy $A(0)/V(0)$ coupon bonds. At time t we cash the coupon and sell the bond for $V(t)$, receiving the total sum $C + V(t) = V(0)e^{rt}$ (see Example 2.25). Because the interest rate is constant, this sum of money is certain. In this way we have effectively created a zero-coupon bond with face value $V(0)e^{rt}$ maturing at time t. It means that the scheme worked out above for zero-coupon bonds applies to coupon bonds as well, resulting in the same formula (2.15) for $A(t)$.

Exercise 2.39

The sum of $1,000 is invested in five-year bonds with face value $100 and $8 coupons paid annually. All coupons are reinvested in bonds of the same kind. Assuming that the bonds are trading at par and the interest rate remains constant throughout the period to maturity, compute the number of bonds held during each consecutive year of the investment.

As we have seen, under the assumption that the interest rate is constant, the function $A(t)$ does not depend on the way the money market account is constructed, that is, it neither depends on the types of bonds selected for investment nor on the method of extending the investment beyond the maturity of the bonds.

Throughout much of this book we shall assume $A(t)$ to be deterministic and known. Indeed, we assume that $A(t) = e^{rt}$, where r is a constant interest rate. Variable interest rates and a random money market account will be studied in Chapter 9.

Case 2: Discussion

It is clear that we have to save some money on a regular basis. The simplest method would be to put away the same amount each year. However, since some growth is assumed, this fixed amount might be relatively large as compared to the salary income in the early years. So we formulate the following question: what fixed percentage of your salary should you be paying into this pension fund?

For simplicity we assume that by salary we mean the annual salary. (The reader is encouraged to analyse the version with monthly payments.) We also make the bold assumption that the interest rate will remain constant. It will be variable, certainly, but during such a long term the fluctuations will have an averaging effect, lending some credibility to the result of our calculations (which, nevertheless, have to be treated as crude estimates).

In the solution to Exercise 2.17, formula (2.9) for the present value of an annuity is extended to the case of payments growing at a constant rate. For simplicity, suppose that all payments are made annually at the end of each year. Let S be the initial salary and let x denote the percentage of the salary to be invested in the pension fund. Then the present value of the savings will be

$$V(0) = \sum_{n=1}^{40} \frac{xS(1+g)^n}{(1+r)^n},$$

where $g = 2\%$ is the growth rate and $r = 5\%$ is the interest rate. Employ (2.8) with $q = \frac{1+g}{1+r}$ to get

$$V(0) = xS(0)\text{GAF}(r, g, N)$$

with the *growing annuity factor* given by

$$\text{GAF}(r, g, N) = \frac{1+g}{r-g}\left(1 - \frac{(1+g)^N}{(1+r)^N}\right),$$

where N is the number of years. In the case in hand

$$\text{GAF}(5\%, 2\%, 40) = 23.34.$$

After 40 years you will accumulate the amount $V(40) = V(0)(1+r)^{40}$, which then becomes the starting capital for the retirement period.

The final salary will be $S(1+g)^{40}$ after 40 years, and we want the initial pension to be 50% of that, growing at rate g in the following years. The pension will also be a growing annuity with present value (the 'present' here being the end of year 40) $\frac{1}{2}S(1+g)^{40}\text{GAF}(r, g, 20)$, which must be equal to the accumulated capital $V(40)$. This gives an equation for x,

$$xS(1+r)^{40}\text{GAF}(r, g, 40) = \frac{1}{2}S(1+g)^{40}\text{GAF}(r, g, 20),$$

with solution $x = 10.05\%$. It would be sufficient to pay just over 10% of the salary into this pension scheme to achieve its objectives.

However, in the course of 60 years the rates r and g will certainly fluctuate. It is therefore interesting to discuss the sensitivity of the solution to changes in those rates. The values of x for a range of values of r and g are shown in Table 2.1.

	$r = 4\%$	$r = 5\%$	$r = 6\%$
$g = 1\%$	9.96%	7.24%	5.24%
$g = 2\%$	13.70%	10.05%	7.34%
$g = 3\%$	18.62%	13.78%	10.14%

Table 2.1 Percentage x of salary paid into the pension scheme

Observe that x depends quite strongly on the premium rate of interest $r - g$ above the growth rate g, but much less so on g (or r) itself when the value of $r-g$ is fixed. (For example, examine the values of x on the diagonal, corresponding to $r-g = 3\%$.) While r and g will tend to increase or decrease together with the rate of inflation, the difference $r - g$ can be expected to remain relatively stable over the years, lending some justification to the numerical results obtained under the manifestly false assumption of constant rates.

3
Portfolio Management

Case 3

The case in Chapter 2 involved a pension fund invested in a risk-free bank account. However, you may be willing to accept some reasonable risk in the hope that this could lead to a more profitable pension scheme. Consider investing in some risky securities in addition to the bank account.

3.1 Risk and Return

An investment in a risky security always carries the burden of possible losses or poor performance. In this chapter we analyse the advantages of spreading the investment among several securities to keep the inevitable risk under control.

In Chapter 1 we considered one risky security in a single time step. Here, we shall assume a similar simple single time step setup but consider many risky securities. The length of the time step can be arbitrary, and our typical example will be one year. We shall keep the same notation as in Chapter 1, namely 0 for the beginning and T for the end of the time period.

M. Capiński, T. Zastawniak, *Mathematics for Finance*,
Springer Undergraduate Mathematics Series,
© Springer-Verlag London Limited 2011

3.1.1 Expected Return

Suppose we make a single-period investment in some stock with known current price $S(0)$. The future price $S(T)$ is unknown, hence assumed to be a random variable

$$S(T) : \Omega \to [0, +\infty)$$

on a probability space Ω. The return

$$K = \frac{S(T) - S(0)}{S(0)}$$

is a random variable with expected value $\mathbb{E}(K) = \mu_K = \mu$. We introduce the convention of using the Greek letter μ for expectations of various random returns, with subscripts indicating the context, if necessary. By the linearity of mathematical expectation

$$\mu = \frac{\mathbb{E}(S(T)) - S(0)}{S(0)}.$$

The relationships between the prices and returns can be written as

$$S(T) = S(0)(1 + K),$$
$$\mathbb{E}(S(T)) = S(0)(1 + \mu),$$

which indicates the possibility of reversing the approach: given the returns we can find the prices.

Example 3.1

Suppose that the probability space Ω is finite. We can consider the elements of Ω to be the possible economic scenarios, writing $\Omega = \{\omega_1, \ldots, \omega_N\}$. The possible values of $S(T)$ form a vector $(S^{\omega_1}(T), \ldots, S^{\omega_N}(T))$ in $\mathbb{R}^N$. It is natural to equip Ω with the σ-field $\mathcal{F} = 2^\Omega$ of all subsets. To define a probability measure $P : \mathcal{F} \to [0, 1]$ it is sufficient to prescribe its values on single element sets, $P(\{\omega_i\}) = p_i$, by choosing $p_i \in [0, 1]$ such that $\sum_{i=1}^{N} p_i = 1$. Denoting the vector of probabilities by $p = (p_1, \ldots, p_N)$, we can write the expected price at the end of the period as

$$\mathbb{E}(S(T)) = \sum_{i=1}^{N} S^{\omega_i}(T) p_i = \langle S(T), p \rangle,$$

where $\langle \cdot, \cdot \rangle$ denotes the inner product of vectors in $\mathbb{R}^N$. The return

$$K^{\omega_i} = \frac{S^{\omega_i}(T) - S(0)}{S(0)}, \quad K = (K^{\omega_1}, \ldots, K^{\omega_N})$$

has expected value

$$\mathbb{E}(K) = \langle K, p \rangle = \sum_{i=1}^{N} K^{\omega_i} p_i.$$

Example 3.2

Now we consider some special kinds of investment, which can be cast into the general framework of portfolios. The following considerations are motivated by the practical regulations encountered in some financial markets:

1. Margin purchases (leveraged purchases). Suppose an investment in stock is partially financed by a loan obtained at an interest rate $r_l \geq r$, higher that the rate r for risk-free investments. That is, we buy one share for $S(0)$, investing $wS(0)$ of our own money and borrowing $(1-w)S(0)$, $w \in [0,1]$.

2. Short selling. We borrow one share, paying some percentage of its value as collateral, say $cS(0)$, where $c \in [0,1]$. This collateral attracts interest at a lower rate r_c than that available for risk-free investments, $0 \leq r_c \leq r$. We sell the stock and can invest the proceeds at rate r. At the end of the period we repurchase and return the stock, and recover the collateral. (For mathematical convenience we often take $c = 0$, a valid assumption for some large investors.)

Exercise 3.1

Write down the formulae for the return and expected return on a) a leveraged purchase, and b) a short position in stock.

3.1.2 Standard Deviation as Risk Measure

First of all, we need to identify a suitable quantity to measure risk. An investment in zero-coupon bonds held to maturity gives a return free of risk, 6% say, in which case the measure of risk should be equal to zero. If the return on an investment is, say 11% or 13%, depending on the market scenario, then the risk is clearly lower as compared with an investment returning 2% or 22%, respectively.

However, the spread of return values can hardly be used to measure risk because it ignores the probabilities. If the return rate is 22% with probability 0.99 and 2% with probability 0.01, the risk can be considered quite small, whereas the same rates of return occurring with probability 0.5 each would indicate a rather more risky investment.

The uncertainty is understood as the scatter of returns around some reference point. A natural candidate for the reference value is the expected return. The extent of scatter can be conveniently measured by standard deviation. This notion takes care of two aspects of risk:

1) the distances between possible values and the reference point;
2) the probabilities of attaining these values.

Definition 3.3

By (the measure of) *risk* we mean the standard deviation

$$\sigma_K = \sqrt{\text{Var}(K)}$$

of the return K,

$$\text{Var}(K) = \mathbb{E}(K - \mu)^2 = \mathbb{E}(K^2) - \mu^2$$

being the variance of the return.

Exercise 3.2

Compute the risk as measured by the standard deviations $\sigma_{K_1}, \sigma_{K_2}, \sigma_{K_3}$ for each of the following three investment projects, where the returns K_1, K_2, K_3 depend on the market scenario:

Scenario	Probability	Return K_1	Return K_2	Return K_3
ω_1	0.25	12%	11%	2%
ω_2	0.75	12%	13%	22%

Which of these is the most risky and the least risky project?

Exercise 3.3

Consider two scenarios, ω_1 with probability $\frac{1}{4}$ and ω_2 with probability $\frac{3}{4}$. Suppose that the return on some security is $K_1^{\omega_1} = -2\%$ in the first scenario and $K_1^{\omega_2} = 8\%$ in the second scenario. If the return on another security is $K_2^{\omega_1} = -4\%$ in the first scenario, find the return $K_2^{\omega_2}$ in the other scenario such that the two securities have the same risk.

Example 3.4

Let the return on an investment be $K = 3\%$ or -1%, both with probability 0.5. Then the risk is

$$\sigma_K = 0.02.$$

Now suppose that the return on another investment is double that on the first investment, being equal to $2K = 6\%$ or -2%, also with probability 0.5 each. Then the risk of the second investment will be

$$\sigma_{2K} = 0.04.$$

The risk as measured by the standard deviation is doubled.

This illustrates the following general rule:

$$\sigma_{aK} = |a|\, \sigma_K,$$
$$\mathrm{Var}(aK) = a^2 \mathrm{Var}(K),$$

for any real number a.

3.2 Two Securities

We begin a detailed discussion of the relationship between risk and expected return in the simple situation of a portfolio with just two risky securities.

Example 3.5

Suppose that the prices of two stocks behave as follows:

Scenario	Probability	Return K_1	Return K_2
ω_1	0.5	10%	-5%
ω_2	0.5	-5%	10%

If we split our money equally between these two stocks, then we shall earn 5% in each scenario (losing 5% on one stock, but gaining 10% on the other). Even though an investment in either stock separately involves risk, we have reduced the overall risk to nil by splitting the investment between the two stocks. This is a simple example of diversification, which is particularly effective here because the returns have perfect negative correlation.

We introduce weights to describe the allocation of funds between the securities as a convenient alternative to specifying portfolios in terms of the number of shares of each security. The *weights* are defined by

$$w_1 = \frac{x_1 S_1(0)}{V(0)}, \quad w_2 = \frac{x_2 S_2(0)}{V(0)},$$

where x_k denotes the number of shares of kind $k = 1, 2$ in the portfolio. This means that w_k is the percentage of the initial value of the portfolio invested in security k. Observe that the weights always add up to 100%,

$$w_1 + w_2 = \frac{x_1 S_1(0) + x_2 S_2(0)}{V(0)} = \frac{V(0)}{V(0)} = 1. \tag{3.1}$$

If short selling is allowed, then one of the weights may be negative and the other one greater than 100%.

Example 3.6

Suppose that the prices of two kinds of stock are $S_1(0) = 30$ and $S_2(0) = 40$ dollars. We prepare a portfolio worth $V(0) = 1,000$ dollars by purchasing $x_1 = 20$ shares of stock 1 and $x_2 = 10$ shares of stock 2. The allocation of funds between the two securities is

$$w_1 = \frac{30 \times 20}{1,000} = 60\%, \quad w_2 = \frac{10 \times 40}{1,000} = 40\%.$$

These are the weights in the portfolio. If the stock prices change to $S_1(T) = 35$ and $S_2(T) = 39$ dollars, then the portfolio will be worth $V(T) = 20 \times 35 + 10 \times 39 = 1,090$ dollars. Observe that this amount is no longer split between the two securities as 60% to 40%, but as follows:

$$\frac{20 \times 35}{1,090} \cong 64.22\%, \quad \frac{10 \times 39}{1,090} \cong 35.78\%,$$

even though the actual number of shares of each stock in the portfolio remains unchanged.

Remark 3.7

In reality the number of shares has to be an integer, which places a constraint on possible weights. To simplify matters, as in Chapter 1 we shall assume divisibility of assets. This means that the weights can be any real numbers that add up to one. Since not all real markets allow short selling, sometimes we need to distinguish a special case when the weights are non-negative.

Example 3.8

Suppose that a portfolio worth $V(0) = 1,000$ dollars is constructed by taking a long position in stock 1 and a short position in stock 2 from Example 3.6 with weights $w_1 = 120\%$ and $w_2 = -20\%$. The portfolio will consist of

$$x_1 = w_1 \frac{V(0)}{S_1(0)} = 120\% \times \frac{1,000}{30} = 40,$$

$$x_2 = w_2 \frac{V(0)}{S_2(0)} = -20\% \times \frac{1,000}{40} = -5$$

shares of type 1 and 2. If the stock prices change as in Example 3.6, then this portfolio will be worth

$$V(T) = x_1 S_1(T) + x_2 S_2(T) = V(0) \left(w_1 \frac{S_1(T)}{S_1(0)} + w_2 \frac{S_2(T)}{S_2(0)} \right)$$

$$= 1,000 \left(120\% \times \frac{35}{30} - 20\% \times \frac{39}{40} \right) = 1,205$$

dollars, benefiting from both the rise in the price of stock 1 and the fall of stock 2. However, a small investor might face some restrictions on short selling. For example, it may be necessary to provide a cash collateral equal to 50% of the amount raised by shorting stock 2. The investor would then have to forego the interest that could otherwise have been earned on the collateral, which will need to be subtracted from the final value $V(T)$ of the portfolio.

Exercise 3.4

Compute the value $V(T)$ of a portfolio worth initially $V(0) = 100$ dollars that consists of two securities with weights $w_1 = 25\%$ and $w_2 = 75\%$, given that the security prices are $S_1(0) = 45$ and $S_2(0) = 33$ dollars initially, changing to $S_1(T) = 48$ and $S_2(T) = 32$ dollars.

We can see in Example 3.8 and Exercise 3.4 that $V(T)/V(0)$ depends on the security prices only through the ratios $S_1(T)/S_1(0) = 1 + K_1$ and $S_2(T)/S_2(0) = 1 + K_2$, and on the positions in the securities only through the weights w_1, w_2. This indicates that the return on the portfolio should depend only on the weights w_1, w_2 and the returns K_1, K_2 on the two securities.

Proposition 3.9

The return K_V on a portfolio consisting of two securities is the weighted average

$$K_V = w_1 K_1 + w_2 K_2, \tag{3.2}$$

where w_1 and w_2 are the weights and K_1 and K_2 are the returns on the two components.

Proof

Suppose that the portfolio consists of x_1 shares of security 1 and x_2 shares of security 2. Then the initial and final values of the portfolio are

$$V(0) = x_1 S_1(0) + x_2 S_2(0),$$
$$\begin{aligned} V(T) &= x_1 S_1(0)(1 + K_1) + x_2 S_2(0)(1 + K_2) \\ &= V(0)\left(w_1(1 + K_1) + w_2(1 + K_2)\right) \\ &= V(0)(1 + w_1 K_1 + w_2 K_2). \end{aligned}$$

As a result, the return on the portfolio is

$$K_V = \frac{V(T) - V(0)}{V(0)} = w_1 K_1 + w_2 K_2.$$

$\square$

Exercise 3.5

Find the return on a portfolio consisting of two kinds of stock with weights $w_1 = 30\%$ and $w_2 = 70\%$ if the returns on the components are as follows:

Scenario	Return K_1	Return K_2
ω_1	12%	−4%
ω_2	−10%	7%

3.2.1 Risk and Expected Return on a Portfolio

The expected return on a portfolio consisting of two securities can be expressed in terms of the weights and the expected returns on the components as

$$\mathbb{E}(K_V) = w_1 \mathbb{E}(K_1) + w_2 \mathbb{E}(K_2). \tag{3.3}$$

This follows at once from (3.2) by the additivity of mathematical expectation.

Example 3.10

Consider three scenarios with the probabilities given below (a trinomial model). Let the returns on two different stocks in these scenarios be as follows:

Scenario	Probability	Return K_1	Return K_2
ω_1 (recession)	0.2	−10%	−30%
ω_2 (stagnation)	0.5	0%	20%
ω_3 (boom)	0.3	10%	50%

The expected returns are

$$\mathbb{E}(K_1) = -0.2 \times 10\% + 0.5 \times 0\% + 0.3 \times 10\% = 1\%,$$
$$\mathbb{E}(K_2) = -0.2 \times 30\% + 0.5 \times 20\% + 0.3 \times 50\% = 19\%.$$

Suppose that $w_1 = 60\%$ of available funds is invested in stock 1 and 40% in stock 2. The expected return on such a portfolio will be

$$\mathbb{E}(K_V) = w_1\mathbb{E}(K_1) + w_2\mathbb{E}(K_2)$$
$$= 0.6 \times 1\% + 0.4 \times 19\% = 8.2\%.$$

Exercise 3.6

Compute the weights in a portfolio consisting of two kinds of stock if the expected return on the portfolio is to be $\mathbb{E}(K_V) = 10\%$, given the following information on the returns on stock 1 and 2:

Scenario	Probability	Return K_1	Return K_2
w_1	0.1	-10%	10%
w_2	0.3	0%	-5%
w_3	0.6	15%	20%

To compute the variance of K_V we need to know not only the variances of the returns K_1 and K_2 on the components in the portfolio, but also the covariance between the two returns.

Theorem 3.11

The variance of the return on a portfolio is given by

$$\text{Var}(K_V) = w_1^2\text{Var}(K_1) + w_2^2\text{Var}(K_2) + 2w_1w_2\text{Cov}(K_1, K_2). \qquad (3.4)$$

Proof

Substituting $K_V = w_1K_1 + w_2K_2$ and collecting the terms with w_1^2, w_2^2 and w_1w_2, we compute

$$\text{Var}(K_V) = \mathbb{E}(K_V^2) - \mathbb{E}(K_V)^2$$
$$= w_1^2[\mathbb{E}(K_1^2) - \mathbb{E}(K_1)^2] + w_2^2[\mathbb{E}(K_2^2) - \mathbb{E}(K_2)^2]$$
$$+ 2w_1w_2[\mathbb{E}(K_1K_2) - \mathbb{E}(K_1)\mathbb{E}(K_2)]$$
$$= w_1^2\text{Var}(K_1) + w_2^2\text{Var}(K_2) + 2w_1w_2\text{Cov}(K_1, K_2).$$

$\square$

To avoid clutter, we introduce the following notation:

$$\mu_V = \mathbb{E}(K_V), \qquad \sigma_V = \sqrt{\text{Var}(K_V)},$$
$$\mu_1 = \mathbb{E}(K_1), \qquad \sigma_1 = \sqrt{\text{Var}(K_1)},$$
$$\mu_2 = \mathbb{E}(K_2), \qquad \sigma_2 = \sqrt{\text{Var}(K_2)}.$$
$$c_{12} = \text{Cov}(K_1, K_2).$$

Formulae (3.3) and (3.4) can be written as

$$\mu_V = w_1\mu_1 + w_2\mu_2, \tag{3.5}$$
$$\sigma_V^2 = w_1^2\sigma_1^2 + w_2^2\sigma_2^2 + 2w_1w_2c_{12}. \tag{3.6}$$

We shall also use the correlation coefficient

$$\rho_{12} = \frac{c_{12}}{\sigma_1\sigma_2}. \tag{3.7}$$

Note that the correlation coefficient is undefined when $\sigma_1\sigma_2 = 0$. This condition means that at least one of the assets is risk free.

Remark 3.12

For risky securities the returns K_1 and K_2 are non-constant random variables. Because of this $\sigma_1\sigma_2 > 0$ and ρ_{12} is well defined.

Example 3.13

We use the following data:

Scenario	Probability	Return K_1	Return K_2
ω_1	0.2	-10%	5%
ω_2	0.4	0%	30%
ω_3	0.4	20%	-5%

We want to compare the risk of a portfolio such that $w_1 = 40\%$ and $w_2 = 60\%$ with the risk of each of its components. Direct computations give

$$\sigma_1^2 \cong 0.0144, \quad \sigma_2^2 \cong 0.0254, \quad \rho_{12} \cong -0.6065.$$

By (3.6)

$$\sigma_V^2 \cong (0.4)^2 \times 0.0144 + (0.6)^2 \times 0.0254$$
$$+2 \times 0.4 \times 0.6 \times (-0.6065) \times \sqrt{0.0144} \times \sqrt{0.0254}$$
$$\cong 0.00588.$$

Observe that the variance σ_V^2 is smaller than both σ_1^2 and σ_2^2.

Example 3.14

Consider another portfolio with weights $w_1 = 10\%$ and $w_2 = 90\%$, all other things being the same as in Example 3.13. Then

$$\sigma_V^2 \cong (0.1)^2 \times 0.0144 + (0.9)^2 \times 0.0254$$
$$+ 2 \times 0.1 \times 0.9 \times (-0.6065) \times \sqrt{0.0144} \times \sqrt{0.0254}$$
$$\cong 0.01863,$$

which is between σ_1^2 and σ_2^2.

Proposition 3.15

The variance σ_V^2 of a portfolio cannot exceed the greater of the variances σ_1^2 and σ_2^2 of the components,

$$\sigma_V^2 \le \max\{\sigma_1^2, \sigma_2^2\},$$

if short sales are not allowed.

Proof

Let us assume that $\sigma_1^2 \le \sigma_2^2$. If short sales are not allowed, then $w_1, w_2 \ge 0$ and

$$w_1\sigma_1 + w_2\sigma_2 \le (w_1 + w_2)\sigma_2 = \sigma_2.$$

Since the correlation coefficient satisfies $-1 \le \rho_{12} \le 1$, it follows that

$$\sigma_V^2 = w_1^2\sigma_1^2 + w_2^2\sigma_2^2 + 2w_1w_2\rho_{12}\sigma_1\sigma_2$$
$$\le w_1^2\sigma_1^2 + w_2^2\sigma_2^2 + 2w_1w_2\sigma_1\sigma_2$$
$$= (w_1\sigma_1 + w_2\sigma_2)^2 \le \sigma_2^2.$$

If $\sigma_1^2 \ge \sigma_2^2$, the proof is analogous. □

Example 3.16

Now consider a portfolio with weights $w_1 = -50\%$ and $w_2 = 150\%$ (allowing short sales of security 1), all the remaining data being the same as in Example 3.13. The variance of this portfolio is

$$\sigma_V^2 \cong (-0.5)^2 \times 0.0144 + (1.5)^2 \times 0.0254$$
$$+ 2 \times (-0.5) \times 1.5 \times (-0.6065) \times \sqrt{0.0144} \times \sqrt{0.0254}$$
$$\cong 0.2795,$$

which is greater than both σ_1^2 and σ_2^2.

Exercise 3.7

Using the data in Example 3.13, find the weights in a portfolio with expected return $\mu_V = 50\%$ and compute the risk σ_V of this portfolio.

3.2.2 Feasible Set

The collection of all portfolios that can be manufactured by investing in two given assets is called the *feasible* (or *attainable*) *set*. Each portfolio can be represented by a point with coordinates σ_V and μ_V in the σ, μ plane.

We want to investigate the shape of the feasible set in this plane, which will allow valuable insights. The set consists of all points with coordinates

$$\mu_V = w_1\mu_1 + w_2\mu_2,$$
$$\sigma_V^2 = w_1^2\sigma_1^2 + w_2^2\sigma_2^2 + 2w_1w_2c_{12},$$

where $w_1, w_2 \in \mathbb{R}$ and

$$1 = w_1 + w_2.$$

Portfolios in the feasible set can be parameterised by one of the weights. Here we shall use $s = w_1$ as a parameter. Then $1 - s = w_2$, and the above expressions for σ_V^2 and μ_V can be written as

$$\mu_V = s\mu_1 + (1 - s)\mu_2, \tag{3.8}$$
$$\sigma_V^2 = s^2\sigma_1^2 + (1 - s)^2\sigma_2^2 + 2s(1 - s)c_{12}, \tag{3.9}$$

where $s \in \mathbb{R}$.

We begin with finding the portfolio with the smallest variance (that is, smallest risk) among all feasible portfolios.

Proposition 3.17

If $\rho_{12} < 1$ or $\sigma_1 \neq \sigma_2$, then σ_V^2 as a function of s attains its minimum value at

$$s_0 = \frac{\sigma_2^2 - c_{12}}{\sigma_1^2 + \sigma_2^2 - 2c_{12}}. \tag{3.10}$$

The corresponding values of the expected return μ_V and variance σ_V^2 are

$$\mu_0 = \frac{\mu_1\sigma_2^2 + \mu_2\sigma_1^2 - (\mu_1 + \mu_2)c_{12}}{\sigma_1^2 + \sigma_2^2 - 2c_{12}}, \tag{3.11}$$

$$\sigma_0^2 = \frac{\sigma_1^2\sigma_2^2 - c_{12}}{\sigma_1^2 + \sigma_2^2 - 2c_{12}}. \tag{3.12}$$

If $\rho_{12} = 1$ and $\sigma_1 = \sigma_2$, then all feasible portfolios have the same variance equal to $\sigma_1^2 = \sigma_2^2$.

Proof

To find the value of s for which σ_V^2 in (3.9) attains a minimum we differentiate σ_V^2 with respect to s and equate the derivative to zero. This gives an equation for s,

$$\frac{d(\sigma_V^2)}{ds} = 2s\left(\sigma_1^2 + \sigma_2^2 - 2c_{12}\right) - 2\left(\sigma_2^2 - c_{12}\right) = 0,$$

which has a solution s_0 given by (3.10) as long as the denominator in (3.10) is non-zero. This is guaranteed by the condition that $\rho_{12} < 1$ or $\sigma_1 \neq \sigma_2$. Indeed, if $\rho_{12} < 1$, then

$$\sigma_1^2 + \sigma_2^2 - 2c_{12} > \sigma_1^2 + \sigma_2^2 - 2\sigma_1\sigma_2 = (\sigma_1 - \sigma_2)^2 \geq 0,$$

and if $\sigma_1 \neq \sigma_2$, then

$$\sigma_1^2 + \sigma_2^2 - 2c_{12} \geq \sigma_1^2 + \sigma_2^2 - 2\sigma_1\sigma_2 = (\sigma_1 - \sigma_2)^2 > 0.$$

For the second derivative this implies that

$$\frac{d^2(\sigma_V^2)}{ds^2} = 2\left(\sigma_1^2 + \sigma_2^2 - 2c_{12}\right) > 0.$$

We can conclude that σ_V^2 attains its minimum at s_0. The expressions for μ_0 and σ_0^2 follow by substituting s_0 for s in (3.8) and (3.9). If $\rho_{12} = 1$ and $\sigma_1 = \sigma_2$, then $c_{12} = \sigma_1\sigma_2$, and for any $s \in \mathbb{R}$

$$\sigma_V^2 = s^2\sigma_1^2 + (1-s)^2\sigma_2^2 + 2s(1-s)c_{12} = \left(s\sigma_1 + (1-s)\sigma_2\right)^2 = \sigma_1^2 = \sigma_2^2.$$

$\square$

The curve described by the parametric equations (3.8), (3.9), which represents the feasible set on the σ, μ plane, turns out to be a branch of a hyperbola. This is shown in Figure 3.1 along with the asymptotes of the hyperbola. The bold segment of the hyperbola corresponds to portfolios without short selling.

Proposition 3.18

Assume that $-1 < \rho_{12} < 1$ and $\mu_1 \neq \mu_2$. Then for each portfolio V in the feasible set, $x = \sigma_V$ and $y = \mu_V$ satisfy the equation of a hyperbola

$$x^2 - A^2\left(y - \mu_0\right)^2 = \sigma_0^2 \tag{3.13}$$

with μ_0 and $\sigma_0^2 > 0$ given in Proposition 3.17 and with

$$A^2 = \frac{\sigma_1^2 + \sigma_2^2 - 2c_{12}}{(\mu_1 - \mu_2)^2} > 0.$$

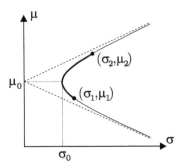

Figure 3.1 Hyperbola representing feasible portfolios on the σ, μ plane

The two asymptotes of the hyperbola are

$$y = \mu_0 \pm \frac{1}{A}x.$$

Proof

From (3.8) we have $s = \frac{\mu_V - \mu_2}{\mu_1 - \mu_2}$. On substituting this expression for s into (3.9), it follows by simple (if slightly tedious) transformations that $x = \sigma_V$ and $y = \mu_V$ satisfy (3.13). Since $-1 < \rho_{12} < 1$ and $\mu_1 \neq \mu_2$, it follows that $A^2 > 0$ and $\sigma_0^2 > 0$, so that (3.13) indeed describes a hyperbola. The equation for the asymptotes follows directly from (3.13).   □

The correlation coefficient always satisfies $-1 \leq \rho_{12} \leq 1$. The next proposition is concerned with the two special cases when ρ_{12} assumes one of the extreme values 1 or -1. This means perfect positive or negative correlation between the securities in the portfolio. In these cases the feasible set will have a particularly simple shape, see Figure 3.2.

Proposition 3.19

Suppose that $\rho_{12} = 1$ and $\sigma_1 \neq \sigma_2$. Then $\sigma_V = 0$ if and only if

$$w_1 = -\frac{\sigma_2}{\sigma_1 - \sigma_2}, \quad w_2 = \frac{\sigma_1}{\sigma_1 - \sigma_2}. \tag{3.14}$$

This involves short selling, since either w_1 or w_2 is negative.

Suppose that $\rho_{12} = -1$. Then $\sigma_V = 0$ if and only if

$$w_1 = \frac{\sigma_2}{\sigma_1 + \sigma_2}, \quad w_2 = \frac{\sigma_1}{\sigma_1 + \sigma_2}. \tag{3.15}$$

No short selling is necessary, since both w_1 and w_2 are positive.

Proof

Let $\rho_{12} = 1$ and $\sigma_1 \neq \sigma_2$. Then (3.6) takes the form

$$\sigma_V^2 = w_1^2\sigma_1^2 + w_2^2\sigma_2^2 + 2w_1w_2\sigma_1\sigma_2 = (w_1\sigma_1 + w_2\sigma_2)^2$$

and $\sigma_V^2 = 0$ if and only if $w_1\sigma_1 + w_2\sigma_2 = 0$. This is equivalent to (3.14) because $w_1 + w_2 = 1$.

Now let $\rho_{12} = -1$. Then (3.6) becomes

$$\sigma_V^2 = w_1^2\sigma_1^2 + w_2^2\sigma_2^2 - 2w_1w_2\sigma_1\sigma_2 = (w_1\sigma_1 - w_2\sigma_2)^2$$

and $\sigma_V^2 = 0$ if and only if $w_1\sigma_1 - w_2\sigma_2 = 0$. The last equality is equivalent to (3.15) because $w_1 + w_2 = 1$. (We know that $\sigma_1 + \sigma_2 > 0$ because both securities are risky.) $\qquad\square$

Figure 3.2 shows two typical lines representing portfolios with $\rho_{12} = -1$ (left) and $\rho_{12} = 1$ (right). The bold segments correspond to portfolios without short selling.

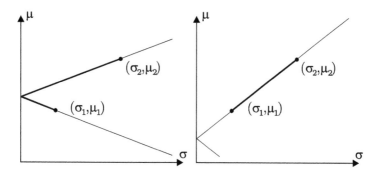

Figure 3.2 Typical portfolio lines with $\rho_{12} = -1$ and 1

Suppose that $\rho_{12} = -1$. It follows from the proof of Proposition 3.19 that $\sigma_V = |w_1\sigma_1 - w_2\sigma_2|$. In addition, $\mu_V = w_1\mu_1 + w_2\mu_2$ by (3.5) and $w_1 + w_2 = 1$ by (3.1). We can choose $s = w_1$ as a parameter. Then $1 - s = w_2$ and

$$\sigma_V = |s\sigma_1 - (1 - s)\sigma_2|,$$
$$\mu_V = s\mu_1 + (1 - s)\mu_2.$$

These parametric equations describe the line in Figure 3.2 with a broken segment between (σ_1, μ_1) and (σ_2, μ_2). As s decreases, the point (σ_V, μ_V) moves along the line in the direction from (σ_1, μ_1) to (σ_2, μ_2).

If $\rho_{12} = 1$, then $\sigma_V = |w_1\sigma_1 + w_2\sigma_2|$. We choose $s = w_1$ as a parameter once again, and obtain the parametric equations

$$\sigma_V = |s\sigma_1 + (1-s)\sigma_2|,$$
$$\mu_V = s\mu_1 + (1-s)\mu_2$$

of the line in Figure 3.2 with a straight segment between (σ_1, μ_1) and (σ_2, μ_2).

Exercise 3.8

Suppose that there are just two scenarios ω_1 and ω_2 and consider two risky securities with returns K_1 and K_2. Show that $K_1 = aK_2 + b$ for some numbers $a \neq 0$ and b, and deduce that $\rho_{12} = 1$ or -1.

Figure 3.3 illustrates the following corollary, in which we study the shape of the feasible set depending on the correlation coefficient ρ_{12}.

Corollary 3.20

Suppose that $\sigma_1 \leq \sigma_2$. The following five cases are possible:

1) If $\rho_{12} = 1$, then there is a feasible portfolio V with short selling such that $\sigma_V = 0$ (line 1 in Figure 3.3) whenever $\sigma_1 < \sigma_2$. Each portfolio V in the feasible set has the same σ_V whenever $\sigma_1 = \sigma_2$.

2) If $\frac{\sigma_1}{\sigma_2} < \rho_{12} < 1$, then there is a feasible portfolio V with short selling such that $\sigma_V < \sigma_1$, but for each portfolio without short selling $\sigma_V \geq \sigma_1$ (line 2 in Figure 3.3).

3) If $\rho_{12} = \frac{\sigma_1}{\sigma_2}$, then $\sigma_V \geq \sigma_1$ for each feasible portfolio V (line 3 in Figure 3.3).

4) If $-1 < \rho_{12} < \frac{\sigma_1}{\sigma_2}$, then there is a feasible portfolio V without short selling such that $\sigma_V < \sigma_1$ (line 4 in Figure 3.3).

5) If $\rho_{12} = -1$, then there is a feasible portfolio V without short selling such that $\sigma_V = 0$ (line 5 in Figure 3.3).

Proof

1) The case when $\rho_{12} = 1$ is covered in Proposition 3.19.

2) If $\frac{\sigma_1}{\sigma_2} < \rho_{12} < 1$, then $s_0 > 1$. In this case the portfolio V with minimum variance that corresponds to the parameter s_0 involves short selling of security 2 and satisfies $\sigma_V < \sigma_1$. For $s \geq s_0$ the variance σ_V is an increasing function of s, which means that $\sigma_V \geq \sigma_1$ for every portfolio without short selling.

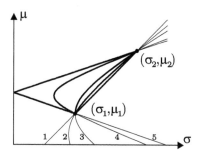

Figure 3.3 Portfolio lines for various values of ρ_{12}

3) If $\rho_{12} = \frac{\sigma_1}{\sigma_2}$, then $s_0 = 1$. As a result, $\sigma_V \geq \sigma_1$ for every portfolio V because σ_1^2 is the minimum variance.

4) If $-1 < \rho_{12} < \frac{\sigma_1}{\sigma_2}$, then $0 < s_0 < 1$, which means that the portfolio V with minimum variance, which corresponds to s_0, involves no short selling and satisfies $\sigma_V < \sigma_1$.

5) The case when $\rho_{12} = -1$ has been covered in Proposition 3.19.

$\square$

The above corollary is important because it shows when it is possible to construct a portfolio with risk lower than that of either of its components. In cases 4) and 5) this is possible without short selling. In cases 1) and 2) this is also possible, but only if short selling is allowed. In case 3) it is impossible to construct such a portfolio.

Example 3.21

Suppose that

$$\sigma_1^2 = 0.0041, \quad \sigma_2^2 = 0.0121, \quad \rho_{12} = 0.9796.$$

Clearly, $\sigma_1 < \sigma_2$ and $\frac{\sigma_1}{\sigma_2} < \rho_{12} < 1$, so this is case 2) in Corollary 3.20. Our task will be to find the portfolio with minimum risk with and without short selling. Using Proposition 3.17, we compute

$$s_0 \cong 2.1663.$$

It follows that in the portfolio with minimum risk the weights should be $w_1 \cong 2.1663$ and $w_2 \cong -1.1663$ if short selling is allowed. Without short selling $w_1 = 1$ and $w_2 = 0$.

Exercise 3.9

Compute the weights in the portfolio with minimum risk for the data in Example 3.13. Does this portfolio involve short selling?

We conclude this section with a brief discussion of portfolios in which one of the securities is risk free. The variance of the risky security is positive, whereas that of the risk-free component is zero.

Proposition 3.22

The standard deviation σ_V of a portfolio consisting of a risky security with expected return μ_1 and standard deviation $\sigma_1 > 0$, and a risk-free security with return R and standard deviation zero depends on the weight w_1 of the risky security as follows:

$$\sigma_V = |w_1| \sigma_1.$$

Proof

Let $\sigma_1 > 0$ and $\sigma_2 = 0$. Then (3.6) reduces to $\sigma_V^2 = w_1^2 \sigma_1^2$, and the formula for σ_V follows by taking the square root.   □

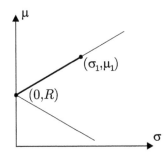

Figure 3.4 Portfolio line for one risky and one risk-free security

The line on the σ, μ plane representing portfolios constructed from one risky and one risk-free security is shown in Figure 3.4. As before, the bold line segment corresponds to portfolios without short selling.

3.3 Several Securities

3.3.1 Risk and Expected Return on a Portfolio

A portfolio constructed from n different securities can be described in terms of their weights

$$w_i = \frac{x_i S_i(0)}{V(0)}, \quad i = 1, \ldots, n,$$

where x_i is the number of shares of type i in the portfolio, $S_i(0)$ is the initial price of security i, and $V(0)$ is the amount initially invested in the portfolio. It will prove convenient to arrange the weights into a one-row matrix

$$\mathbf{w} = \begin{bmatrix} w_1 & w_2 & \cdots & w_n \end{bmatrix}.$$

Just like for two securities, the weights add up to one, which can be written in matrix form as

$$1 = \mathbf{wu}^{\mathrm{T}}, \tag{3.16}$$

where

$$\mathbf{u} = \begin{bmatrix} 1 & 1 & \cdots & 1 \end{bmatrix}$$

is a one-row matrix with all n entries equal to 1, $\mathbf{u}^{\mathrm{T}}$ is a one-column matrix, the transpose of $\mathbf{u}$, and the usual matrix multiplication rules apply. The *feasible* (or *attainable*) *set* consists of all portfolios with weights $\mathbf{w}$ satisfying (3.16), called the *feasible* (or *attainable*) *portfolios*.

Suppose that the returns on the securities are $K_1, \ldots, K_n$. The expected returns $\mu_i = \mathbb{E}(K_i)$ for $i = 1, \ldots, n$ will also be arranged into a one-row matrix

$$\mathbf{m} = \begin{bmatrix} \mu_1 & \mu_2 & \cdots & \mu_n \end{bmatrix}.$$

The covariances between returns will be denoted by $c_{ij} = \mathrm{Cov}(K_i, K_j)$. They are the entries of the $n \times n$ *covariance matrix*

$$\mathbf{C} = \begin{bmatrix} c_{11} & c_{12} & \cdots & c_{1n} \\ c_{21} & c_{22} & \cdots & c_{2n} \\ \vdots & \vdots & \ddots & \vdots \\ c_{n1} & c_{n2} & \cdots & c_{nn} \end{bmatrix}.$$

The diagonal elements of $\mathbf{C}$ are simply the variances of returns, $c_{ii} = \sigma_i = \mathrm{Var}(K_i)$.

The covariance matrix is symmetric and non-negative definite, see Appendix 10.4. In what follows, we shall assume that $\det \mathbf{C} \neq 0$, which implies that $\mathbf{C}$ has an inverse $\mathbf{C}^{-1}$.

Formula (3.2) for the return on a portfolio extends to the case of n securities:

$$K_V = w_1 K_1 + \cdots + w_n K_n.$$

Proposition 3.23

The expected return $\mu_V = \mathbb{E}(K_V)$ and variance $\sigma_V^2 = \text{Var}(K_V)$ of the return $K_V = w_1 K_1 + \cdots + w_n K_n$ on a portfolio V with weights $\mathbf{w} = [w_1, \ldots, w_n]$ are given by

$$\mu_V = \mathbf{wm}^{\mathrm{T}}, \tag{3.17}$$
$$\sigma_V^2 = \mathbf{wCw}^{\mathrm{T}}. \tag{3.18}$$

Proof

The formula for μ_V follows by the linearity of expectation,

$$\mu_V = \mathbb{E}(K_V) = \mathbb{E}\left(\sum_{i=1}^{n} w_i K_i\right) = \sum_{i=1}^{n} w_i \mu_i = \mathbf{wm}^{\mathrm{T}}.$$

For σ_V^2 we use the linearity of covariance with respect to each of its arguments,

$$\sigma_V^2 = \text{Var}(K_V) = \text{Var}\left(\sum_{i=1}^{n} w_i K_i\right)$$

$$= \text{Cov}\left(\sum_{i=1}^{n} w_i K_i, \sum_{j=1}^{n} w_j K_j\right) = \sum_{i,j=1}^{n} w_i w_j c_{ij}$$

$$= \mathbf{wCw}^{\mathrm{T}}.$$

$\square$

Exercise 3.10

Compute the expected return μ_V and standard deviation σ_V of a portfolio consisting of three securities with weights $w_1 = 40\%$, $w_2 = -20\%$, $w_3 = 80\%$, given that the securities have expected returns $\mu_1 = 8\%$, $\mu_2 = 10\%$, $\mu_3 = 6\%$, standard deviations $\sigma_1 = 0.15$, $\sigma_2 = 0.05$, $\sigma_3 = 0.12$ and correlations $\rho_{12} = 0.3$, $\rho_{23} = 0.0$, $\rho_{31} = -0.2$.

3.3.2 Minimum Variance Portfolio

In Proposition 3.17 we found the portfolio with the smallest variance (that is, smallest risk) among all feasible portfolios constructed from two risky assets. Here we shall extend this result to the case of n risky assets.

The portfolio with the smallest variance among all feasible portfolios will be called the *minimum variance portfolio* (MVP). To find this portfolio we need to minimise the variance $\sigma_V^2 = \mathbf{wCw}^T$ over all weights $\mathbf{w}$. Because the weights must add up to 1 this leads to a constrained minimum problem:

$$\min \mathbf{wCw}^T,$$

where the minimum is taken over all vectors $\mathbf{w} \in \mathbb{R}$ that satisfy the condition

$$\mathbf{wu}^T = 1.$$

To compute this constrained minimum we can use the method of Lagrange multipliers, see Appendix 10.1.2.

Proposition 3.24

Assume that $\det \mathbf{C} \neq 0$. Then the minimum variance portfolio has weights

$$\mathbf{w}_{\text{MVP}} = \frac{\mathbf{uC}^{-1}}{\mathbf{uC}^{-1}\mathbf{u}^T}. \tag{3.19}$$

Proof

We need to find the minimum of $\mathbf{wCw}^T$ subject to the constraint $\mathbf{wu}^T = 1$. According to the method of Lagrange multipliers, we put

$$F(\mathbf{w}, \lambda) = \mathbf{wCw}^T - \lambda(\mathbf{wu}^T - 1).$$

The first order necessary condition gives $2\mathbf{wC} - \lambda \mathbf{u} = 0$, hence

$$\mathbf{w} = \frac{\lambda}{2}\mathbf{uC}^{-1}.$$

Substituting this into the constraint, we obtain $\frac{\lambda}{2}\mathbf{uC}^{-1}\mathbf{u}^T = 1$. The properties of the covariance matrix guarantee that $\mathbf{uC}^{-1}\mathbf{u}^T \neq 0$, see Appendix 10.4. Therefore, we can compute the Lagrange multiplier λ and substitute the result into the expression for $\mathbf{w}$ to obtain the formula for $\mathbf{w}_{\text{MVP}}$.

We have verified a necessary condition for a minimum. Observe that $\mathbf{wCw}^T$ is a quadratic function of the weights, bounded below by 0. It must therefore have a minimum, which completes the proof. □

Exercise 3.11

Among all feasible portfolios constructed using three securities with expected returns $\mu_1 = 0.20$, $\mu_2 = 0.13$, $\mu_3 = 0.17$, standard deviations of returns $\sigma_1 = 0.25$, $\sigma_2 = 0.28$, $\sigma_3 = 0.20$, and correlations between returns $\rho_{12} = 0.30$, $\rho_{23} = 0.00$, $\rho_{31} = 0.15$, find the minimum variance portfolio. What are the weights in this portfolio? Also compute the expected return and standard deviation of this portfolio.

3.3.3 Efficient Frontier

Given the choice between two securities a rational investor will, if possible, choose that with the higher expected return and lower standard deviation, that is, lower risk. This motivates the following definition.

Definition 3.25

We say that a security with expected return μ_1 and standard deviation σ_1 *dominates* another security with expected return μ_2 and standard deviation σ_2 whenever

$$\mu_1 \geq \mu_2 \quad \text{and} \quad \sigma_1 \leq \sigma_2.$$

This definition readily extends to portfolios, which can be considered as securities in their own right.

The ordering is only partial. For example, for two securities with $\mu_1 = 10\%, \sigma_1 = 20\%$ and $\mu_2 = 15\%, \sigma_2 = 22\%$, neither of them dominates the other.

Definition 3.26

A portfolio is called *efficient* if there is no other portfolio, except itself, that dominates it. The subset of efficient portfolios among all feasible portfolios is called the *efficient frontier*.

We assume that every rational investor prefers dominating portfolios to dominated ones. However, different investors may select different portfolios on the efficient frontier, depending on their individual preferences. Given two efficient portfolios with $\mu_1 \leq \mu_2$ and $\sigma_1 \leq \sigma_2$, a cautious person may prefer that with lower risk σ_1 and lower expected return μ_1, while others may choose a portfolio with higher risk σ_2, regarding the higher expected return μ_2 as compensation for increased risk.

To find the efficient frontier we have to recognise and eliminate the dominated portfolios. To this end, for any $\mu \in \mathbb{R}$ consider all feasible portfolios whose expected return is μ. The set of such portfolios is called an *isoexpected line*. All portfolios on an isoexpected line are dominated by that with the smallest variance among them.

Definition 3.27

The family of portfolios V, parameterised by $\mu \in \mathbb{R}$, such that $\mu_V = \mu$ and $\sigma_V^2 \leq \sigma_{V'}^2$ for each portfolio V' with $\mu_{V'} = \mu$ is called the *minimum variance line* (MVL).

To compute the weights of the portfolio on the minimum variance line for any $\mu \in \mathbb{R}$ we shall solve the constrained minimum problem

$$\min \mathbf{w}\mathbf{C}\mathbf{w}^\mathrm{T},$$

where the minimum is taken over all vectors $\mathbf{w} \in \mathbb{R}^n$ such that

$$\mathbf{w}\mathbf{m}^\mathrm{T} = \mu, \quad \mathbf{w}\mathbf{u}^\mathrm{T} = 1.$$

We introduce Lagrange multipliers λ_1, λ_2 and minimise the function

$$G(\mathbf{w}, \lambda_1, \lambda_2) = \mathbf{w}\mathbf{C}\mathbf{w}^\mathrm{T} - \lambda_1(\mathbf{w}\mathbf{m}^\mathrm{T} - \mu) - \lambda_2(\mathbf{w}\mathbf{u}^\mathrm{T} - 1).$$

The gradient of G with respect to $\mathbf{w}$ and the partial derivatives with respect to λ_1, λ_2 give necessary conditions for a minimum, see Appendix 10.1.2:

$$2\mathbf{w}\mathbf{C} - \lambda_1\mathbf{m} - \lambda_2\mathbf{u} = \mathbf{0},$$
$$\mathbf{w}\mathbf{m}^\mathrm{T} - \mu = 0,$$
$$\mathbf{w}\mathbf{u}^\mathrm{T} - 1 = 0.$$

The first of these is a system of n equations, $\mathbf{0}$ representing a vector of zeros. Assuming invertibility of $\mathbf{C}$, we solve this system for $\mathbf{w}$

$$2\mathbf{w} = \lambda_1\mathbf{m}\mathbf{C}^{-1} + \lambda_2\mathbf{u}\mathbf{C}^{-1}. \tag{3.20}$$

Multiplying the last equality on the right by $\mathbf{m}^\mathrm{T}$ and, respectively, by $\mathbf{u}^\mathrm{T}$, we obtain a system of linear equations for the Lagrange multipliers:

$$\lambda_1\mathbf{m}\mathbf{C}^{-1}\mathbf{m}^\mathrm{T} + \lambda_2\mathbf{u}\mathbf{C}^{-1}\mathbf{m}^\mathrm{T} = 2\mu,$$
$$\lambda_1\mathbf{m}\mathbf{C}^{-1}\mathbf{u}^\mathrm{T} + \lambda_2\mathbf{u}\mathbf{C}^{-1}\mathbf{u}^\mathrm{T} = 2.$$

The matrix of coefficients of this system will be denoted by

$$M = \begin{bmatrix} \mathbf{m}\mathbf{C}^{-1}\mathbf{m}^\mathrm{T} & \mathbf{u}\mathbf{C}^{-1}\mathbf{m}^\mathrm{T} \\ \mathbf{m}\mathbf{C}^{-1}\mathbf{u}^\mathrm{T} & \mathbf{u}\mathbf{C}^{-1}\mathbf{u}^\mathrm{T} \end{bmatrix}.$$

If the vectors $\mathbf{u}, \mathbf{m}$ are linearly independent, then M is an invertible matrix, see Appendix 10.4. We can solve the system of linear equations for λ_1 and λ_2, obtaining a linear expression in μ:

$$\begin{bmatrix} \lambda_1 \\ \lambda_2 \end{bmatrix} = 2M^{-1}\begin{bmatrix} \mu \\ 1 \end{bmatrix}. \tag{3.21}$$

Inserting this into (3.20), we get

$$\mathbf{w} = \mu\mathbf{a} + \mathbf{b} \tag{3.22}$$

for some vectors $\mathbf{a}, \mathbf{b} \in \mathbb{R}^n$. The important point here is that the vectors $\mathbf{a}, \mathbf{b}$ are the same for each portfolio on the minimum variance line. They can be expressed in terms of $\mathbf{C}$, $\mathbf{m}$ and $\mathbf{u}$.

We can summarise the above results as a theorem.

Theorem 3.28

Assume that $\det \mathbf{C} \neq 0$ and $\mathbf{m}, \mathbf{u}$ are linearly independent vectors. Then a vector of weights $\mathbf{w}$ represents a portfolio V on the minimum variance line if and only if $\mathbf{w}$ is of the form (3.20) with λ_1, λ_2 given by (3.21) and with $\mu = \mu_V$.

Proof

The above argument shows that (3.20) with (3.21) is a necessary condition for minimising the risk at a given level μ of expected return. Sufficiency follows from the fact that the minimisation problem has a unique solution for each μ since the function minimised is quadratic in $\mathbf{w}$ and bounded below by 0. $\square$

Example 3.29

Consider three securities with expected returns, standard deviations of returns and correlations between returns

$$\begin{aligned} \mu_1 &= 0.10, & \sigma_1 &= 0.28, & \rho_{12} = \rho_{21} &= -0.10, \\ \mu_2 &= 0.15, & \sigma_2 &= 0.24, & \rho_{23} = \rho_{32} &= 0.20, \\ \mu_3 &= 0.20, & \sigma_3 &= 0.25, & \rho_{31} = \rho_{13} &= 0.25. \end{aligned}$$

We arrange the μ_i's into a one-row matrix $\mathbf{m}$ and 1's into a one-row matrix $\mathbf{u}$,

$$\mathbf{m} = \begin{bmatrix} 0.10 & 0.15 & 0.20 \end{bmatrix}, \quad \mathbf{u} = \begin{bmatrix} 1 & 1 & 1 \end{bmatrix}.$$

Next we compute the entries $c_{ij} = \rho_{ij}\sigma_i\sigma_j$ of the covariance matrix $\mathbf{C}$, and find the inverse matrix $\mathbf{C}^{-1}$,

$$\mathbf{C} \cong \begin{bmatrix} 0.0784 & -0.0067 & 0.0175 \\ -0.0067 & 0.0576 & 0.0120 \\ 0.0175 & 0.0120 & 0.0625 \end{bmatrix}, \quad \mathbf{C}^{-1} \cong \begin{bmatrix} 13.954 & 2.544 & -4.396 \\ 2.544 & 18.548 & -4.274 \\ -4.396 & -4.274 & 18.051 \end{bmatrix}.$$

This gives

$$\mathbf{u}\mathbf{C}^{-1} \cong \begin{bmatrix} 12.102 & 16.818 & 9.382 \end{bmatrix},$$
$$\mathbf{m}\mathbf{C}^{-1} \cong \begin{bmatrix} 0.898 & 2.182 & 2.530 \end{bmatrix},$$

and

$$M = \left[\begin{array}{cc} \mathbf{mC}^{-1}\mathbf{m}^{\mathrm{T}} & \mathbf{uC}^{-1}\mathbf{m}^{\mathrm{T}} \\ \mathbf{mC}^{-1}\mathbf{u}^{\mathrm{T}} & \mathbf{uC}^{-1}\mathbf{u}^{\mathrm{T}} \end{array} \right] \cong \left[\begin{array}{cc} 0.923 & 5.609 \\ 5.609 & 38.302 \end{array} \right],$$

$$M^{-1} \cong \left[\begin{array}{cc} 9.842 & -1.441 \\ -1.441 & 0.237 \end{array} \right].$$

The weights in the minimum variance portfolio can be found using (3.19):

$$\mathbf{w} = \frac{\mathbf{uC}^{-1}}{\mathbf{uC}^{-1}\mathbf{u}^{\mathrm{T}}} \cong \left[\begin{array}{ccc} 0.316 & 0.439 & 0.245 \end{array} \right].$$

The expected return and standard deviation of this portfolio are

$$\mu_V = \mathbf{mw}^{\mathrm{T}} \cong 0.146, \quad \sigma_V = \sqrt{\mathbf{wCw}^{\mathrm{T}}} \cong 0.162.$$

To obtain the weights $\mathbf{w}$ of the portfolio on the minimum variance line with expected return μ_V we apply formula (3.22):

$$\mathbf{w} \cong \mu_V \mathbf{a} + \mathbf{b},$$

where

$$\mathbf{a} = \left[\begin{array}{ccc} -8.614 & -2.769 & 11.384 \end{array} \right], \quad \mathbf{b} = \left[\begin{array}{ccc} 1.578 & 0.845 & -1.422 \end{array} \right].$$

The standard deviation of this portfolio is

$$\sigma_V = \sqrt{\mathbf{wCw}^{\mathrm{T}}} \cong \sqrt{0.237 - 2.885\mu_V + 9.850\mu_V^2}.$$

Exercise 3.12

Among all feasible portfolios with expected return $\mu_V = 20\%$ constructed using the three securities in Exercise 3.11 find the portfolio with the smallest variance. Compute the weights and the standard deviation of this portfolio.

The following result is an important consequence of the linear dependence (3.22) of the weights of portfolios belonging to the minimum variance line on the expected return.

Theorem 3.30 (Two-Fund Theorem)

Under the assumptions of Theorem 3.28, let w_1, w_2 be the weights of any two portfolios V_1, V_2 on the minimum variance line with different expected returns

$\mu_{V_1} \neq \mu_{V_2}$. Then each portfolio V on the minimum variance line can be obtained as a linear combination of these two, that is,

$$\mathbf{w} = \alpha \mathbf{w}_1 + (1 - \alpha)\mathbf{w}_2$$

for some $\alpha \in \mathbb{R}$.

Proof

We find α so that

$$\mu_V = \alpha \mu_{V_1} + (1 - \alpha)\mu_{V_2}.$$

This is possible since the returns are different:

$$\alpha = \frac{\mu_V - \mu_{V_2}}{\mu_{V_1} - \mu_{V_2}}.$$

Since the two portfolios V_1, V_2 belong to the minimum variance line, they satisfy

$$\mathbf{w}_1 = \mu_{V_1}\mathbf{a} + \mathbf{b}, \quad \mathbf{w}_2 = \mu_{V_2}\mathbf{a} + \mathbf{b}.$$

As a result, because V also belongs to the minimum variance line

$$\alpha \mathbf{w}_1 + (1 - \alpha)\mathbf{w}_2 = (\alpha \mu_{V_1} + (1 - \alpha)\mu_{V_2})\mathbf{a} + \mathbf{b} = \mu_V \mathbf{a} + \mathbf{b} = \mathbf{w}.$$

$\square$

The practical significance of the two-fund theorem stems from the fact that any portfolio on the minimum variance line can be realised by splitting available wealth between just two different investment funds (two different portfolios) instead of investing in individual assets. Trading in the units of only two investment funds can significantly reduce the costs and simplify the procedures as compared with simultaneous transactions in n risky assets.

Another consequence of the two-fund theorem is that the minimum variance line can be viewed as if it were a two-asset line constructed from the portfolios V_1 and V_2 (regarded for this purpose as if they were individual risky assets). From the results of Section 3.2 we can therefore conclude that the minimum variance line is a hyperbola of the kind described in Proposition 3.18. We can therefore conclude that the efficient frontier consists of portfolios with weights of the form $\mathbf{w} = \mu \mathbf{a} + \mathbf{b}$ for μ greater than or equal to expected return on the minimal variance portfolio.

Example 3.31

There are two convenient ways to visualise all portfolios that can be constructed from the three securities in Example 3.29. One is presented in Figure 3.5. Here

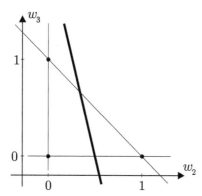

Figure 3.5 Feasible portfolios on the w_2, w_3 plane

two of the three weights, namely w_2 and w_3, are used as parameters. The remaining weight is given by $w_1 = 1 - w_2 - w_3$. (Of course any other two weights can also be used as parameters.) Each point on the w_2, w_3 plane represents a different portfolio. The vertices of the triangle represent the portfolios consisting of only one of the three securities. For example, the vertex with coordinates $(1, 0)$ corresponds to weights $w_1 = 0$, $w_2 = 1$ and $w_3 = 0$, that is, represents a portfolio with all money invested in security 2. The straight lines through the vertices correspond to portfolios consisting of two securities only. For example, the line through $(1, 0)$ and $(0, 1)$ corresponds to portfolios containing securities 2 and 3 only. Points inside the triangle, including the boundaries, correspond to portfolios without short selling. For example, $(\frac{2}{5}, \frac{1}{2})$ represents a portfolio with 10% of the initial funds invested in security 1, 40% in security 2, and 50% in security 3. Points outside the triangle correspond to portfolios with one or two of the three securities shorted. The minimum variance line is a straight line because of the linear dependence of the weights on the expected return. It is represented by the bold line in Figure 3.5.

Figure 3.6 shows another way to visualise feasible portfolios by plotting the expected return of a portfolio against the standard deviation. This is sometimes called the risk–expected return graph. The three points indicated in this picture correspond to portfolios consisting of only one of the three securities. For instance, the portfolio with all funds invested in security 2 is represented by the point $(0.24, 0.15)$. The hyperbolas passing through a pair of these three points correspond to portfolios consisting of just two securities. These are the two-security lines studied in detail in Section 3.2. For example, all portfolios containing securities 2 and 3 only lie on the line through $(0.24, 0.15)$ and $(0.25, 0.20)$. The three points and the lines passing through them correspond to the vertices of the triangle and the straight lines passing through them in Figure 3.5. The shaded area (both dark and light), including the boundary, represents portfolios that can be constructed from the three securities, that is, all feasible portfolios. The boundary, shown as a bold line, is the minimum variance line. The shape of it is known as the *Markowitz bullet*. The darker part of the shaded area corresponds to the interior of the triangle in Figure 3.5, that is, it represents portfolios without short selling.

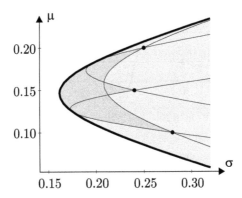

Figure 3.6 Feasible portfolios on the σ, μ plane

It is instructive to imagine how the whole w_2, w_3 plane in Figure 3.5 is mapped onto the shaded area representing all feasible portfolios in Figure 3.6. Namely, the w_2, w_3 plane is folded along the minimum variance line, being simultaneously warped and stretched to attain the shape of the Markowitz bullet. This means, in particular, that pairs of points on opposite sides of the minimum variance line on the w_2, w_3 plane are mapped into single points on the σ, μ plane. In other words, each point inside the shaded area in Figure 3.6 corresponds to two different portfolios. However, each point on the minimum variance line corresponds to a single portfolio.

Example 3.32

For the same three securities as in Examples 3.29 and 3.31, Figure 3.7 shows what happens if no short selling is allowed. All portfolios without short selling are represented by the interior and boundary of the triangle on the w_1, w_2 plane and by the shaded area with boundary on the σ, μ plane. The minimum variance line without short selling is shown as a bold line in both plots. For comparison, the minimum variance line with short selling is shown as a broken line.

Exercise 3.13

For portfolios constructed with and without short selling from the three securities in Exercise 3.11 compute the minimum variance line parame-

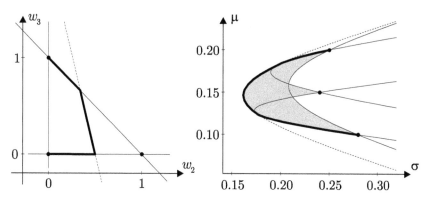

Figure 3.7 Portfolios without short selling

terised by the expected return and sketch it a) on the w_2, w_3 plane and
b) on the σ, μ plane. Also sketch the set of all feasible portfolios with
and without short selling.

3.3.4 Market Portfolio

From now on we shall assume that a risk-free security is available in addition
to n risky securities. The return on the risk-free security will be denoted by R.
The standard deviation of the return is of course zero for the risk-free security.

Consider a portfolio consisting of the risk-free security and a specified risky
security (or a portfolio of risky securities) V with expected return μ_V and
standard deviation $\sigma_V > 0$. By Proposition 3.22 all such portfolios form a line
consisting of two rectilinear half-lines meeting at the point $(0, R)$ on the vertical
axis on the σ, μ plane, represented by the broken line in Figure 3.8. By taking
portfolios containing the risk-free security and a risky security V with σ_V, μ_V
anywhere in the feasible set on the σ, μ plane, represented by the shaded area in
Figure 3.8, we can construct any portfolio between the two solid half-lines. The
area between these two half-lines serves as a new set of admissible portfolios,
which can now include the risk-free security.

The efficient frontier of this new feasible set is the bold half-line passing
through $(0, R)$ and tangent to the hyperbola representing the minimum variance
line constructed from risky securities. Every rational investor who respects the
dominance relation between portfolios will select his or her portfolio on this
half-line, called the *capital market line* (CML). This argument works as long
as the risk-free return R is not too high, namely R is less than the expected
return on the minimum variance portfolio (this is because the asymptotes of
the hyperbola intersect on the μ axis, see Proposition 3.18).

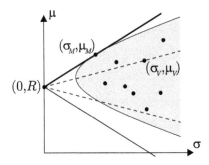

Figure 3.8 Efficient frontier for portfolios with a risk-free security

The portfolio M corresponding to the tangency point (σ_M, μ_M) plays a special role. It is called the *market portfolio*. In order to compute the weights in the market portfolio we observe that the straight line passing though $(0, R)$ and (σ_M, μ_M) has the steepest slope (highest gradient) among all lines through $(0, R)$ and (σ_V, μ_V) for any portfolio V in the feasible set constructed from risky assets. The slope of such a line can be written as

$$\frac{\mu_V - R}{\sigma_V} = \frac{\mathbf{w}\mathbf{m}^{\mathrm{T}} - R}{\sqrt{\mathbf{w}\mathbf{C}\mathbf{w}^{\mathrm{T}}}},$$

where $\mathbf{w}$ are the weights in portfolio V. We shall therefore find

$$\max \frac{\mathbf{w}\mathbf{m}^{\mathrm{T}} - R}{\sqrt{\mathbf{w}\mathbf{C}\mathbf{w}^{\mathrm{T}}}}$$

over all vectors $\mathbf{w} \in \mathbb{R}^n$ subject to the usual constraint

$$\mathbf{w}\mathbf{u}^{\mathrm{T}} = 1.$$

To solve this maximisation problem we form the Lagrange function

$$F(\mathbf{w}, \lambda) = \frac{\mathbf{w}\mathbf{m}^{\mathrm{T}} - R}{\sqrt{\mathbf{w}\mathbf{C}\mathbf{w}^{\mathrm{T}}}} - \lambda(\mathbf{w}\mathbf{u}^{\mathrm{T}} - 1),$$

and equate its gradient to zero:

$$\nabla_{\mathbf{w}} F(\mathbf{w}, \lambda) = \frac{\sqrt{\mathbf{w}\mathbf{C}\mathbf{w}^{\mathrm{T}}}\,\mathbf{m} - \frac{\mathbf{w}\mathbf{m}^{\mathrm{T}} - R}{\sqrt{\mathbf{w}\mathbf{C}\mathbf{w}^{\mathrm{T}}}}\mathbf{w}\mathbf{C}}{\mathbf{w}\mathbf{C}\mathbf{w}^{\mathrm{T}}} - \lambda\mathbf{u} = 0.$$

This can be written as

$$\frac{\mu_V - R}{\sigma_V^2}\mathbf{w}\mathbf{C} = \mathbf{m} - \lambda\sigma_V\mathbf{u}.$$

Multiplying by $\mathbf{w}^{\mathrm{T}}$ on the right we get

$$\frac{\mu_V - R}{\sigma_V^2}\mathbf{w}\mathbf{C}\mathbf{w}^{\mathrm{T}} = \mu_V - \lambda\sigma_V$$

so

$$\lambda = \frac{R}{\sigma_V}.$$

Therefore we have the equation

$$\gamma\mathbf{w}\mathbf{C} = \mathbf{m} - R\mathbf{u} \qquad (3.23)$$

where $\gamma = \frac{\mu_V - R}{\sigma_V^2}$, so that

$$\gamma\mathbf{w} = (\mathbf{m} - R\mathbf{u})\mathbf{C}^{-1}.$$

The difficulty in solving this for $\mathbf{w}$ lies in the fact that γ depends on $\mathbf{w}$. To compute γ we multiply by $\mathbf{u}^{\mathrm{T}}$ on the right and use the constraint once again, which gives

$$\gamma = (\mathbf{m}-R\mathbf{u})\mathbf{C}^{-1}\mathbf{u}^{\mathrm{T}}.$$

This leads to the following result.

Theorem 3.33

If $\det\mathbf{C} \neq 0$ and the risk-free rate R is lower than the expected return μ_{MVP} of the minimum variance portfolio, then the market portfolio M exists and its weights are given by

$$\mathbf{w}_M = \frac{(\mathbf{m} - R\mathbf{u})\mathbf{C}^{-1}}{(\mathbf{m}-R\mathbf{u})\mathbf{C}^{-1}\mathbf{u}^{\mathrm{T}}}.$$

Proof

It only remains to make sure that $\gamma \neq 0$. By (3.19) the expected return of the minimum variance portfolio can be written as

$$\mu_{\mathrm{MVP}} = \mathbf{w}_{\mathrm{MVP}}\mathbf{m}^{\mathrm{T}} = \frac{\mathbf{u}\mathbf{C}^{-1}\mathbf{m}^{\mathrm{T}}}{\mathbf{u}\mathbf{C}^{-1}\mathbf{u}^{\mathrm{T}}}.$$

Since, by assumption, $R < \mu_{\mathrm{MVP}}$ and $\mathbf{u}\mathbf{C}^{-1}\mathbf{u}^{\mathrm{T}} > 0$, it follows that

$$\gamma = \mathbf{u}\mathbf{C}^{-1}\mathbf{m}^{\mathrm{T}} - R\mathbf{u}\mathbf{C}^{-1}\mathbf{u}^{\mathrm{T}} > 0.$$

$\square$

Exercise 3.14

Suppose that the risk-free return is $R = 5\%$. Compute the weights in the market portfolio constructed from the three securities in Exercise 3.10. Also compute the expected return and standard deviation of the market portfolio.

Exercise 3.15

In a market consisting of the three securities in Exercise 3.11, consider the portfolio V on the efficient frontier with expected return $\mu_V = 21\%$. Compute the values of γ and R such that the weights $\mathbf{w}$ in this portfolio satisfy $\gamma \mathbf{w} \mathbf{C} = \mathbf{m} - R\mathbf{u}$.

The capital market line (CML), which joins the risk-free security represented by $(0, R)$ and the market portfolio with coordinates (σ_M, μ_M), satisfies the equation

$$\mu = R + \frac{\mu_M - R}{\sigma_M}\sigma. \tag{3.24}$$

For a portfolio on the CML with risk σ the term $\frac{\mu_M - R}{\sigma_M}\sigma$ is called the *risk premium*. It can be viewed as additional return above the risk-free return R which compensates for exposure to risk.

If each investor uses the same values of the model parameters (the expected returns on the basic assets and the entries of the covariance matrix) and chooses a portfolio that is not dominated by another one, then each investor will construct their portfolio on the CML. In effect, everyone will combine an investment in the risk-free security with just one risky portfolio, namely the market portfolio M. Consequently, the relative proportions of risky assets held in the portfolio will be the same for all investors. This can only happen if the weights in the market portfolio are proportional to the total value (called *market capitalisation*) of each of the risky assets available in the market.

In practice, the market portfolio can be represented by a market index. An empirical test of the theory is therefore to see if the market index lies on the efficient frontier and determines the tangent line when combined with the risk-free asset.

3.4 Capital Asset Pricing Model (CAPM)

Paradoxically, in a model where decisions are based on risk expressed as standard deviation, if we look at a portfolio of assets to assess its risk, the standard deviations of individual assets may turn out not as relevant as the covariances between them. This can be illustrated by the following example.

Example 3.34

Suppose that the weights in a portfolio are of the form $w_j = \frac{1}{n}$ where n is the number of risky assets. We shall investigate the risk of this portfolio depending on n. Assume that the variances of all assets traded in the market are uniformly bounded, $\sigma_j^2 \leq L$ for each j. Then

$$\sigma_V^2 = \sum_{j,k} w_j w_k c_{jk}$$

$$= \sum_j w_j^2 \sigma_j^2 + \sum_{j \neq k} w_j w_k c_{jk} \leq \frac{1}{n} L + \frac{1}{n^2} \sum_{j \neq k} c_{jk}.$$

Assume further that the off-diagonal elements of the covariance matrix are all equal, $c_{jk} = c > 0$ for each $j \neq k$. Then

$$\sigma_V^2 \leq \frac{L}{n} + \frac{1}{n^2} n(n-1)c.$$

The upper bound converges to c as $n \to \infty$. The risk of a portfolio containing many such assets is determined by the covariances. The variances of the individual assets become irrelevant for large n.

This example motivates the following distinction between two kinds of risk: *diversifiable*, or *specific risk*, which can be reduced to zero by expanding the portfolio, and *undiversifiable*, *systematic*, or *market risk*, which cannot be avoided because the securities are to some extent linked to the market.

As we know, the CML is tangent to the efficient frontier at the point (σ_M, μ_M) representing the market portfolio M with weights $\mathbf{w}_M$. Consider any other portfolio (or an individual asset) V, represented by a point (σ_V, μ_V) on the σ, μ plane. Now consider all portfolios constructed from M and V. They form a hyperbola, which we claim to be tangent to the CML at (σ_M, μ_M). If this hyperbola were to intersect the CML, this would contradict the fact that the slope of CML is maximal, see Figure 3.9.

We shall compute the slope of the tangent line to the hyperbola at M, and then use the fact that the slope of the CML must be the same. Denote the weights of V and M in a portfolio P on the hyperbola by $(x, 1-x)$. The risk and return of P are

$$\sigma_P = (x^2 \sigma_V^2 + (1-x)^2 \sigma_M^2 + 2x(1-x)\text{Cov}(K_V, K_M))^{\frac{1}{2}},$$

$$\mu_P = x\mu_V + (1-x)\mu_M.$$

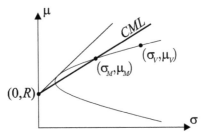

Figure 3.9 The CML cannot both have the steepest slope and intersect the two-asset line through M and V

We compute the derivatives with respect to x at $x = 0$ to get

$$\left.\frac{\partial \sigma_P}{\partial x}\right|_{x=0} = \frac{\mathrm{Cov}(K_V, K_M) - \sigma_M^2}{\sigma_M},$$

$$\left.\frac{\partial \mu_P}{\partial x}\right|_{x=0} = \mu_V - \mu_M,$$

The slope of the tangent line is the ratio of these derivatives, which we equate to the slope of the CML:

$$\frac{\mu_V - \mu_M}{\frac{\mathrm{Cov}(K_V, K_M) - \sigma_M^2}{\sigma_M}} = \frac{\mu_M - R}{\sigma_M}.$$

Solving for μ_V, we get

$$\mu_V = R + \frac{\mathrm{Cov}(K_V, K_M)}{\sigma_M^2}(\mu_M - R).$$

Definition 3.35

We call

$$\beta_V = \frac{\mathrm{Cov}(K_V, K_M)}{\sigma_M^2}$$

the *beta factor* of the portfolio V.

We have proved the following theorem.

Theorem 3.36 (CAPM)

Suppose that the risk-free rate R is lower than the expected return μ_{MVP} of the minimum variance portfolio (so that the market portfolio M exists). Then the expected return μ_V on any feasible portfolio V is given by

$$\mu_V = R + \beta_V(\mu_M - R). \tag{3.25}$$

The term $\beta_V(\mu_M - R)$ is called the *risk premium*. It represents the additional return required by an investor who faces risk due to the link between the portfolio and the market. This is similar to the risk premium in (3.24). However, (3.24) applies only to portfolios on the capital market line, whereas (3.25) is much more general in that it applies to all feasible portfolios and individual securities. Of course, for a portfolio V on the capital market line the risk premium in (3.25) and that in (3.24) are the same:

$$\beta_V(\mu_M - R) = \frac{\mu_M - R}{\sigma_M}\sigma_V.$$

Exercise 3.16

Show that the beta factor β_V of a portfolio consisting of n securities with weights $w_1, \ldots, w_n$ is given by $\beta_V = w_1\beta_1 + \cdots + w_n\beta_n$, where $\beta_1, \ldots, \beta_n$ are the beta factors of the securities.

The CAPM formula (3.25) is concerned with expected returns. Our next step is to consider the returns themselves. Suppose that we want to approximate the return K_V on a feasible portfolio V by a linear function $\beta K_M + \alpha$ of the return K_M on the market portfolio M. The error of this approximation

$$\varepsilon = K_V - (\beta K_M + \alpha)$$

is called the *residual error*. The best approximation in the sense of minimising the variance $\mathrm{Var}(\varepsilon)$, known as *linear regression*, is achieved for

$$\beta = \beta_V, \tag{3.26}$$
$$\alpha = \mu_V - \beta_V\mu_M = R(1 - \beta_V), \tag{3.27}$$

see Appendix 10.5. This provides an alternative interpretation of β_V as the gradient of the regression line, which minimises the variance of the residual error.

We shall denote the residual error by ε_V when β, α are given by (3.26) and (3.27). This means that

$$\varepsilon_V = (K_V - \mu_V) - \beta_V(K_M - \mu_M) \tag{3.28}$$
$$= (K_V - R) - \beta_V(K_M - R).$$

The second equality follows from the CAPM formula (3.25).

Exercise 3.17

Verify that

$$\mathbb{E}(\varepsilon_V) = 0,$$
$$\mathrm{Cov}(K_M, \varepsilon_V) = 0.$$

Formula (3.28) can be written as $K_V - \mu_V = \varepsilon_V + \beta_V(K_M - \mu_M)$. Applying variance to both sides of this equality and using the results of Exercise 3.17, we obtain

$$\sigma_V^2 = \mathrm{Var}(\varepsilon_V) + \beta_V^2 \sigma_M^2.$$

This sheds more light on the distinction between diversifiable and systematic risk. The left-hand side is directly related to the risk of the portfolio V. The second term on the right represents the systematic risk that is due to the link between the portfolio and the market and is measured by β_V. The first term corresponds to the diversifiable part of the risk. In particular, if $V = M$, then $\mathrm{Var}(\varepsilon_M) = 0$. There is no diversifiable risk if we invest in the market portfolio (or in a portfolio on the CML, which is a combination of the market portfolio and the risk-free asset).

Remark 3.37

Drawing on the results established in Section 3.3.4, we can give an alternative proof of the CAPM formula. Consider an arbitrary portfolio V with weights $\mathbf{w}$. The weights $\mathbf{w}_M$ in the market portfolio satisfy

$$\gamma \mathbf{w}_M \mathbf{C} = \mathbf{m} - R\mathbf{u}$$

for some number $\gamma > 0$, see (3.23). The beta factor of portfolio V with weights $\mathbf{w}$ can, therefore, be written as

$$\beta_V = \frac{\mathrm{Cov}(K_V, K_M)}{\sigma_M^2} = \frac{\gamma \mathbf{w} \mathbf{C} \mathbf{w}_M^{\mathrm{T}}}{\gamma \mathbf{w}_M \mathbf{C} \mathbf{w}_M^{\mathrm{T}}} = \frac{\mathbf{w} \frac{1}{\gamma}(\mathbf{m} - R\mathbf{u})^{\mathrm{T}}}{\mathbf{w}_M \frac{1}{\gamma}(\mathbf{m} - R\mathbf{u})^{\mathrm{T}}} = \frac{\mu_V - R}{\mu_M - R}.$$

Solving this for μ_V, we obtain the CAPM formula (3.25) once again.

According to (3.25), the expected return on a portfolio is an affine function of the beta factor. The graph of this function in the β, μ plane is called the *security market line* (SML). This straight line is shown in Figure 3.10, in which the CML is also plotted for comparison.

The CAPM describes a state of equilibrium in the market. Everyone is holding a portfolio of risky securities with the same weights as the market

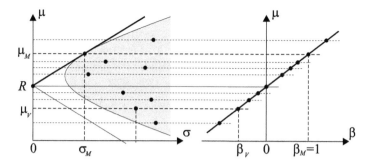

Figure 3.10 Capital market line and security market line

portfolio. Any trades that may be executed by investors will only affect their split of funds between the risk-free security and the market portfolio. As a result, the demand and supply of all securities will be balanced.

The market will remain in equilibrium as long as the estimates of expected returns and beta factors satisfy (3.25). Any new information about a particular security V can affect the expected return μ_V or the beta factor β_V, so that the CAPM formula (3.25) may no longer be valid. Suppose, for example, that

$$\mu_V > R + \beta_V(\mu_M - R).$$

In this case investors will want to increase their relative position in this security, which offers higher expected return than that required as compensation for systematic risk. Demand will exceed supply, the price of the security will begin to rise and the expected return will decline. On the other hand, if the reverse inequality

$$\mu_V < R + \beta_V(\mu_M - R)$$

holds, then investors will want to sell or even short sell the security, causing the price to fall because of the excess supply, so that the expected return will increase. In both cases we should observe price adjustments restoring the CAPM formula and the state of equilibrium.

Case 3: Discussion

Let us set similar targets for the pension scheme and make similar assumptions as in Case 2. When you retire after 40 years you want to receive a pension for 20 years. Your salary is expected to grow at 2% annually, and you want the pension payments to start at 50% of your final salary and to grow at the same 2% rate. The interest rate offered by the risk-free bank account is 5%.

Moreover, you consider risk of up to 10% (as measured by standard deviation) to be reasonable, and select three stocks A, B, C to be included in your portfolio in addition to the bank account. You estimate the expected returns on the stocks to be, respectively, 8%, 10% and 14% per annum, and the standard deviations of the annual returns to be 15%, 22% and 26%. The estimates for the correlation coefficients are 0.5 between stocks A, B, -0.3 between A, C, and 0.3 between B, C.

The three chosen stocks constitute a mini-market, to which we can apply the theory developed in this chapter. This requires selecting a suitable portfolio on the capital market line. To this end, using the formula in Theorem 3.33, we can compute the weights $(0.6773, -0.1518, 0.4745)$ in the market portfolio M.

However, as a small investor, you are unlikely to be able to do short selling and will need to restrict your portfolios to those with non-negative weights only. We no longer have an analytic formula for the weights, and need to resort to a numerical solution. Using Solver in Excel to maximise the gradient of the straight line through the risk-free security and feasible portfolios without short selling, we find the weights in the market portfolio M' with no short selling to be $(0.5646, 0.0000, 0.4354)$. Portfolio M' has expected return 10.61% (as compared to 10.54% with short selling) and standard deviation 11.93% (11.52% with short selling).

We can now combine M' with the risk-free investment into a bank account at 5% to construct a portfolio P on the capital market line. The choice of how to split your investment between M' and the bank account will depend on your attitude to risk. To construct a portfolio P with 10% risk, you will need to invest 16.19% of your funds in the bank account and the rest in portfolio M'. The expected return on P will then be 9.70%. Going back to Case 2, we can see that at this level of return, the premium you need to pay into the pension scheme will be just 2.20% of your salary, as compared to 10.05% when investing into the bank account only.

If risk at 10% is considered too high for a pension scheme, it can be reduced by increasing the proportion of funds invested in the bank account. This is shown in Table 3.1.

risk	0.00%	2.50%	5.00%	7.50%	10.00%
expected return	5.00%	6.18%	7.35%	8.53%	9.70%
proportion of salary paid	10.05%	6.94%	4.76%	3.24%	2.20%
of which invested risk free	100.00%	79.05%	58.09%	37.14%	16.19%

Table 3.1 Pension schemes based on different attitudes to risk

4
Forward and Futures Contracts

Case 4

In Cases 2 and 3 we considered a personal pension scheme. Including risky assets in the portfolio improves the expected return in the long term, helping to reduce pension contributions. An inevitable drawback is the risk involved. You would like to explore some instruments available in financial markets to reduce the risk while sustaining the advantages of your pension scheme.

4.1 Forward Contracts

In Section 1.5 we introduced the notion of a forward contract. Let us just recall that this is an agreement to sell or buy the underlying asset at a specified time for a price determined at the initiation of the contract.

4.1.1 Underlying Asset

The market price of the underlying asset at time $t \in [0, T]$ will be denoted by $S(t)$. It is assumed to be strictly positive for all t. We take $t = 0$ to be the present time, $S(0)$ being the spot price of the underlying, known to all investors. The future prices $S(t)$ for $t > 0$ remain unknown, in general.

M. Capiński, T. Zastawniak, *Mathematics for Finance*,
Springer Undergraduate Mathematics Series,
© Springer-Verlag London Limited 2011

Mathematically, $S(t)$ can be represented as a positive random variable on a probability space Ω, that is,

$$S(t) : \Omega \rightarrow (0, \infty).$$

The probability space Ω consists of all feasible price movement 'scenarios' $\omega \in \Omega$. We shall write $S^\omega(t)$ to denote the price at time t if the market follows scenario $\omega \in \Omega$. The spot price $S(0)$ is simply a positive number, but it can be thought of as a constant random variable. The unknown future prices $S(t)$ for $t > 0$ are non-constant random variables. This means that for each $t > 0$ there are at least two scenarios $\omega, \omega' \in \Omega$ such that $S^\omega(t) \neq S^{\omega'}(t)$.

The probability space and the probability distributions of all $S(t)$ are the main ingredients of a market model. In this chapter the discussion does not require specifying any particular model, being robust enough to be valid in any model. The results concerning forward and futures contracts are *model-independent*, which is an important feature.

Holding the underlying asset may involve some intermediate cash flow. For example, if $S(t)$ represents a stock, a dividend may be paid to the shareholders. This is also the case with currencies, which generate interest if deposited at the corresponding rate. On the other hand, in the case of commodities, storage costs may apply. In general, we apply the term *cost of carry* for the cash flow involved in maintaining a long or short position in the underlying asset.

Consider a forward contract exchanged at time 0 with delivery time T and forward price $F(0, T)$. No payment is made by either party at time 0. At delivery, the party with a long forward position will receive (or pay if negative) the amount of $S(T) - F(0, T)$ by buying the underlying asset for $F(0, T)$ and selling it for the market price $S(T)$. The payoff for the short forward position will be $F(0, T) - S(T)$, see Figure 4.1.

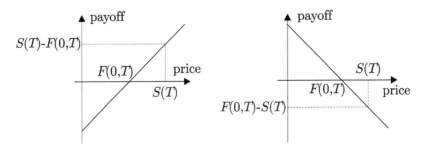

Figure 4.1 Payoff for long and short forward positions at delivery

If the contract is initiated at time $t < T$ rather than 0, then we shall write $F(t, T)$ for the forward price, the payoff at the delivery being $S(T) - F(t, T)$ for a long forward position and $F(t, T) - S(T)$ for a short position.

4.1.2 Forward Price

The No-Arbitrage Principle (Assumption 1.9) makes it possible to obtain formulae for the forward prices of assets of various kinds. Recall Proposition 1.10, where we established the formula

$$F(0,T) = S(0)\frac{A(T)}{A(0)}$$

under the assumption of zero cost of carry.

We also assumed that the future value of the money market account is known at time 0. This amounts to assuming that the interest rate is constant during the whole time period. For example, using continuous compounding we can write $A(t) = A(0)e^{rt}$, so the above formula takes the intuitively clear form

$$F(0,T) = S(0)e^{rT}.$$

Note that $e^{rT} = \frac{1}{B(0,T)}$, where $B(0,T)$ denotes the price of a unit zero-coupon bond maturing at time T. This motivates the following version of Proposition 1.10, where is no longer necessary to assume a constant rate of interest. In addition, we take an arbitrary time t prior to delivery time T as the time when the forward contract is initiated, with the corresponding forward price $F(t,T)$.

Theorem 4.1

For an underlying asset with no cost of carry, if a forward contract is initiated at time $t \le T$, then the forward price must be

$$F(t,T) = \frac{S(t)}{B(t,T)}. \qquad (4.1)$$

Proof

Suppose that $F(t,T) > S(t)/B(t,T)$. In this case, at time t:

- sell $S(t)/B(t,T)$ zero-coupon bonds, receiving the amount $S(t)$;
- buy one share for $S(t)$;
- take a short forward position, that is, agree to sell one share for $F(t,T)$ at time T.

Then, at time T:

- sell the stock for $F(t,T)$ using the forward contract;
- pay the face value $1 for each unit bond sold.

This will bring a risk-free profit of

$$F(t,T) - \frac{S(t)}{B(t,T)} > 0,$$

contrary to the No-Arbitrage Principle.

Next, suppose that $F(t,T) < S(t)/B(t,T)$. In this case construct the opposite strategy to the one above. Namely, at time t:

- sell short one share for $S(t)$;
- buy $S(t)/B(t,T)$ zero-coupon bonds;
- enter into a long forward contract with forward price $F(t,T)$.

Then, at time T:

- receive the face value \$1 for each unit bond held, collecting $S(t)/B(t,T)$ dollars;
- buy the stock for $F(t,T)$ using the forward contract;
- close out the short position in stock by returning it to the owner.

You will end up with a positive amount

$$\frac{S(t)}{B(t,T)} - F(t,T) > 0,$$

again a contradiction with the No-Arbitrage Principle.   □

Remark 4.2

Consider an S-maturity bond $B(t,S)$ as an underlying asset. The forward price of the bond is $F(t,T) = B(t,S)/B(t,T)$ for delivery at time T, where $t \le T \le S$. In particular, $F(0,S) = 1$, which agrees with the fact that the bond will have value 1 with certainty at time S.

In a market with restrictions on short sales the inequality $F(t,T) < S(t)/B(t,T)$ does not necessarily lead to arbitrage opportunities.

Exercise 4.1

Suppose that $S(0) = 17$ dollars, $F(0,1) = 18$ dollars, $r = 8\%$, and short selling requires a 30% security deposit attracting interest at $d = 4\%$. Is there an arbitrage opportunity? Find the highest rate d for which there is no arbitrage opportunity.

Exercise 4.2

Suppose that the price of stock on 1 April 2000 turns out to be 10% lower than it was on 1 January 2000. Assuming that the risk-free rate is constant at $r = 6\%$, what is the percentage drop of the forward price on 1 April 2000 as compared to that on 1 January 2000 for a forward contract with delivery on 1 October 2000?

Remark 4.3

In the case considered here we always have $F(t,T) = S(t)/B(t,T) > S(t)$ since $B(t,T) < 1$. The difference $F(t,T) - S(t)$, which is called the *basis*, converges to 0 as $t \to T$ since $B(t,T) \to B(T,T) = 1$.

Including Dividends. We shall generalise the formula for the forward price to cover assets that generate income during the lifetime of the forward contract. The income may be in the form of dividends or interest. We shall also cover the case when the asset involves some costs (called the cost of carry), such as storage or insurance, by treating the costs as negative income.

Suppose that the stock will pay a dividend div at an intermediate time t between initiating the forward contract and delivery. At time t the stock price will drop by the amount of the dividend paid. The formula for the forward price, which involves the present stock price, can be modified by subtracting the present value of the dividend.

Theorem 4.4

The forward price of a stock paying dividend div at time t, where $0 < t < T$, is

$$F(0,T) = [S(0) - B(0,t)\text{div}]\frac{1}{B(0,T)}. \tag{4.2}$$

Proof

Suppose that

$$F(0,T) > [S(0) - B(0,t)\text{div}]\frac{1}{B(0,T)}.$$

We shall construct an arbitrage strategy. At time 0:

- enter into a short forward contract with forward price $F(0,T)$ and delivery time T;
- borrow $S(0)$ dollars by issuing $S(0)/B(0,T)$ zero-coupon bonds;
- buy one share;
- enter into div$\frac{B(0,t)}{B(0,T)}$ long forward contracts on a bond maturing at time T with delivery time t and forward price $\frac{B(0,T)}{B(0,t)}$.

At time t:

- cash the dividend div and invest it in bonds $B(t,T)$, buying $\text{div}\frac{B(0,t)}{B(0,T)}$ bonds at the forward price $\frac{B(0,T)}{B(0,t)}$.

At time T:

- sell the share for $F(0,T)$;
- pay $S(0)/B(0,T)$ to the bondholders;
- receive the amount $\text{div}\frac{B(0,t)}{B(0,T)}$ from the bonds purchased at time t.

The final balance will be positive:

$$F(0,T) - \frac{S(0)}{B(0,T)} + \text{div}\frac{B(0,t)}{B(0,T)} > 0,$$

a contradiction with the No-Arbitrage Principle. The case when $F(0,T) < [S(0) - B(0,t)\text{div}]\frac{1}{B(0,T)}$ can be dealt with similarly by inverting all positions.

□

The formula can easily be generalised to the case when dividends are paid more than once:

$$F(0,T) = [S(0) - \text{div}_0]B(0,T), \tag{4.3}$$

where div_0 is the present value of all dividends due during the lifetime of the forward contract.

Exercise 4.3

Consider a stock whose price on 1 January is $120 and which will pay a dividend of $1 on 1 July 2000 and $2 on 1 October 2000. The interest rate is 12%. Is there an arbitrage opportunity if on 1 January 2000 the forward price for delivery of the stock on 1 November 2000 is $131? If so, compute the arbitrage profit.

Exercise 4.4

Suppose that the risk-free rate is 8%. However, as a small investor, you can invest money at 7% only and borrow at 10%. Does either of the strategies in the proof of Theorem 4.4 give an arbitrage profit if $F(0,1) = 89$ and $S(0) = 83$ dollars, and a $2 dividend is paid in the middle of the year, that is, at time $1/2$?

Dividend Yield. Dividends are often assumed to be paid continuously at a specified rate, rather than at discrete time instants. For example, in a case of a highly diversified portfolio of stocks it is better to avoid analysing frequent payments scattered throughout the year. Another example is foreign currency, attracting interest at the corresponding rate.

Recall that in Section 1.7 we gave a formula for the forward price of foreign currency:

$$F(0,T) = S(0)\frac{1+K_\mathrm{h}}{1+K_\mathrm{f}} = S(0)\frac{A_\mathrm{h}(T)/A_\mathrm{h}(0)}{A_\mathrm{f}(T)/A_\mathrm{f}(0)},$$

where $A_\mathrm{h}(t)$, $A_\mathrm{f}(t)$ denote the home and foreign currency money market accounts and where $K_\mathrm{h}, K_\mathrm{f}$ are the corresponding returns. Using the above approach and employing bonds denominated in the home and foreign currencies, we can write this formula as

$$F(0,T) = S(0)\frac{B_\mathrm{f}(0,T)}{B_\mathrm{h}(0,T)}.$$

Next, suppose that a stock pays dividends continuously at a rate $r_\mathrm{div} > 0$, called the (continuous) *dividend yield*. If the dividends are reinvested in the stock, then an investment in one share held at time 0 will increase to become $e^{r_\mathrm{div}T}$ shares at time T. (The situation is similar to continuous compounding.) Consequently, in order to have one share at time T we should begin with $e^{-r_\mathrm{div}T}$ shares at time 0. This observation is used in the arbitrage proof below.

Theorem 4.5

The forward price for stock paying dividends continuously at a rate r_div is

$$F(0,T) = S(0)e^{-r_\mathrm{div}T}\frac{1}{B(0,T)}. \tag{4.4}$$

Proof

Suppose that

$$F(0,T) > S(0)e^{-r_\mathrm{div}T}\frac{1}{B(0,T)}.$$

In this case, at time 0:

- enter into a short forward contract;
- borrow the amount $S(0)e^{-r_\mathrm{div}T}$ to buy $e^{-r_\mathrm{div}T}$ shares.

Between time 0 and T collect the dividends paid continuously, reinvesting them in the stock. At time T you will have 1 share. At that time:

- sell the share for $F(0,T)$, closing out the short forward position;
- pay $S(0)e^{-r_\mathrm{div}T}/B(0,T)$ to clear the loan with interest.

The final balance $F(0,T) - S(0)e^{-r_{\text{div}}T}/B(0,T) > 0$ will be your arbitrage profit. The argument in the case when

$$F(0,T) < S(0)e^{-r_{\text{div}}T}\frac{1}{B(0,T)}$$

is similar and involves the opposite positions. □

In general, if the contract is initiated at time $t < T$, then

$$F(t,T) = S(t)e^{-r_{\text{div}}(T-t)}\frac{1}{B(t,T)}. \qquad (4.5)$$

Exercise 4.5

A US importer of German cars wants to arrange a forward contract to buy euros in half a year. The interest rates for investments in US dollars and euros are $r_\$ = 4\%$ and $r_{\text{€}} = 3\%$, respectively, the current exchange rate being 0.9834 euros to a dollar. What is the forward price of euros expressed in dollars (that is, the forward exchange rate)?

4.1.3 Value of Forward Contract

Every forward contract has value zero when initiated. As time goes by, the price of the underlying asset may change. Along with it, the value of the forward contract will vary and will no longer be zero, in general. In particular, the value of a long forward contract will be $S(T) - F(0,T)$ at delivery, which may turn out to be positive, zero or negative. We shall derive formulae to capture the changes in the value of a forward contract.

Suppose that the forward price $F(t,T)$ for a forward contract initiated at time t, where $0 < t < T$, is higher than $F(0,T)$. This is good news for an investor with a long forward position initiated at time 0. At time T such an investor will gain $F(t,T) - F(0,T)$ as compared to an investor entering into a new long forward contract at time t with the same delivery date T. To find the value of the original forward position at time t all we have to do is to discount this gain back to time t. This discounted amount would be received (or paid, if negative) by the investor with a long position should the forward contract initiated at time 0 be closed out at time t, that is, prior to delivery T. This intuitive argument needs to be supported by a rigorous arbitrage proof.

Theorem 4.6

For any t such that $0 \le t \le T$ the time t value of a long forward contract with forward price $F(0,T)$ is given by

$$V(t) = [F(t,T) - F(0,T)]B(t,T). \qquad (4.6)$$

Proof

At time t the holder of the long forward position may simultaneously enter the short position at no cost, which is done with the forward price prevailing at that time, namely $F(t,T)$. The final payoff of the combined position is

$$S(T) - F(0,T) - [S(T) - F(t,T)] = F(t,T) - F(0,T)$$

which is non-random, that is, risk free. Therefore its time t value is the discounted T value, with $B(t,T)$ as the discount factor.   □

For a stock paying no dividends formula (4.6) gives

$$V(t) = \left[\frac{S(t)}{B(t,T)} - \frac{S(0)}{B(0,T)}\right]B(t,T) = S(t) - S(0)\frac{B(t,T)}{B(0,T)}. \qquad (4.7)$$

The message is this: if the stock price grows at the same rate as a risk-free investment, then the value of the forward contract will be zero. The growth above the risk-free rate results in a gain for the holder of a long forward position.

Remark 4.7

Consider a contract with delivery price X rather than $F(0,T)$. The value of this contract at time t will be given by (4.6) with $F(0,T)$ replaced by X,

$$V_X(t) = [F(t,T) - X]B(t,T).$$

Such a contract may have non-zero value initially. In the case of a stock paying no dividends

$$V_X(0) = [F(0,T) - X]B(0,T) = S(0) - XB(0,T). \qquad (4.8)$$

Exercise 4.6

Suppose that the price of a stock is \$45 at the beginning of the year, the risk-free rate is 6%, assumed constant, and a \$2 dividend is to be paid after half a year. For a long forward position with delivery in one year, find its value after 9 months if at that time the stock price turns out to be a) \$49, b) \$51.

4.2 Futures

One of the two parties to a forward contract will be losing money. There is always some risk of default by the party suffering a loss. Futures contracts are designed to eliminate such risk.

Just like a forward contract, a *futures contract* involves an underlying asset, let it be a stock with price $S(t)$, and a specified time of delivery T. In addition to the usual stock prices, the market dictates the so-called *futures prices* $f(t, T)$ for each $t \leq T$. These prices are unknown at time 0, except for $f(0, T)$, and we shall treat them as random variables.

As in the case of a forward contract, it costs nothing to initiate a futures position. The difference lies in the cash flow during the lifetime of the contract. A long forward contract involves just a single payment $S(T) - F(0, T)$ at delivery. A futures contract involves a random cash flow, known as *marking to market*. Namely, for prescribed time instants $t_1 < t_2 < \cdots < t_N = T$, typically these will be consecutive trading days, the holder of a long futures position will receive at time t_n, $n = 1, \ldots, N$ the amount

$$f(t_n, T) - f(t_{n-1}, T)$$

if positive, or will have to pay it if negative. The opposite payments apply for a short futures position. The following two conditions are imposed:

1) the futures price at delivery is $f(T, T) = S(T)$;
2) for each time $t_n \leq T$, $n = 1, \ldots, N$ the futures price $f(t_n, T)$ is such that it costs nothing to enter a futures contract at that time.

4.2.1 Pricing

We shall show that in some circumstances the forward and the futures prices are the same. Let r be the risk-free rate under continuous compounding.

Theorem 4.8

If the interest rate is constant, then $f(0, T) = F(0, T)$.

Proof

Suppose for simplicity that marking to market is performed at just one intermediate time instant $0 < t < T$. The argument below can easily be extended to cover more frequent marking to market.

Suppose $f(0, T) > F(0, T)$. If so, then at time 0:

- take a long forward position (at no cost);
- open a fraction $e^{-r(T-t)}$ of a short futures position (at no cost).

At time t:

- pay the amount $e^{-r(T-t)}[f(t, T) - f(0, T)]$ as a result of marking to market;
- borrow $e^{-r(T-t)}[f(t, T) - f(0, T)]$ risk free;
- increase the short futures position to 1 contract (at no cost).

Note that the difference $f(t, T) - f(0, T)$ can be negative in which case 'pay' means 'receive' and 'borrow' means 'invest'. Then at time T:

- close the risk-free investment paying $f(t, T) - f(0, T)$;
- close the short futures position paying $S(T) - f(t, T)$ (since $S(T) = f(T, T)$);
- close the long forward position receiving $S(T) - F(0, T)$.

The final balance will be $f(0, T) - F(0, T) > 0$, which means that the strategy is an arbitrage opportunity, a contradiction.

The case when $f(0, T) < F(0, T)$ can be dealt with similarly by taking the opposite positions. ☐

The above argument applies to any time t so that $f(t, T) = F(t, T)$. The construction cannot be performed if the interest rate changes unpredictably. However if interest rate movements are known in advance, then the argument can be suitably modified and the equality between the futures and forward prices remains valid.

In an economy with constant interest rates r we obtain a simple structure of futures prices,

$$f(t, T) = S(t)e^{r(T-t)} \qquad (4.9)$$

if the stock pays no dividends. The futures prices are random, but this is caused entirely by the randomness of the underlying asset price. If the futures prices depart from the values given by (4.9), this reflects the market's view of future interest rate changes.

Example 4.9

Consider three scenarios with the same stock price at maturity of the contract. Let $T = 2/365$ be two days from now with marking-to-market days $t_1 = 1/365$ and $t_2 = T$.

1. The stock price goes up at the rate r. Then $S(t) = S(0)e^{rt}$,

$$f(t, T) = S(0)e^{rt}e^{r(T-t)} = S(0)e^{rT}$$

for all t, and marking-to-market payments are zero.

2. The price goes up more than in scenario 1, so on day one $S(t_1) > S(0)e^{rt_1}$.
   Then
   $$f(t_1, T) = S(t_1)e^{r(T-t_1)} > f(0, T) = S(0)e^{rT},$$
   with positive cash flow to the holder of a long futures position. Next, $S(T) = S(0)e^{rT} = f(T, T)$, which means that a negative cash flow on day two
   $$f(T, T) - f(t_1, T) = S(0)e^{rT} - S(t_1)e^{r(T-t_1)} = -(f(t_1, T) - f(0, 0)),$$
   exactly offsetting the payment received on day one.
3. The price on day one is below $S(0)e^{rt_1}$ so first we have a negative payment
   followed by a positive payment of the same size given by the above formulae.

The difference in the timing of the balancing payments is crucial. Due to the
time value of money the long position in scenario 2 is better than in scenario 3.

   This example also shows an important benchmark for the profitability of
a futures position. An investor who wants to take advantage of a predicted
increase in the price of stock above the risk-free rate should enter into a long
futures position. A short futures position will bring a profit should the stock
price go down or increase below the risk-free rate.

4.2.2 Margins

To ensure that the obligations involved in a futures position are fulfilled, certain
practical regulations are enforced. Each investor entering into a futures contract
has to pay a deposit, called the *initial margin*, which is kept by the clearing
house as collateral. In the case of a long futures position the amount $f(t_n, T) - f(t_{n-1}, T)$ is added to the deposit if positive or subtracted if negative at each
time step n, typically once a day. (The opposite amount is added or subtracted
for a short futures position.) Any excess that builds up above the initial margin
can be withdrawn by the investor. On the other hand, if the deposit drops below
a certain level, called the *maintenance margin*, the clearing house will issue a
margin call, requesting the investor to make a payment and restore the deposit
to the level of the initial margin. A futures position can be closed at any time,
in which case the deposit will be returned to the investor. In particular, the
futures position will be closed immediately by the clearing house if the investor
fails to respond to a margin call. As a result, the risk of default is eliminated.

Example 4.10

Suppose that the initial margin is set at 10% and the maintenance margin at
5% of the futures price. The table below shows a scenario with futures prices

$f(t,T)$. The columns labelled 'margin 1' and 'margin 2' show the deposit at the beginning and at the end of each day, respectively. The 'payment' column contains the amounts paid to top up the deposit (negative numbers) or withdrawn (positive numbers).

t-days	$f(t,T)$	cash flow	margin 1	payment	margin 2
0	140	opening:	0	-14	14
1	138	-2	12	0	12
2	130	-8	4	-9	13
3	140	$+10$	23	$+9$	14
4	150	$+10$	24	$+9$	15
		closing:	15	$+15$	0
			total:	10	

On day 0 a futures position is opened and a 10% deposit paid. On day 1 the futures price drops by $2, which is subtracted from the deposit. On day 2 the futures price drops further by $8, triggering a margin call because the deposit falls below 5%. The investor has to pay $9 to restore the deposit to the 10% level. On day 3 the forward price increases and $9 is withdrawn, leaving a 10% margin. On day 4 the forward price goes up again, allowing the investor to withdraw another $9. At the end of the day the investor decides to close the position, collecting the balance of the deposit. The total of all payments is $10, the increase in the futures price between day 0 and 4.

Remark 4.11

An important feature of the futures market is liquidity. This is possible due to standardisation and the presence of a clearing house. Only futures contracts with particular delivery dates are traded. Moreover, futures contracts on commodities such as gold or timber specify standardised delivery arrangements as well as standardised physical properties of the assets. The clearing house acts as an intermediary, matching the total of a large number of short and long futures positions of various sizes. The clearing house also maintains the margin deposit for each investor to eliminate the risk of default. This is in contrast to forward contracts negotiated directly between two parties.

An investor believing that the price of certain stock will go up can invest in the stock directly but can also take a long futures position. The investment required for the latter is the initial margin which is just a percentage of the stock price. If the prediction is correct, since the money engaged is smaller, the result is more impressive. This is best illustrated by an example.

Example 4.12

Let $S(0) = 100$ dollars, $r = 6\%$, $T = 0.5$ and let marking to market be restricted to just one instant $t_1 = 0.25$, equal to the investment horizon. If the initial deposit is 20% of the price, the investor can open 5 futures positions. Suppose $S(t_1) = 108$ dollars and then $f(t_1, T) \cong 106.39$ dollars, which together with $f(0, T) \cong 103.05$ dollars gives a marking-to-market gain of about $16.73. This is more than double the gain of $8 from a direct investment in stock.

4.2.3 Hedging with Futures

One relatively simple way to reduce an exposure to stock price variations is to enter a forward contract. However, a contract of this kind may not be readily available, not to mention the risk of default. Another possibility is to use the futures market, which is well-developed, liquid and protected from the risk of default.

Example 4.13

Let $S(0) = 100$ dollars and let the risk-free rate be constant at $r = 8\%$. Assume that marking to market takes place once a month. Suppose that we wish to sell the stock after 3 months. To hedge the exposure to stock price variations we enter into one short futures contract on the stock with the same maturity. The payments resulting from marking to market are invested (or borrowed), attracting interest at the risk-free rate. The results for two different stock price scenarios are displayed below. The column labelled 'm2m' represents the payments due to marking to market and the last column shows the interest accrued up to the delivery date.

Scenario 1

t-months	$S(t)$	$f(t, 3/12)$	m2m	interest
0	100	102.02		
1	102	103.37	-1.35	-0.02
2	101	101.67	$+1.69$	$+0.01$
3	105	105.00	-3.32	0.00
		total:	-2.98	-0.01

In this scenario we sell the stock for $105.00, but marking to market brings losses, reducing the sum to $105.00 - 2.98 - 0.01 = 102.01$ dollars. Note that if the marking-to-market payments were not invested at the risk-free rate, then the realised sum would be $105.00 - 2.98 = 102.02$ dollars, that is, exactly equal to the futures price $f(0, 3/12)$.

Scenario 2

t-months	$S(t)$	$f(t, 3/12)$	m2m	interest
0	100	102.02		
1	98	99.31	+2.70	+0.04
2	97	97.65	+1.67	+0.01
3	92	92.00	+5.65	0.00
		total:	+10.02	+0.05

In this case we sell the stock for $92.00 and benefit from marking to market along with the interest earned, bringing the final sum to $92.00 + 10.02 + 0.05 = 102.07$ dollars. Without the interest the final sum would be $92.00 + 10.02 = 102.02$ dollars, once again exactly the futures price $f(0, 3/12)$.

In reality the calculations in Example 4.13 are slightly more complicated because of the presence of the initial margin, which we have neglected for simplicity. Some limitations come from the standardisation of futures contracts. As a result, a difficulty may arise in matching the terms of the contract to our needs. The exercise dates for futures are typically certain fixed days in four specified months in a year, for example, the third Friday in March, June, September and December. If we want to close out our investment at the end of April, we will need to hedge with futures contracts with delivery date beyond the end of April.

Example 4.14

Suppose we wish to sell stock after 2 months and we hedge using futures with delivery in 3 months (we work in the same scenarios as in Example 4.13).

Scenario 1

t-months	$S(t)$	$f(t, 3/12)$	m2m	interest
0	100	102.02		
1	102	103.37	−1.35	−0.01
2	101	101.67	+1.69	0.00
		total:	+0.34	−0.01

We sell the stock for $101.00, which together with marking to market and interest will give $101.33.

Scenario 2

t-months	$S(t)$	$f(t, 3/12)$	m2m	interest
0	100	102.02		
1	98	99.31	+2.70	+0.02
2	97	97.65	+1.67	0.00
		total:	+4.37	+0.02

In this case we sell the stock for $97.00, and together with marking to market and interest obtain $101.39.

We almost hit the target, which is the futures price $f(0,2) \cong 101.34$ dollars, that is, the value of $100 compounded at the risk-free rate.

Remark 4.15

The difference between the spot and futures prices is called the *basis* (as for forward contracts):

$$b(t, T) = S(t) - f(t, T).$$

(Sometimes the basis is defined as $f(t, T) - S(t)$.) The basis converges to zero as $t \to T$, since $f(T, T) = S(T)$. In a market with constant interest rates it is given explicitly by

$$b(t, T) = S(t)(1 - e^{r(T-t)}),$$

being negative for $t < T$. If the asset pays dividends at a rate $r_{\text{div}} > r$, then:

$$b(t, T) = S(t)(1 - e^{(r-r_{\text{div}})(T-t)}).$$

Going back to the problem of designing a hedge, suppose that we wish to sell an asset at time $t < T$. To protect ourselves against a decrease in the asset price, at time 0 we can short a futures contract with futures price $f(0, T)$. At time t we shall receive $S(t)$ from selling the asset plus the cash flow $f(0, T) - f(t, T)$ due to marking to market (for simplicity, we neglect any intermediate cash flow, assuming that t is the first instance when marking to market takes place), that is, we obtain

$$f(0, T) + S(t) - f(t, T) = f(0, T) + b(t, T).$$

The price $f(0, T)$ is known at time 0, so the risk involved in the hedging position will be related to the unknown level of the basis. This uncertainty is mainly concerned with unknown future interest rates.

If the goal of a hedger is to minimise risk, it may be best to use a certain optimal hedge ratio, that is, to enter into N futures contracts, with N not necessarily equal to the number of shares of the underlying asset held. To see this compute the risk as measured by the variance of the basis $b_N(t, T) = S(t) - Nf(t, T)$:

$$\text{Var}(b_N(t, T)) = \sigma^2_{S(t)} + N^2 \sigma^2_{f(t,T)} - 2N\sigma_{S(t)}\sigma_{f(t,T)}\rho_{S(t)f(t,T)},$$

where $\rho_{S(t)f(t,T)}$ is the correlation coefficient between the spot and futures prices, and $\sigma_{S(t)}, \sigma_{f(t,T)}$ are the standard deviations. The variance is a quadratic function in N and has a minimum at

$$N = \rho_{S(t)f(t,T)} \frac{\sigma_{S(t)}}{\sigma_{f(t,T)}},$$

which is the optimal hedge ratio.

Exercise 4.7

Find the optimal hedge ratio if the interest rates are constant.

4.2.4 Index Futures

A stock exchange index is a weighted average of a selection of stock prices with weights proportional to the market capitalisation of stocks. An index of this kind will be approximately proportional to the value of the market portfolio (see Chapter 3) if the chosen set of stocks is large enough. For example, the Standard and Poor Index S&P500 is computed using 500 stocks, representing about 80% of trade at the New York Stock Exchange. For the purposes of futures markets the index can be treated as a security. This is because the index can be identified with a portfolio, even though in practice transaction costs would impede trading in this portfolio. The futures prices $f(t, T)$, expressed in index points, are assumed to satisfy the same conditions as before. Marking to market is given by the difference $f(t_n, T) - f(t_{n-1}, T)$ multiplied by a fixed amount ($500 for futures on S&P500). If the number of stocks included in the index is large, it is possible and convenient to assume that the index is an asset with dividends paid continuously.

Exercise 4.8

Suppose that the value of a stock exchange index is $13, 500$, the futures price for delivery in 9 months is $14, 100$ index points, and the interest rate is 8%. Find the dividend yield.

Our goal in this section is to study applications of index futures for hedging based on the Capital Asset Pricing Model introduced in Chapter 3. As we know, see (3.25), the expected return on a portfolio over a time step of length t is given by

$$\mu_V = R + \beta_V(\mu_M - R),$$

where β_V is the beta factor of the portfolio, μ_M is the expected return on the market portfolio and R is the risk-free return over a single time step τ. By $V(0)$ we denote the initial value of the portfolio. We assume for simplicity that the index is equal to the value of the market portfolio, so that the futures prices are given by

$$f(t, T) = M(t)e^{r(T-t)},$$

$M(t)$ being the value of the market portfolio and r the continuously compounded rate such that $e^{rt} = 1 + R$.

We can form a new portfolio with value $\widetilde{V}(0)$ by supplementing the original portfolio with N short futures contracts on the index with delivery time T. The initial value of the new portfolio is the same as the value $V(0)$ of the original portfolio, since it costs nothing to initiate a futures contract. At the first marking-to-market instant t the value of the new portfolio will be

$$\widetilde{V}(t) = V(t) - N(f(t, T) - f(0, T)).$$

The return on the new portfolio over the period $[0, t]$ will be

$$K_{\widetilde{V}} = \frac{\widetilde{V}(t) - \widetilde{V}(0)}{\widetilde{V}(0)} = \frac{V(t) - N(f(t, T) - f(0, T)) - V(0)}{V(0)}.$$

We shall show that the beta factor $\beta_{\widetilde{V}}$ of the new portfolio can be modified arbitrarily by a suitable choice of the futures position N.

Proposition 4.16

For any given number a, if

$$N = (\beta_V - a)\frac{V(0)e^{tr}}{f(0, T)},$$

then $\beta_{\widetilde{V}} = a$.

Proof

We shall compute the beta factor from the definition:

$$\beta_{\widetilde{V}} = \mathrm{Cov}(K_{\widetilde{V}}, K_M)/\sigma_M^2$$

$$= \mathrm{Cov}(K_V, K_M)/\sigma_M^2 - \frac{1}{V(0)}\mathrm{Cov}(N(f(t, T) - f(0, T)), K_M)/\sigma_M^2,$$

where K_M is the return on the market portfolio and K_V the return on the portfolio without futures. Since $\text{Cov}(f(0,T),K_M) = 0$ and covariance is linear with respect to each argument,

$$\text{Cov}(N(f(t,T) - f(0,T)), K_M) = N\text{Cov}(f(t,T), K_M).$$

Inserting the futures price $f(t,T) = M(t)e^{r(T-t)}$, we have

$$\text{Cov}(f(t,T), K_M) = e^{r(T-t)}\text{Cov}(M(t), K_M).$$

Again by the linearity of covariance in each argument

$$\text{Cov}(M(t), K_M) = M(0)\text{Cov}\left(\frac{M(t) - M(0)}{M(0)}, K_M\right) = M(0)\sigma_M^2.$$

Subsequent substitutions give

$$\beta_{\widetilde{V}} = \beta_V - \frac{NM(0)e^{r(T-t)}}{V(0)} = \beta_V - N\frac{f(0,T)}{V(0)e^{rt}},$$

which implies the asserted property. □

Corollary 4.17

If $a = 0$, then $\mu_{\widetilde{V}} = R$.

Example 4.18

Suppose that the index drops from $M(0) = 890$ down to $M(t) = 850$, that is, by 4.49% within one time step. Suppose further that the risk-free rate for the period of length t is $r = 1\%$, that is, $1 + R = e^{tr} = 1.1$. This means that the futures prices on the index (with delivery after 3 steps of length t) are

$$f(0, 3t) = M(0)(1 + R)^3 = 890 \times 1.01^3 \cong 916.97,$$
$$f(t, 3t) = M(t)(1 + R)^2 = 850 \times 1.01^2 \cong 867.09.$$

Consider a portfolio with $\beta_V = 1.5$ and initial value $V(0) = 100$ dollars. This portfolio will have negative expected return

$$\mu_V = R + (\mu_M - R)\beta_V$$
$$\cong 1\% + (-4.49\% - 1\%)1.5 \cong -7.24\%.$$

To construct a new portfolio with $\beta_{\widetilde{V}} = 0$ we can supplement the original portfolio with

$$N = \beta_V \frac{(1 + R)V(0)}{f(0, 3t)} \cong 1.5 \times \frac{1.01 \times 100}{916.97} \cong 0.1652$$

short forward contracts on the index with delivery after 3 steps.

Suppose that the actual return on the original portfolio during the first time step happens to be equal to the expected return. This gives $V(t) \cong 92.76$ dollars. Marking to market gives a payment of

$$-N\left(f(t,3t) - f(0,3t)\right) \cong -0.1652 \times (867.09 - 916.97) \cong 8.24$$

dollars due to the holder of $N \cong 0.1652$ short forward contracts. This makes the new portfolio worth

$$\widetilde{V}(t) = V(t) - N\left(f(t,3t) - f(0,3t)\right) \cong 92.76 + 8.24 = 101.00$$

dollars at time t, matching the risk-free growth exactly.

Exercise 4.9

Perform the same calculations in the case when the index increases from 890 to 920.

Remark 4.19

The ability to adjust the beta factor of a portfolio is valuable to investors who may wish either to reduce or to magnify the systematic risk. For example, suppose that an investor is able to design a portfolio with superior average performance to that of the market. By entering into a futures position such that the beta factor of the resulting portfolio is zero, the investor will be hedged against adverse movements of the market. This is crucial in the event of recession, so that the superior performance of the portfolio as compared to the market can be turned into a profit despite a decline in the market. On the other hand, should the market show some growth, the expected return on the hedged portfolio will be reduced by comparison because the futures position will result in a loss.

It needs to be emphasised that this type of hedging with futures works only on average. In particular, setting the beta factor to zero will not make the investment risk free.

Let us conclude this chapter with a surprising application of index futures.

Example 4.20

In emerging markets short sales are rarely available. This was the case in Poland in the late 1990's. However, index futures were traded. Due to the fact that one of the indices (WIG20) was composed of 20 stocks only, it was possible to manufacture a short sale of any stock among those 20 by entering into a short futures position on the index, combined with purchasing a suitable portfolio of the remaining 19 stocks. With a larger number of stocks comprising the index the transaction costs would have been too high to make such a construction practical.

Case 4: Discussion

First of all, we should discuss the choice between forward and futures contracts. Futures are more likely to be available, but margin maintenance may be causing difficulties. In addition to the initial deposit, we might face short term liquidity problems, so we consider forward contracts.

In the first phase we are building up our portfolio by purchasing assets. Hence, we should be worried about price increases. To hedge this risk we should take long positions. For instance, consider some stock, which we wish to include in our portfolio, with current price $S(0) = 100$ dollars. Taking a long forward position with maturity T corresponding to the nearest time when we will be buying more shares, we commit ourselves to buying the stock for $S(0)e^{rT}$. Let $T = 0.5$, $r = 5\%$, so the forward price is 102.53 dollars. This is less than the price related to our average growth assumptions, which for $\mu = 10\%$ (with continuous compounding) is $S(0)e^{\mu T} = 105.13$, approximately. Thus, from the funds devoted to the stock investment we would be able to buy more units at the forward price or save some money. Namely, if we plan to invest 1000 dollars in stock, using the prices based on the expected growth we would buy 9.51 shares, but these at the forward price would cost just 975.31 dollars.

Unfortunately, if the price $S(T)$ is lower than $S(0)e^{rT}$, which may happen even if the long term expected growth is at the assumed level, we would not benefit from this, since the forward contract is binding. To have more flexibility we have to learn about more sophisticated instruments.

In the second phase, when we will be gradually liquidating our portfolio by paying ourselves a pension, a short position in forwards or futures would preserve our return at the level of the risk-free rate. Given a share worth $S(0)$ (we moved our clock, and 0 is now some time in the future), we can sell it for the forward price $S(0)e^{rT}$. This is favourable should the market price drop, unfavourable otherwise, but on the whole makes little sense. If we are to earn the risk-free rate, we can invest risk free.

5
Options: General Properties

Case 5

European call and put options with strike price 98 euros on a German company G with current stock price 99 euros cost 4.145 and 2.738 euros, respectively. The stock of a British company B costs 66 pounds, and European call and put options on B with strike 65 pounds and exercise time in one month cost 2.55 and 1.34 pounds, respectively. A European call option on the pound with exercise price equal to the current rate, 1.5 euros to a pound, costs 14.40 euros and a put costs 13.16 euros for 1000 units. All these options have exercise time in one month. The world is ideal here, no transaction costs, long and short positions can be taken at the same prices since we assume that the reader is employed by a large financial institution for the purpose of spotting any instances of mispricing.

5.1 Definitions

In Chapter 1 we have seen simple examples of call and put options in a one-step discrete time setting. Here we give the definitions and then establish some fundamental properties of options. These properties are universal in the sense that they are valid in any market model, and the arguments are based on the No-Arbitrage Principle alone.

M. Capiński, T. Zastawniak, *Mathematics for Finance*,
Springer Undergraduate Mathematics Series,
© Springer-Verlag London Limited 2011

A *European call option* is a contract giving the holder the right to buy an asset, called the *underlying*, for a price X fixed in advance, known as the *exercise price* or *strike price*, at a specified future time T, called the *exercise* or *expiry time*. A *European put option* gives the right to sell the underlying asset for the strike price X at the exercise time T.

An *American call* or *put option* gives the right to buy or, respectively, to sell the underlying asset for the strike price X at any time between now and a specified future time T, called the *expiry time*. In other words, an American option can be exercised at any time up to and including expiry.

The term 'underlying asset' has quite general scope. Apart from typical assets such as stocks, commodities or foreign currency, there are options on stock indices, interest rates, or even on the snow level at a ski resort. Some underlying assets may be impossible to buy or sell. The option is then cleared in cash in a fashion which resembles settling a bet. For example, the holder of a European call option on the Standard and Poor Index (see page 107) with strike price 800 will gain if the index turns out to be 815 on the exercise date. The writer of the option will have to pay the holder an amount equal to the difference $815 - 800 = 15$ multiplied by a fixed sum of money, say by \$100. No payment will be due if the index turns out to be lower than 800 on the exercise date.

An option is determined by its payoff, which for a European call is

$$
\begin{cases}
S(T) - X & \text{if } S(T) > X, \\
0 & \text{otherwise.}
\end{cases}
$$

This payoff is a random variable, contingent on the price $S(T)$ of the underlying on the exercise date T. (This explains why options are often referred to as *contingent claims*.) It is convenient to use the notation

$$
x^{+} = \begin{cases}
x & \text{if } x > 0 \\
0 & \text{otherwise}
\end{cases}
$$

for the *positive part* of a real number x. Then the payoff of a European call option can be written as $(S(T) - X)^{+}$. For a put option the payoff is $(X - S(T))^{+}$.

Since the payoffs are always non-negative and with positive probability they are strictly positive, a premium must be paid to buy an option. If no premium had to be paid, an investor purchasing an option could under no circumstances lose money and would in fact make a profit whenever the payoff turned out to be positive. This would be contrary to common sense and in particular to the No-Arbitrage Principle. The premium is the market price of the option.

Establishing bounds and some general properties for option prices is the primary goal of the present chapter. The prices of calls and puts will be denoted by C_E, P_E for European options and C_A, P_A for American options, respectively. The same constant interest rate r will apply for lending and borrowing money without risk, and continuous compounding will be used.

Example 5.1

On 20 October 2007 European calls on Cadbury Schweppes PLC stock with strike price 640 pence to be exercised on 21 December 2007 traded at 22.5 pence at the Euronext London International Financial Futures Exchange (LIFFE). Suppose that the purchase of such an option was financed by a loan at 5.23% compounded continuously, so that $22.5e^{0.0523 \times \frac{2}{12}} \cong 22.7$ pence would have to be paid back on the exercise date. The investment would bring a profit if the stock price turned out to be higher than $640 + 22.7 = 662.7$ pence on the exercise date.

Exercise 5.1

Find the stock price on the exercise date for a European put option with strike price $36 and exercise date in three months to produce a profit of $3 if the option is bought for $4.50, financed by a loan at 12% compounded continuously.

The gain of an option holder (writer) is the payoff modified by the premium C_E or P_E paid (received) for the option. At time T the gain of the holder of a European call is $(S(T) - X)^+ - C_E e^{rT}$, where the time value of the premium is taken into account. In other words, we include the opportunity cost of an alternative risk-free investment. For the holder of a European put the gain is $(X - S(T))^+ - P_E e^{rT}$. These gains are illustrated in Figure 5.1. For the writer of an option the gains are $C_E e^{rT} - (S(T) - X)^+$ for a call and $P_E e^{rT} - (X - S(T))^+$ for a put option. Note that the potential loss for the holder of a call or put is always limited to the premium paid. For the writer the loss can be much higher, even unbounded in the case of a call option.

Exercise 5.2

Find the expected gain (or loss) for a holder of a European call option with strike price $90 to be exercised in 6 months if the stock price on the exercise date may turn out to be $87, $92 or $97 with probability $\frac{1}{3}$ each, given that the option is bought for $8, financed by a loan at 9% compounded continuously.

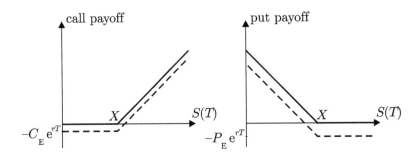

Figure 5.1 Payoffs (solid lines) and gains (broken lines) for the holder of a European call or put option

5.2 Put-Call Parity

In this section we shall make an important link between the prices of European call and put options.

Consider a portfolio constructed by writing and selling one put and buying one call option, each with the same strike price X and exercise date T. Adding the payoffs of the long position in the call and the short position in the put, we obtain the payoff of a long forward contract with forward price X and delivery time T. Indeed, if $S(T) \geq X$, then the call will pay $S(T) - X$ and the put will be worthless. If $S(T) < X$, then the call will be worth nothing and the writer of the put will need to pay $X - S(T)$. In either case, the value of the portfolio will be $S(T) - X$ at expiry, the same as for the long forward position, see Figure 5.2. As a result, the current value of such a portfolio of options should be that of the forward contract, which is $S(0) - Xe^{-rT}$ if the interest rate r is constant, see Remark 4.7.

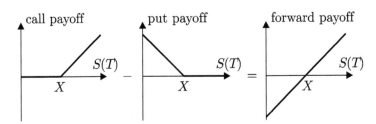

Figure 5.2 Long forward payoff constructed from a call and a put

This motivates the theorem below. Even though the theorem follows from the above intuitive argument, we shall give another proof with a view to possible generalisations.

Theorem 5.2 (Put-Call Parity)

For a stock that pays no dividends the following relation holds between the prices of European call and put options, each with exercise price X and exercise time T:

$$C_E - P_E = S(0) - Xe^{-rT}. \tag{5.1}$$

Proof

Suppose that

$$C_E - P_E > S(0) - Xe^{-rT}. \tag{5.2}$$

In this case an arbitrage strategy can be constructed as follows. At time 0:

- buy one share for $S(0)$;
- buy one put option for P_E;
- write and sell one call option for C_E;
- invest the sum $C_E - P_E - S(0)$ (borrow, if negative) in the money market at the interest rate r.

The balance of these transactions is 0. Then, at time T:

- close out the money market position, collecting (paying, if negative) the sum $(C_E - P_E - S(0))e^{rT}$;
- sell the share for X either by exercising the put if $S(T) \leq X$ or settling the short position in calls if $S(T) > X$.

The balance will be $(C_E - P_E - S(0))e^{rT} + X$, which is positive by (5.2), contradicting the No-Arbitrage Principle.

Now suppose that

$$C_E - P_E < S(0) - Xe^{-rT}. \tag{5.3}$$

Then the following reverse strategy will result in arbitrage. At time 0:

- sell short one share for $S(0)$;
- write and sell a put option for P_E;
- buy one call option for C_E;
- invest the sum $S(0) - C_E + P_E$ (borrow, if negative) in the money market at the interest rate r.

The balance of these transactions is 0. At time T:

- close out the money market position, collecting (paying, if negative) the sum $(S(0) - C_E + P_E)e^{rT}$;
- buy one share for X either by exercising the call if $S(T) > X$ or settling the short position in puts if $S(T) \leq X$, and close the short position in stock.

The balance will be $(S(0) - C_E + P_E)e^{rT} - X$, positive by (5.3), once again contradicting the No-Arbitrage Principle. $\qquad\square$

Exercise 5.3

Suppose that a stock paying no dividends is trading at $15.60 a share. European calls on the stock with strike price $15 and exercise date in three months are trading at $2.83. The interest rate is $r = 6.72\%$, compounded continuously. What is the price of a European put with the same strike price and exercise date?

Exercise 5.4

European call and put options with strike price $24 and exercise date in six months are trading at $5.09 and $7.78. The price of the underlying stock is $20.37 and the interest rate is 7.48%. Find an arbitrage opportunity.

Remark 5.3

We can make a simple but powerful observation based on Theorem 5.2: the difference between European calls and puts prices does not depend on any variables that are absent in the put-call parity relation (5.1). As an example, consider the expected return on stock. If the price of a call were to increase with the expected return, which on first sight seems consistent with intuition because higher stock prices mean higher payoffs on calls, then the price of a put should also increase. The latter, however, contradicts common sense because higher stock prices mean lower payoffs on puts. Because of this, one could conjecture that put and call prices should be independent of the expected return on stock. We shall see that this is indeed the case once pricing formulae are developed for calls and puts.

Following the argument presented at the beginning of this section, we can reformulate put-call parity as follows:

$$C_E - P_E = V_X(0), \tag{5.4}$$

where $V_X(0)$ is the value of a long forward contract, see (4.8). Note that if X is equal to the theoretical forward price $S(0)e^{rT}$ of the asset, then the value of the forward contract is zero, $V_X(0) = 0$, and so $C_E = P_E$. Formula (5.4) allows us to generalise put-call parity by drawing on the relationships established in Remark 4.7. Namely, if the stock pays a dividend between times 0 and T, then

$V_X(0) = S(0) - \text{div}_0 - X e^{-rT}$, where div_0 is the present value of the dividend. It follows that

$$C_E - P_E = S(0) - \text{div}_0 - X e^{-rT}. \qquad (5.5)$$

If dividends are paid continuously at a rate r_{div}, then

$$V_X(0) = S(0) e^{-r_{\text{div}}T} - X e^{-rT},$$

so

$$C_E - P_E = S(0) e^{-r_{\text{div}}T} - X e^{-rT}. \qquad (5.6)$$

Exercise 5.5

Outline an arbitrage proof of (5.5).

Exercise 5.6

Outline an arbitrage proof of (5.6).

Exercise 5.7

For the data in Exercise 4.5, find the strike price for European calls and puts to be exercised in six months such that $C_E = P_E$.

For American options put-call parity gives only some estimates, rather than a strict equality, involving put and call prices.

Theorem 5.4 (American Put-Call Parity Estimates)

The prices of American put and call options with the same strike price X and expiry time T on a stock that pays no dividends satisfy

$$S(0) - X e^{-rT} \geq C_A - P_A \geq S(0) - X.$$

Proof

Suppose that the first inequality fails to hold, that is,

$$C_A - P_A - S(0) + X e^{-rT} > 0.$$

Then we can write and sell a call, and buy a put and a share, financing the transactions in the money market. If the holder of the American call chooses to exercise it at time $t \leq T$, then we shall receive X for the share and settle the money market position, ending up with the put and a positive amount

$$
\begin{aligned}
X + (C_A - P_A - S(0))e^{rt} &= (Xe^{-rt} + C_A - P_A - S(0))e^{rt} \\
&\geq (Xe^{-rT} + C_A - P_A - S(0))e^{rt} > 0.
\end{aligned}
$$

If the call option is not exercised at all, we can sell the share for X by exercising the put at time T and close the money market position, also ending up with a positive amount

$$
X + (C_A - P_A - S(0))e^{rT} > 0.
$$

Now suppose that

$$
C_A - P_A - S(0) + X < 0.
$$

In this case we can write and sell a put, buy a call and sell short one share, investing the balance in the money market. If the American put is exercised at time $t \leq T$, then we can withdraw X from the money market to buy a share and close the short sale. We shall be left with the call option and a positive amount

$$
(-C_A + P_A + S(0))e^{rt} - X > Xe^{rt} - X \geq 0.
$$

If the put is not exercised at all, then we can buy a share for X by exercising the call at time T and close the short position in stock. On closing the money market position, we shall also end up with a positive amount

$$
(-C_A + P_A + S(0))e^{rT} - X > Xe^{rT} - X > 0.
$$

The theorem, therefore, holds by the No-Arbitrage Principle.   □

Exercise 5.8

Modify the proof of Theorem 5.4 to show that

$$
S(0) - Xe^{-rT} \geq C_A - P_A \geq S(0) - \text{div}_0 - X
$$

for a stock paying a dividend between time 0 and the expiry time T, where div_0 is the value of the dividend discounted to time 0.

Exercise 5.9

Modify the proof of Theorem 5.4 to show that

$$
S(0) - Xe^{-rT} \geq C_A - P_A \geq S(0)e^{-r_{\text{div}}T} - X
$$

for a stock paying dividends continuously at a rate r_{div}.

5.3 Bounds on Option Prices

First of all, we note the obvious inequalities

$$C_{\mathrm{E}} \leq C_{\mathrm{A}}, \quad P_{\mathrm{E}} \leq P_{\mathrm{A}}, \tag{5.7}$$

for European and American options with the same strike price X and expiry time T. They hold because an American option gives at least the same rights as the corresponding European option.

Figure 5.3 shows a scenario of stock prices in which the payoff of a European call is zero at the exercise time T, whereas that of an American call will be positive if the option is exercised at an earlier time $t < T$ when the stock price $S(t)$ is higher than X. Nevertheless, it does not necessarily follow that the inequalities in (5.7) can be replaced by strict ones; see Section 5.3.2, where it is shown that $C_{\mathrm{E}} = C_{\mathrm{A}}$ for call options on an asset that pays no dividends.

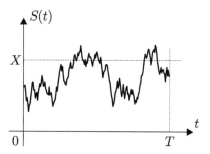

Figure 5.3 Scenario in which an American call can bring a positive payoff, but a European call cannot

Exercise 5.10

Prove (5.7) by an arbitrage argument.

It is also obvious that the price of a call or put option has to be non-negative because an option of this kind offers the possibility of a future gain with no liability. Therefore,

$$C_{\mathrm{E}} \geq 0, \quad P_{\mathrm{E}} \geq 0.$$

Similar inequalities are of course valid for the more valuable American options. In what follows we are going to discuss some further simple bounds for the prices of European and American options.

5.3.1 European Options

We shall establish some upper and lower bounds on the prices of European call and put options.

On the one hand, observe that

$$C_E < S(0).$$

If the reverse inequality were satisfied, that is, if $C_E \geq S(0)$, then we could write and sell the option and buy the stock, investing the balance on the money market. On the exercise date T we could then sell the stock for $\min(S(T), X)$, settling the call option. Our arbitrage profit would be $(C_E - S(0))e^{rT} + \min(S(T), X) > 0$. This proves that $C_E < S(0)$. On the other hand, we have the lower bound

$$S(0) - Xe^{-rT} \leq C_E,$$

which follows immediately by put-call parity, since $P_E \geq 0$. Moreover, put-call parity implies that

$$P_E < Xe^{-rT},$$

since $C_E < S(0)$, and

$$-S(0) + Xe^{-rT} \leq P_E,$$

since $C_E \geq 0$.

These results are summarised in the following proposition and illustrated in Figure 5.4. The shaded areas correspond to option prices that satisfy the bounds.

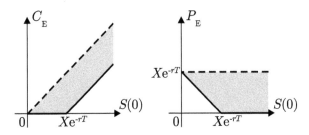

Figure 5.4 Bounds on European call and put prices

Proposition 5.5

The prices of European call and put options on a stock paying no dividends satisfy the inequalities

$$\max\{0, S(0) - Xe^{-rT}\} \leq C_E < S(0),$$
$$\max\{0, -S(0) + Xe^{-rT}\} \leq P_E < Xe^{-rT}.$$

For dividend-paying stock the bounds are

$$\max\{0, S(0) - \mathrm{div}_0 - Xe^{-rT}\} \leq C_E < S(0) - \mathrm{div}_0,$$
$$\max\{0, -S(0) + \mathrm{div}_0 + Xe^{-rT}\} \leq P_E < Xe^{-rT}.$$

Exercise 5.11

Prove the above bounds on option prices for dividend-paying stock.

Exercise 5.12

For dividend-paying stock, sketch the regions of call and put prices determined by the bounds.

5.3.2 Calls on Non-Dividend-Paying Stock

Consider European and American call options with the same strike price X and expiry time T. We know that $C_A \geq C_E$, since the American option gives more rights than its European counterpart. If the underlying stock pays no dividends, then $C_E \geq S(0) - Xe^{-rT}$ by Proposition 5.5. It follows that $C_A > S(0) - X$ if $r > 0$. Because the price of the American option is greater than the payoff, the option will sooner be sold than exercised at time 0.

The choice of 0 as the starting time is of course arbitrary. Replacing 0 by any given $t < T$, we can show by the same argument that the American option will not be exercised at time t. This means that the American option will in fact never be exercised prior to expiry. This being so, it should be equivalent to the European option. In particular, their prices should be equal, leading to the following theorem.

Theorem 5.6

The prices of European and American call options on a stock that pays no dividends are equal, $C_A = C_E$, whenever the strike price X and expiry time T are the same for both options.

Proof

We already know that $C_A \geq C_E$, see (5.7) and Exercise 5.10. If $C_A > C_E$, then write and sell an American call and buy a European call, investing the balance $C_A - C_E$ at the interest rate r. If the American call is exercised at time $t \leq T$, then borrow a share and sell it for X to settle your obligation as writer of the call option, investing X at rate r. Then, at time T you can use the European call to buy a share for X and close your short position in stock. Your arbitrage profit will be $(C_A - C_E)e^{rT} + Xe^{r(T-t)} - X > 0$. If the American option is not exercised at all, you will end up with the European option and an arbitrage profit of $(C_A - C_E)e^{rT} > 0$. This proves that $C_A = C_E$. $\qquad\square$

Theorem 5.6 may seem counter-intuitive at first sight. While it is possible to gain $S(t) - X$ by exercising an American call option if $S(t) > X$ at time $t < T$, this is not so with a European option, which cannot be exercised at time $t < T$. It might, therefore, appear that the American call option should be more valuable than the European one. Nevertheless, there is no contradiction. Even though a European call option cannot be exercised at time $t < T$, it can be sold for at least $S(t) - X$.

The situation is different for dividend-paying stock. Example 6.21 in the next chapter shows a case in which an American call option is worth more than its European counterpart and should be exercised prior to expiry, at least in some scenarios.

On the other hand, it often happens that an American put should be exercised prematurely even if the underlying stock pays no dividends, as in the following example.

Example 5.7

Suppose that the stock price is \$10, the strike price of an American put expiring in one year is \$80, and the interest rate is 16%. Exercising the option now, we can gain \$70, which can be invested at 16% to become \$81.20 after one year. As will be shown in (5.8), the value of a put option cannot possibly exceed the strike price, so we are definitely better off by exercising the option early.

5.3.3 American Options

First we consider options on non-dividend-paying stock. In this case, the price of an American call is equal to that of a European call, $C_A = C_E$, see Theorem 5.6, so it must satisfy the same bounds as in Proposition 5.5. For an American put we have

$$-S(0) + X \le P_A$$

because P_A cannot be less than the payoff of the option at time 0. This gives a sharper lower bound than that for a European put. However, the upper bound has to be relaxed as compared to a European put. Namely,

$$P_A < X. \tag{5.8}$$

Indeed, if $P_A \geq X$, then the following arbitrage strategy could be constructed. Write and sell an American put for P_A and invest this amount at the interest rate r. If the put is exercised at time $t \leq T$, then a share of the underlying stock will have to be bought for X and can then be sold for $S(t)$. The final cash balance will be positive, $P_A e^{rt} - X + S(t) > 0$. If the option is not exercised at all, the final balance will also be positive, $P_A e^{rT} > 0$, at expiry. These results can be summarised as follows.

Proposition 5.8

The prices of American call and put options on a stock paying no dividends satisfy the inequalities

$$\max\{0, S(0) - Xe^{-rT}\} \leq C_A < S(0),$$
$$\max\{0, -S(0) + X\} \leq P_A < X.$$

Next we consider options on dividend-paying stock. The lower bounds for European options imply

$$S(0) - \mathrm{div}_0 - Xe^{-rT} \leq C_E \leq C_A,$$

$$-S(0) + \mathrm{div}_0 + Xe^{-rT} \leq P_E \leq P_A.$$

But because the price of an American option cannot be less than its payoff at any time, we also have $S(0) - X \leq C_A$ and $X - S(0) \leq P_A$. Moreover, the upper bounds $C_A < S(0)$ and $P_A < X$ can be established in a similar manner as for non-dividend-paying stock. All these inequalities can be summarised as follows: for dividend-paying stock

$$\max\{0, S(0) - \mathrm{div}_0 - Xe^{-rT}, S(0) - X\} \leq C_A < S(0),$$
$$\max\{0, -S(0) + \mathrm{div}_0 + Xe^{-rT}, -S(0) + X\} \leq P_A < X.$$

Exercise 5.13

Prove by an arbitrage argument that $C_A < S(0)$ for an American call on dividend-paying stock.

5.4 Variables Determining Option Prices

The option price depends on a number of variables. These can be variables characterising the option, such as the strike price X or expiry time T, variables describing the underlying asset, for example, the current price $S(0)$ or dividend rate r_{div}, variables connected with the market as a whole such as the risk-free rate r, and of course the running time t.

We shall analyse option prices as functions of one of the variables, keeping the remaining variables constant. This is a significant simplification because usually a change in one variable is accompanied by changes in some or all of the other variables. Nevertheless, even the simplified situation will provide interesting insights.

5.4.1 European Options

Dependence on the Strike Price. We shall consider options on the same underlying asset and with the same exercise time T, but with different values of the strike price X. The call and put prices will be denoted by $C_{\text{E}}(X)$ and, respectively, $P_{\text{E}}(X)$ to emphasise their dependence on X. All remaining variables such as the exercise time T, running time t and the underlying asset price $S(0)$ will be kept fixed for the time being.

Proposition 5.9

If $X' < X''$, then

$$C_{\text{E}}(X') \geq C_{\text{E}}(X''),$$
$$P_{\text{E}}(X') \leq P_{\text{E}}(X'').$$

This means that $C_{\text{E}}(X)$ is a decreasing and $P_{\text{E}}(X)$ an increasing function of X.

These inequalities are obvious. The right to buy at a lower price is more valuable than the right to buy at a higher price. Similarly, it is better to sell an asset at a higher price than at a lower one. However, at the exercise time the prices are equal to the payoffs and can be zero for various exercise prices. We shall see that the same may hold prior to exercise in some models.

Exercise 5.14

Give a rigorous arbitrage argument to prove the inequalities in Proposition 5.9.

Proposition 5.10

If $X' < X''$, then

$$C_{\mathrm{E}}(X') - C_{\mathrm{E}}(X'') \le e^{-rT}\,(X'' - X')\,,$$
$$P_{\mathrm{E}}(X'') - P_{\mathrm{E}}(X') \le e^{-rT}\,(X'' - X')\,.$$

Proof

By put-call parity (5.1)

$$C_{\mathrm{E}}(X') - P_{\mathrm{E}}(X') = S(0) - X'e^{-rT},$$
$$C_{\mathrm{E}}(X'') - P_{\mathrm{E}}(X'') = S(0) - X''e^{-rT}.$$

Subtracting, we get

$$(C_{\mathrm{E}}(X') - C_{\mathrm{E}}(X'')) + (P_{\mathrm{E}}(X'') - P_{\mathrm{E}}(X')) = (X'' - X')e^{-rT}.$$

Since, by Proposition 5.9, both terms on the left-hand side are non-negative, neither of them can exceed the right-hand side. $\qquad\square$

Remark 5.11

The inequalities in Proposition 5.10 mean that the call and put prices as functions of the strike price satisfy the Lipschitz condition with constant $e^{-rT} < 1$,

$$|C_{\mathrm{E}}(X'') - C_{\mathrm{E}}(X')| \le e^{-rT}|X'' - X'|,$$
$$|P_{\mathrm{E}}(X'') - P_{\mathrm{E}}(X')| \le e^{-rT}|X'' - X'|.$$

In particular, the slope of the graph of the option price as a function of the strike price is less than $45°$. This is illustrated in Figure 5.5 for a call option.

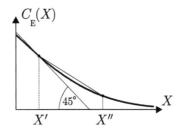

Figure 5.5 Lipschitz property of call prices $C_{\mathrm{E}}(X)$

The reader may suspect that in Proposition 5.10 we should have strict inequalities. However, at exercise the option price is the same as the payoff, and if $S(T) \ge X''$, then we will have equality (with $T = 0$). Later we shall see that the equality may also hold for small T in some models.

Proposition 5.12

For any $\alpha \in [0,1]$ and any X', X''

$$C_E(\alpha X' + (1-\alpha)X'') \le \alpha C_E(X') + (1-\alpha)C_E(X''),$$
$$P_E(\alpha X' + (1-\alpha)X'') \le \alpha P_E(X') + (1-\alpha)P_E(X'').$$

In other words, $C_E(X)$ and $P_E(X)$ are convex functions of X.

Proof

For brevity, we put $X = \alpha X' + (1-\alpha)X''$. Suppose that

$$C_E(X) > \alpha C_E(X') + (1-\alpha)C_E(X'').$$

We can write and sell an option with strike price X, and purchase α options with strike price X' and $1 - \alpha$ options with strike price X'', investing the balance $C_E(X) - (\alpha C_E(X') + (1-\alpha)C_E(X'')) > 0$ on the money market. If the option with strike price X is exercised at expiry, then we shall have to pay $(S(T) - X)^+$. We can raise the amount $\alpha (S(T) - X')^+ + (1-\alpha)(S(T) - X'')^+$ by exercising α calls with strike X' and $1 - \alpha$ calls with strike X''. In this way we will realise an arbitrage profit because of the following inequality, which is easy to verify (the details are left to the reader, see Exercise 5.15):

$$(S(T) - X)^+ \le \alpha (S(T) - X')^+ + (1-\alpha)(S(T) - X'')^+. \qquad (5.9)$$

Convexity for put options follows from that for calls by put-call parity (5.1). Alternatively, an arbitrage argument can be given along similar lines as for call options.   □

Exercise 5.15

Verify inequality (5.9).

Remark 5.13

According to Proposition 5.12, $C_E(X)$ and $P_E(X)$ are convex functions of X. Geometrically, this means that if two points on the graph of the function are joined with a straight line, then the graph of the function between the two points will lie below the line. This is illustrated in Figure 5.6 for call prices.

Dependence on the Underlying Asset Price. The current price $S(0)$ of the underlying asset is given by the market and cannot be altered. However, we can consider an option on a portfolio consisting of x shares, worth $S = xS(0)$.

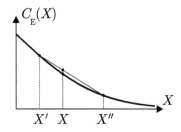

Figure 5.6 Convexity of call prices $C_{\mathrm{E}}(X)$

(Typically, options are traded with packages of shares as the underlying assets.) The payoff of a European call with strike price X on such a portfolio to be exercised at time T is $(xS(T) - X)^+$ and for a put it is $(X - xS(T))^+$. We shall study the dependence of option prices on S. Assuming that all remaining variables are fixed, we shall denote the call and put prices by $C_{\mathrm{E}}(S)$ and $P_{\mathrm{E}}(S)$.

Remark 5.14

The functions $C_{\mathrm{E}}(S)$ and $P_{\mathrm{E}}(S)$ are important because they should reflect the dependence of option prices on very sudden changes of the price of the underlying such that the remaining variables remain almost unaltered. The claim of the next proposition is obvious from the economic point of view. For instance, a call becomes more valuable if the stock goes up because a positive final payoff becomes more likely.

Proposition 5.15

If $S' < S''$, then

$$C_{\mathrm{E}}(S') \leq C_{\mathrm{E}}(S''),$$
$$P_{\mathrm{E}}(S') \geq P_{\mathrm{E}}(S''),$$

that is, $C_{\mathrm{E}}(S)$ is a non-decreasing function and $P_{\mathrm{E}}(S)$ a non-increasing function of S.

Proof

Suppose that $C_{\mathrm{E}}(S') > C_{\mathrm{E}}(S'')$ for some $S' < S''$, where $S' = x'S(0)$ and $S'' = x''S(0)$. We can write and sell a call on a portfolio with x' shares and buy a call on a portfolio with x'' shares, the two options sharing the same strike price X and exercise time T, and we can invest the balance $C_{\mathrm{E}}(S') - C_{\mathrm{E}}(S'') > 0$ on the money market. Since $x' < x''$, the payoffs satisfy $(x'S(T) - X)^+ \leq$

$(x''S(T) - X)^+$. The payoff of the option bought is sufficient to cover that of the option sold. We shall be left with no less than the balance $C_E(S') - C_E(S'') > 0$ plus interest as our arbitrage profit.

The inequality for puts follows by a similar arbitrage argument. □

Exercise 5.16

Prove the inequality in Proposition 5.15 for put options.

Proposition 5.16

Suppose that $S' < S''$. Then

$$C_E(S'') - C_E(S') \le S'' - S',$$
$$P_E(S') - P_E(S'') \le S'' - S'.$$

Proof

We employ put-call parity (5.1):

$$C_E(S'') - P_E(S'') = S'' - Xe^{-rT},$$
$$C_E(S') - P_E(S') = S' - Xe^{-rT}.$$

Subtracting, we get

$$(C_E(S'') - C_E(S')) + (P_E(S') - P_E(S'')) = S'' - S'.$$

Both terms on the left-hand side are non-negative by Proposition 5.15, so each is less than or equal to the right-hand side. □

Remark 5.17

A consequence of Proposition 5.16 is that the slope of the straight line joining two points on the graph of the call or put price as a function of S is no greater than 45°. This is illustrated in Figure 5.7 for call options. In other words, the call and put prices $C_E(S)$ and $P_E(S)$ satisfy the Lipschitz condition with constant 1,

$$|C_E(S'') - C_E(S')| \le |S'' - S'|,$$
$$|P_E(S'') - P_E(S')| \le |S'' - S'|.$$

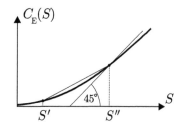

Figure 5.7 Lipschitz property of call prices $C_{\mathrm{E}}(S)$

Proposition 5.18

Let $S' < S''$ and let $\alpha \in [0,1]$. Then

$$C_{\mathrm{E}}(\alpha S' + (1-\alpha)S'') \le \alpha C_{\mathrm{E}}(S') + (1-\alpha)C_{\mathrm{E}}(S''),$$
$$P_{\mathrm{E}}(\alpha S' + (1-\alpha)S'') \le \alpha P_{\mathrm{E}}(S') + (1-\alpha)P_{\mathrm{E}}(S'').$$

This means that the call and put prices are convex functions of S.

Proof

We put $S = \alpha S' + (1-\alpha)S''$ for brevity. Let $S' = x'S(0)$, $S'' = x''S(0)$ and $S = xS(0)$, so $x = \alpha x' + (1-\alpha)\,x''$. Suppose that

$$C_{\mathrm{E}}(S) > \alpha C_{\mathrm{E}}(S') + (1-\alpha)C_{\mathrm{E}}(S'').$$

We write and sell a call on a portfolio with x shares, and purchase α calls on a portfolio with x' shares and $1 - \alpha$ calls on a portfolio with x'' shares, investing the balance $C_{\mathrm{E}}(S) - \alpha C_{\mathrm{E}}(S') - (1-\alpha)C_{\mathrm{E}}(S'')$ on the money market. If the option sold is exercised at time T, then we shall have to pay $(xS(T) - X)^+$. We can cover this liability by the options held since

$$(xS(T) - X)^+ \le \alpha\,(x'S(T) - X)^+ + (1-\alpha)\,(x''S(T) - X)^+,$$

and so this is an arbitrage strategy.

The inequality for put options can be proved by a similar arbitrage argument or using put-call parity. □

5.4.2 American Options

In general, American options have similar properties to their European counterparts. One difficulty is the absence of put-call parity; we only have the weaker estimates in Theorem 5.4. In addition, we have to take into account the possibility of early exercise.

Dependence on the Strike Price. We shall denote the call and put prices by $C_A(X)$ and $P_A(X)$ to emphasise the dependence on X, keeping any other variables fixed.

The following proposition is obvious for the same reasons as for European options: higher strike price makes the right to buy less valuable and the right to sell more valuable.

Proposition 5.19

If $X' < X''$, then

$$C_A(X') \geq C_A(X''),$$
$$P_A(X') \leq P_A(X'').$$

Exercise 5.17

Give a rigorous arbitrage proof of Proposition 5.19.

Proposition 5.20

Suppose that $X' < X''$. Then

$$C_A(X') - C_A(X'') \leq e^{-rT}(X'' - X'),$$
$$P_A(X'') - P_A(X') \leq X'' - X'.$$

Proof

The inequality for calls follows from Theorem 5.6 and Proposition 5.10.

Suppose that $X' < X''$, but $P_A(X'') - P_A(X') > X'' - X'$. Let us write and sell a put with strike X'', buy a put with strike X' and invest the balance $P_A(X'') - P_A(X')$ in the money market. If the written option is exercised at time $t \leq T$, we shall have to pay $X'' - S(t)$. To do so, we can exercise the option with strike X', receiving the payoff $X' - S(t)$. Added to the amount invested in the money market with interest accumulated up to time t, it gives

$$X' - S(t) + (P_A(X'') - P_A(X'))e^{rt} > X' - S(t) + (X'' - X')e^{rt}$$
$$\geq X'' - S(t).$$

The positive surplus above $X'' - S(t)$ would be our arbitrage profit.   □

Proposition 5.21

For any $\alpha \in [0, 1]$ and any X', X''

$$C_A(\alpha X' + (1 - \alpha)X'') \leq \alpha C_A(X') + (1 - \alpha)C_A(X''),$$
$$P_A(\alpha X' + (1 - \alpha)X'') \leq \alpha P_A(X') + (1 - \alpha)P_A(X'').$$

Proof

The inequality for calls follows directly from Theorem 5.6 and Proposition 5.12. For brevity, we put $X = \alpha X' + (1 - \alpha)X''$. Suppose that

$$P_A(X) > \alpha P_A(X') + (1 - \alpha)P_A(X'').$$

We write and sell a put with strike price X and buy α puts with strike price X' and $(1 - \alpha)$ puts with strike price X'', investing the positive balance of these transactions in the money market. If the written option is exercised at time $t \leq T$, then the payoffs of the options held will bring enough money to cover the payment since

$$(X - S(t))^+ \leq \alpha (X' - S(t))^+ + (1 - \alpha) (X'' - S(t))^+ .$$

The investment in the money market will provide an arbitrage profit. □

Dependence on the Underlying Asset Price. Once again, we shall consider options on a portfolio of x shares. The prices of American calls and puts on such a portfolio will be denoted by $C_A(S)$ and $P_A(S)$, where $S = xS(0)$ is the value of the portfolio, all remaining variables being fixed. The payoffs at time t are $(xS(t) - X)^+$ for calls and $(X - xS(t))^+$ for puts.

Proposition 5.22

If $S' < S''$, then

$$C_A(S') \leq C_A(S''),$$
$$P_A(S') \geq P_A(S'').$$

Proof

For calls, this is an immediate consequence of Theorem 5.6 and Proposition 5.15.

Suppose that $P_A(S') < P_A(S'')$ for some $S' < S''$, where $S' = x'S(0)$ and $S'' = x''S(0)$. We can write and sell a put on a portfolio with x'' shares and buy a put on a portfolio with x' shares, both options having the same strike price X and expiry time T. The balance $P_A(S'') - P_A(S')$ of these transactions can be invested in the money market. If the written option is exercised at time $t \leq T$, then we can meet the liability by exercising the other option immediately. Indeed, since $x' < x''$, the payoffs satisfy $(X - x'S(t))^+ \geq (X - x''S(t))^+$. The investment in the money market provides an arbitrage profit. $\qquad\square$

Proposition 5.23

Suppose that $S' < S''$. Then

$$C_A(S'') - C_A(S') \leq S'' - S',$$
$$P_A(S') - P_A(S'') \leq S'' - S'.$$

Proof

The inequality for calls follows from Theorem 5.6 and Proposition 5.16.

For puts, suppose that $P_A(S') - P_A(S'') > S'' - S'$ for some $S' < S''$, where $S' = x'S(0)$ and $S'' = x''S(0)$. We buy $x'' - x' > 0$ shares, buy a put on a portfolio of x'' shares, and sell a put on a portfolio of x' shares, both puts with the same strike price X and expiry T. The cash balance of these transactions is $-(S'' - S') - P_A(S'') + P_A(S') > 0$. Should the holder of the put on x' shares choose to exercise it at time $t \leq T$, we shall need to pay $(X - x'S(t))^+$. By selling $x'' - x'$ shares and exercising the option on x'' shares at time t, we can raise enough cash to settle this liability:

$$(x'' - x')S(t) + (X - x''S(t))^+ \geq (X - x'S(t))^+,$$

since $x'' > x'$. If the put on x' shares is not exercised at all, we need to take no action. In any case, the initial profit plus interest is ours to keep, resulting in an arbitrage opportunity. $\qquad\square$

Proposition 5.24

Let $S' < S''$ and let $\alpha \in [0,1]$. Then

$$C_A(\alpha S' + (1-\alpha)S'') \leq \alpha C_A(S') + (1-\alpha)C_A(S''),$$
$$P_A(\alpha S' + (1-\alpha)S'') \leq \alpha P_A(S') + (1-\alpha)P_A(S'').$$

Proof

The inequality for calls follows from Theorem 5.6 and Proposition 5.18.

Let $S = \alpha S' + (1-\alpha)S''$ and let $S' = x'S(0)$, $S'' = x''S(0)$ and $S = xS(0)$. Suppose that

$$P_A(S) > \alpha P_A(S') + (1-\alpha)P_A(S'').$$

We can write and sell a put on a portfolio with x shares, and purchase α puts on a portfolio with x' shares and $1-\alpha$ puts on a portfolio with x'' shares, all three options sharing the same strike price X and expiry time T. The positive balance $P_A(S) - \alpha P_A(S') - (1-\alpha)P_A(S'')$ of these transactions can be invested in the money market. If the written option is exercised at time $t \leq T$, then we shall have to pay $(X - xS(t))^+$, where $x = \alpha x' + (1-\alpha)x''$. This is an arbitrage strategy because the other two options cover the liability:

$$(X - xS(t))^+ \leq \alpha\left(X - x'S(t)\right)^+ + (1-\alpha)\left(X - x''S(t)\right)^+.$$

$\square$

Dependence on the Expiry Time. For American options we can also formulate a general result on the dependence of their prices on the expiry time T. To emphasise this dependence, we shall now write $C_A(T)$ and $P_A(T)$ for the prices of American calls and puts, assuming that all other variables are fixed.

Proposition 5.25

If $T' < T''$, then

$$C_A(T') \leq C_A(T''),$$
$$P_A(T') \leq P_A(T'').$$

Proof

Suppose that $C_A(T') > C_A(T'')$. We write and sell one option expiring at time T' and buy one with the same strike price but expiring at time T'', investing the balance on the money market. If the written option is exercised at time $t \leq T'$, we can exercise the other option immediately to cover our liability. The positive balance $C_A(T') - C_A(T'') > 0$ invested on the money market will be our arbitrage profit.

The argument is the same for puts. $\square$

5.5 Time Value of Options

The following convenient terminology is often used. We say that at time t a call option with strike price X is

- *in the money* if $S(t) > X$,
- *at the money* if $S(t) = X$,
- *out of the money* if $S(t) < X$.

Similarly, for a put option we say that it is

- *in the money* if $S(t) < X$,
- *at the money* if $S(t) = X$,
- *out of the money* if $S(t) > X$.

Also convenient, though less precise, are the terms *deep in the money* and *deep out of the money*, which mean that the difference between the two sides in the respective inequalities is considerable.

An American option in the money will bring a positive payoff if exercised immediately, whereas an option out of the money will not. We use the same terms for European options, though their meaning is different: even if the option is currently in the money, it may no longer be so on the exercise date, when the payoff may well turn out to be zero. A European option in the money is no more than a promising asset.

Definition 5.26

At time $t \leq T$ the *intrinsic value* of a call option with strike price X is equal to $(S(t) - X)^+$. The intrinsic value of a put option with the same strike price is $(X - S(t))^+$.

We can see that the intrinsic value is zero for options out of the money or at the money. Options in the money have positive intrinsic value. The price of an option at expiry T coincides with the intrinsic value. The price of an American option prior to expiry may be greater than the intrinsic value because of the possibility of future gains. The price of a European option prior to the exercise time may be greater or smaller than the intrinsic value.

Definition 5.27

The *time value* of an option is the difference between the price of the option and its intrinsic value, that is,

$$
\begin{array}{ll}
C_{\mathrm{E}}(t) - (S(t) - X)^+ & \text{for a European call,} \\
P_{\mathrm{E}}(t) - (X - S(t))^+ & \text{for a European put,} \\
C_{\mathrm{A}}(t) - (S(t) - X)^+ & \text{for an American call,} \\
P_{\mathrm{A}}(t) - (X - S(t))^+ & \text{for an American put.}
\end{array}
$$

Example 5.28

Let us examine some typical data. Suppose that the current price of stock is
$125.23 per share. An American call option with strike price $110 is in the
money and has $15.23 intrinsic value. The option price must be at least equal
to the intrinsic value, since the option may be exercised immediately. Typically,
the price will be higher than the intrinsic value because of the possibility of
future gains. On the other hand, a put option with strike price $110 will be out
of the money and its intrinsic value will be zero. The positive price of the put
is entirely due to the possibility of future gains. Similar relationships for other
strike prices can be seen in the table.

Strike Price	Intrinsic Value		Option Price		Time Value	
	Call	Put	Call	Put	Call	Put
110	15.23	0.00	18.40	2.84	3.17	2.84
120	5.23	0.00	12.27	6.46	7.04	6.46
130	0.00	4.77	6.78	9.64	6.78	4.41

The time value of a European call as a function of S is shown in Figure 5.8.
It can never be negative, and for large values of S it exceeds the difference
$X - Xe^{-rT}$. This is because of the inequality $C_E(S) \geq S - Xe^{-rT}$, see Propo-
sition 5.5.

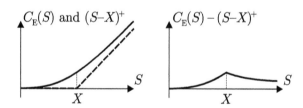

$C_E(S)$ and $(S-X)^+$

$C_E(S) - (S-X)^+$

Figure 5.8   Time value $C_E(S) - (S - X)^+$ of a European call option

The market value of a European put may be lower than its intrinsic value,
that is, the time value may be negative, see Figure 5.9. This may be so only if
the put option is in the money, $S < X$, and it should be deep in the money.
For a European option we have to wait until the exercise time T to realise the
payoff. The risk that the stock price will rise above X in the meantime may be
considerable, which reduces the value of the option.

The time value of an American call option is the same as that of a European
call (if there are no dividends) and Figure 5.8 applies. For an American put, a
typical graph of the time value is shown in Figure 5.10.

Figures 5.8, 5.9 and 5.10 also illustrate the following assertion.

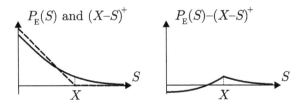

Figure 5.9 Time value $P_{\mathrm{E}}(S) - (X - S)^+$ of a European put option

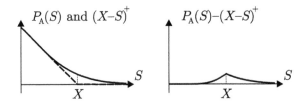

Figure 5.10 Time value $P_{\mathrm{A}}(S) - (X - S)^+$ of an American put option

Proposition 5.29

For any European or American call or put option with strike price X, the time value attains its maximum at $S = X$.

Proof

We shall present an argument for European calls. For $S \leq X$ the intrinsic value of the option is zero. Since $C_{\mathrm{E}}(S)$ is a non-decreasing function of S by Proposition 5.15, this means that the time value is a non-decreasing function of S for $S \leq X$. On the other hand, $C_{\mathrm{E}}(S'') - C_{\mathrm{E}}(S') \leq S'' - S'$ for any $S' < S''$ by Proposition 5.16. It follows that $C_{\mathrm{E}}(S'') - (S'' - X)^+ \leq C_{\mathrm{E}}(S') - (S' - X)^+$ if $X \leq S' < S''$, which means that the time value is a non-increasing function of S for $S \geq X$. As a result, the time value has a maximum at $S = X$.

The proof for other options is similar.   □

Exercise 5.18

Prove Proposition 5.29 for put options.

Case 5: Discussion

If we lock into the strike rate of 1.5 euros to a pound for the currency options available, take a long forward position in stock B with delivery price 65 pounds, and take a short forward position in stock G with delivery price 98 euros, assuming that these forward contracts are available, this will generate a profit at exercise time. We would then be able to sell a share of G for 98 euros, pay 97.50 euros to buy 65 pounds, and purchase B for 65 pounds. We would be left with a profit of 0.50 euros on each share.

However, going forward is free only at the forward prices. Here we have to analyse the costs of taking the above forward positions. These costs, as we know, are the differences between the corresponding option prices. The call price minus the put price is the value of the long forward position. So going long forward in stock B at delivery price 65 pounds will cost $(2.55 - 1.34) \times 1.5 = 1.82$ euros. A short forward contract in G generates $4.145 - 2.738$ euros. To complete the position we have to go long forward in the pound, which costs $65 \times (0.0144 - 0.01316) = 0.0805$ euros. The balance is the cost of 0.4885 euros. This is lower that the future profit, but this cost is paid today, so we face the question of finding the euro interest rate. In practice, this information is readily available, but here it can be extracted from put-call parity on stock G, namely $r = -\frac{1}{T} \ln(\frac{G(0)-C+P}{X}) = 4.994\%$. Borrowing 0.4884 euros at this rate (we live in an ideal world) gives the cost of 0.4905 after a month, so there is an arbitrage profit.

An acute reader may have started the analysis of the case with investigating put-call relationships for all three securities. The pound interest rate generated by stock B is 3.8832%, and then the forward position in the currency should cost 0.00138, as compared to 0.00124 implied by the currency option prices. So call-put parity is violated and arbitrage can be realised in a simpler way following the proof of Theorem 5.2.

6
Binomial Model

Case 6

We are tempted to choose a pension scheme involving risky assets in addition to the risk-free investment. One of the possibilities available within the framework of Chapter 3 would be a portfolio with average annual return of 9% and standard deviation 8.5%. We should be concerned about the accuracy of these parameters. Even if these are correct, there is some danger related to the performance of our investment. Particular scenarios far from expected values may occur. We would like to analyse the behaviour of our portfolio with a time grid more refined than annual. As a first step we could construct a model of quarterly movements; then an extension to monthly steps should be straightforward. Having built such a model, we would like to consider methods of taming the risk.

6.1 Definition of the Model

In the previous chapters we introduced random variables $S(t)$, $0 < t \leq T$ representing future stock prices, with $S(0) > 0$ given. Here we restrict ourselves to evenly spaced discrete time instants $t = 0, h, 2h, \ldots, nh, \ldots$ with a horizon $T = Nh$, where $h > 0$ is the length of a single step. With a slight abuse of notation it is convenient to write $S(n)$ instead of $S(nh)$.

M. Capiński, T. Zastawniak, *Mathematics for Finance*,
Springer Undergraduate Mathematics Series,
© Springer-Verlag London Limited 2011

Our first goal is to build a specific concrete model of random variables $S(n)$. This model is very simple on the one hand, but on the other hand it is quite flexible and very suitable for practical applications.

6.1.1 Single Step

We give a complete exposition of the model already discussed in Chapter 1. To define $S(1)$ we need a probability space. Since we allow $S(1)$ to take only two values, it is sufficient to consider $\Omega = \{u, d\}$ equipped with a field $\mathcal{F}$ of all subsets of Ω and a probability P determined by a single number p such that

$$P(u) = p, \quad P(d) = 1 - p,$$

(see Appendix 10.3). Of course, other choices of Ω are possible.

Now we let

$$S(1) : \Omega \to (0, +\infty)$$

be given by

$$S(1) = S(0)(1 + K),$$

where the rate of return is random and has the form

$$K^\omega = \begin{cases} U & \text{if } \omega = u, \\ D & \text{if } \omega = d, \end{cases}$$

where $-1 < D < U$. Then

$$S^\omega(1) = \begin{cases} S(0)(1 + U) & \text{if } \omega = u, \\ S(0)(1 + D) & \text{if } \omega = d. \end{cases}$$

6.1.2 Two Steps

We assume that the second step has the same form as the first one, the return in the second step being independent of the first return. This requires extending the probability space. To make room for two independent random variables, each with two values U, D, it is convenient to take $\Omega_1 = \Omega_2 = \{u, d\}$ and define $\Omega = \Omega_1 \times \Omega_2 = \{uu, ud, du, ud\}$ (where for better clarity we employ simplified notation writing uu rather than (u, u), for example) with the field $\mathcal{F}$ of all subsets of Ω and probability determined by

$$P(uu) = p^2, \quad P(ud) = P(du) = p(1 - p), \quad P(dd) = (1 - p)^2.$$

Then

$$K^\omega(1) = \begin{cases} U & \text{if } \omega = \text{ud or } \omega = \text{uu}, \\ D & \text{if } \omega = \text{du or } \omega = \text{dd}, \end{cases}$$

$$K^\omega(2) = \begin{cases} U & \text{if } \omega = \text{uu or } \omega = \text{du}, \\ D & \text{if } \omega = \text{ud or } \omega = \text{dd}, \end{cases}$$

are independent. We define

$$S(1) = S(0)(1 + K(1)),$$
$$S(2) = S(1)(1 + K(2)) = S(0)(1 + K(1))(1 + K(2)).$$

At time 2 we thus have three possible stock prices

$$S^\omega(2) = \begin{cases} S(0)(1 + U)^2 & \text{if } \omega = \text{uu}, \\ S(0)(1 + U)(1 + D) & \text{if } \omega = \text{ud or } \omega = \text{du}, \\ S(0)(1 + D)^2 & \text{if } \omega = \text{dd}. \end{cases}$$

The ud and du scenarios give the same result.

6.1.3 General Case

For N steps we take $\Omega = \{\text{u}, \text{d}\}^N$ with the field $\mathcal{F}$ of all subsets of Ω and probability

$$P(\omega) = p^k (1 - p)^{N-k}$$

whenever there are k symbols u (and then $N - k$ symbols d) in the sequence $\omega = \omega_1 \cdots \omega_N$ with $\omega_1, \ldots, \omega_N \in \{\text{u}, \text{d}\}$. The return $K(n)$ is defined by means of the n-th coordinate of the sequence ω:

$$K^\omega(n) = \begin{cases} U & \text{if } \omega_n = \text{u}, \\ D & \text{if } \omega_n = \text{d}. \end{cases}$$

Then, inductively, extending the two-step scheme, we write

$$S(n + 1) = S(n)(1 + K(n + 1)).$$

For example,

$$S^\omega(3) = \begin{cases} S(0)(1 + U)^3 & \text{if } \omega = \text{uuu}, \\ S(0)(1 + U)^2(1 + D) & \text{if } \omega = \text{duu or } \omega = \text{udu or } \omega = \text{uud}, \\ S(0)(1 + U)(1 + D)^2 & \text{if } \omega = \text{udd or } \omega = \text{dud or } \omega = \text{ddu}, \\ S(0)(1 + D)^3 & \text{if } \omega = \text{ddd}. \end{cases}$$

There are fewer stock prices at time n than scenarios. The emerging tree representing the stock prices is called recombining and has the form (here we take $N = 3$)

$S(0)$	$S(1)$	$S(2)$	$S(3)$

$$p \quad S(0)(1+U)^3$$
$$\nearrow$$
$$p \quad S(0)(1+U)^2$$
$$\nearrow \qquad \searrow$$
$$p \quad S(0)(1+U) \qquad {}^{1-p}_{p} \; S(0)(1+U)^2(1+D)$$
$$\nearrow \qquad \searrow \qquad \nearrow$$
$$S(0) \qquad\qquad {}^{1-p}_{p} \; S(0)(1+U)(1+D)$$
$$\searrow \qquad \nearrow \qquad \searrow$$
$${}^{1-p} \; S(0)(1+D) \qquad\qquad {}^{1-p}_{p} \; S(0)(1+U)(1+D)^2$$
$$\searrow \qquad \nearrow$$
$${}^{1-p} \quad S(0)(1+D)^2$$
$$\searrow$$
$${}^{1-p} \quad S(0)(1+D)^3$$

6.1.4 Flow of Information

As time goes by the price movements gradually reveal information about the scenario ω realised in the financial world. We restrict our attention to the case $N = 3$, which is sufficient to explain the ideas. At the beginning each element of

$$\Omega = \{\text{uuu, uud, udu, udd, duu, dud, ddu, ddd}\},$$

can be realised as a future scenario. After one step we shall have more information. If the return on the stock turns out to be U, then the scope is restricted to

$$B_{\text{u}} = \{\text{uuu, uud, udu, udd}\}.$$

Otherwise, if the return is D, then the scenario must be among the elements of

$$B_{\text{d}} = \{\text{duu, dud, ddu, ddd}\}.$$

The knowledge of the first two returns allows us to tell which of the following four sets the scenario belongs to:

$$B_{\text{uu}} = \{\text{uuu, uud}\}, \quad B_{\text{ud}} = \{\text{udu, udd}\},$$
$$B_{\text{du}} = \{\text{duu, dud}\}, \quad B_{\text{dd}} = \{\text{ddu, ddd}\}.$$

After each piece of information is revealed, our uncertainty is reduced.

Mathematically, we have so-called partitions of Ω, meaning that $\Omega = B_u \cup B_d$, or $\Omega = B_{uu} \cup B_{ud} \cup B_{du} \cup B_{dd}$, where the components are pairwise disjoint. This motivates the following general definition.

Definition 6.1

A family $\mathcal{P} = \{B_i\}$ of events is a *partition* of Ω if $B_i \neq \emptyset$, $B_i \cap B_j = \emptyset$ for $i \neq j$ and $\Omega = \bigcup_i B_i$.

Notice that $B_u = B_{uu} \cup B_{ud}$, $B_d = B_{du} \cup B_{dd}$. We say that $\mathcal{P}_2 = \{B_{uu}, B_{ud}, B_{du}, B_{dd}\}$ is a finer partition than $\mathcal{P}_1 = \{B_u, B_d\}$.

Definition 6.2

Given two partitions $\mathcal{P}$, $\mathcal{P}'$, we say that $\mathcal{P}'$ is *finer* than $\mathcal{P}$ if each element of $\mathcal{P}$ can be represented as a union of some elements of $\mathcal{P}'$.

Note that the random variables $S(1)$, $S(2)$ are constant on the components of the corresponding partition $\mathcal{P}_1$, $\mathcal{P}_2$, respectively. For example, $S(1)$ is constant and equal to $S(0)(1+U)$ on B_u. To complete the picture, the same can be said about $S(0)$ being constant on the whole Ω, with $\mathcal{P}_0 = \{\Omega\}$, and $S(3)$ being constant on each of the (single-element) components of $\mathcal{P}_3 = \{B_{uuu}, \ldots, B_{ddd}\}$, where

$$B_{uuu} = \{uuu\}, \quad B_{uud} = \{uud\}, \quad B_{udu} = \{udu\}, \quad B_{udd} = \{udd\},$$
$$B_{duu} = \{duu\}, \quad B_{dud} = \{dud\}, \quad B_{ddu} = \{ddu\}, \quad B_{ddd} = \{ddd\}.$$

Definition 6.3

In general, for any integer $n \leq N$ and any $v_1, \ldots, v_n \in \{u, d\}$ we define $B_{v_1 \cdots v_n}$ to be the set consisting of all $\omega \in \Omega$ such that the first n elements in the sequence $\omega = \omega_1 \cdots \omega_n$ are $\omega_1 = v_1, \ldots, \omega_n = v_n$. The partition $\mathcal{P}_n$ is defined as the family of all such subsets $B_{v_1 \cdots v_n}$ of Ω.

6.1.5 Filtration

A common framework for analysing the flow of information is based on using fields of sets instead of partitions. Given a finite partition $\mathcal{P}$, we consider the family of sets $\mathcal{F}$ consisting of the empty set $\emptyset$ and all possible unions of components in $\mathcal{P}$. The resulting family $\mathcal{F}$ of sets is a field of subsets of Ω (see Appendix 10.3), referred to as the *field extending partition* $\mathcal{P}$.

Remark 6.4

If Ω is infinite, a further extension of the notion of a field is necessary (allowing countable unions), but here we have no need for this.

Definition 6.5

We denote by $\mathcal{F}_n$ the field extending the partition $\mathcal{P}_n$.

For example, for $N = 3$

$$\mathcal{F}_0 = \{\emptyset, \Omega\},$$
$$\mathcal{F}_1 = \{\emptyset, \Omega, B_u, B_d\},$$
$$\mathcal{F}_2 = \{\emptyset, \Omega, B_{uu}, B_{ud}, B_{du}, B_{dd}, B_{uu} \cup B_{ud}, B_{uu} \cup B_{du},$$
$$B_{uu} \cup B_{dd}, B_{ud} \cup B_{du}, B_{ud} \cup B_{dd}, B_{du} \cup B_{dd},$$
$$\Omega \setminus B_{uu}, \Omega \setminus B_{ud}, \Omega \setminus B_{du}, \Omega \setminus B_{dd}\} \quad \text{(16 subsets altogether)},$$
$$\mathcal{F}_3 = 2^\Omega \quad \text{(all subsets of } \Omega\text{)}.$$

We can see that
$$\mathcal{F}_0 \subset \mathcal{F}_1 \subset \mathcal{F}_2 \subset \mathcal{F}_3.$$
This is an example of a *filtration*, which in general is a sequence of fields such that $\mathcal{F}_n \subset \mathcal{F}_{n+1}$.

Remark 6.6

The family $\mathcal{F}_n$ consists of all sets A such that for every scenario $\omega \in \Omega$ it is possible to tell whether or not $\omega \in A$ by knowing the stock prices up to and including time n only (that is, knowing only the first n elements in the sequence $\omega = \omega_1 \cdots \omega_N$).

6.2 Option Pricing

6.2.1 Investment Strategies

We assume that, in addition to trading in stock, which follows the binomial model, we can buy or sell units of so-called *money market account*, a risk-free security whose prices are determined by the interest rate R in the following (deterministic) way:

$$A(n) = A(0)(1 + R)^n.$$

Suppose that $U = 10\%$, $D = -10\%$ and $R = 5\%$ with $S(0) = A(0) = 100$ dollars.

Suppose further that we have at our disposal a certain amount of money, say $V(0) = 300$ dollars, which will be the initial value of our investment. We can use this amount to prepare a portfolio consisting of a position $x(1)$ in stock and $y(1)$ in the money market account. We may decide to buy $x(1) = 2$ shares, and will be left with $V(0) - x(1)S(0)$ to invest risk free, so $y(1) = \frac{V(0) - x(1)S(0)}{A(0)} = 1$. The equality

$$V(0) = x(1)S(0) + y(1)A(0)$$

must hold. The value of the portfolio at time 1 will be

$$\begin{aligned} V^u(1) &= x(1)S^u(1) + y(1)A(1) = 325 \quad \text{on } B_u, \\ V^d(1) &= x(1)S^d(1) + y(1)A(1) = 285 \quad \text{on } B_d. \end{aligned}$$

This is the amount at our disposal for the next step. We assume that there is no withdrawal or injection of external funds, and call such a strategy *self-financing*.

In state u at time 1 we may decide to reduce the position in stock to $x^u(2) = 1$, hence

$$y^u(2) = \frac{V^u(1) - x^u(2) \times S^u(1)}{A(1)} = 2.0476.$$

At time 2 this portfolio will take one of two possible values

$$\begin{aligned} V^{uu}(2) &= x^u(2)S^{uu}(2) + y^u(2)A(2) = 346.75 \quad \text{on } B_{uu}, \\ V^{ud}(2) &= x^u(2)S^{ud}(2) + y^u(2)A(2) = 324.75 \quad \text{on } B_{ud}. \end{aligned}$$

If we find ourselves in state d at time 1, we may choose $x^d(2) = 3$, so

$$y^d(2) = \frac{V^d(1) - x^d(2) \times S^d(1)}{A(1)} = 0.1429.$$

At time 2 we shall have

$$\begin{aligned} V^{du}(2) &= x^d(2)S^{du}(2) + y^d(2)A(2) = 312.75 \quad \text{on } B_{du}, \\ V^{dd}(2) &= x^d(2)S^{dd}(2) + y^d(2)A(2) = 258.75 \quad \text{on } B_{dd}. \end{aligned}$$

We can proceed further in this fashion.

In general, we assume that we build a *strategy*, that is, a sequence of port-folios $(x(n), y(n))$ according to the following principles:

- The strategy is *self-financing*, that is

$$x(n)S(n) + y(n)A(n) = x(n+1)S(n) + y(n+1)A(n),$$

which means that rebalancing the portfolio at time n concerns the choice of $x(n+1)$, and then the value of $y(n+1)$ follows.

- The decision about the position $x(n+1), y(n+1)$ to be held during step $n+1$ is taken on the basis of the information available at time n. Mathematically, this means that $x(n+1)$ and $y(n+1)$ are constant on each of the elements of the partition $\mathcal{P}_n$. We say that such sequences $x(n), y(n)$ are *predictable*.

The value of a strategy is given by

$$V(0) = x(1)S(0) + y(1)A(0),$$
$$V(n) = x(n)S(n) + y(n)A(n) \quad \text{for } n > 0.$$

6.2.2 Single Step

We begin with the case $N = 1$. We shall extend the procedure presented for calls and puts in Section 1.6 to European contingent claim with payoff of the general form $f(S(1))$ for some function f. The value of such an option at time $t = 0, 1$ will be denoted by $H(t)$. At the exercise time

$$H(1) = f(S(1)).$$

As in Section 1.6, the option can be priced by replicating the two possible values $H^{\mathrm{u}}(1), H^{\mathrm{d}}(1)$ of $H(1)$ and computing the time 0 value of the replicating portfolio. This requires solving the system of equations

$$\begin{cases} x(1)S^{\mathrm{u}}(1) + y(1)A(1) = H^{\mathrm{u}}(1), \\ x(1)S^{\mathrm{d}}(1) + y(1)A(1) = H^{\mathrm{d}}(1), \end{cases}$$

so

$$x(1) = \frac{H^{\mathrm{u}}(1) - H^{\mathrm{d}}(1)}{S^{\mathrm{u}}(1) - S^{\mathrm{d}}(1)} = \frac{H^{\mathrm{u}}(1) - H^{\mathrm{d}}(1)}{S(0)(U - D)}$$

which is the replicating position in stock, called the *delta* of the option, and

$$\begin{aligned} y(1) &= \frac{1}{A(1)} \left[H^{\mathrm{u}}(1) - x(1)S^{\mathrm{u}}(1) \right] \\ &= \frac{1}{A(1)} \frac{H^{\mathrm{d}}(1)S^{\mathrm{u}}(1) - H^{\mathrm{u}}(1)S^{\mathrm{d}}(1)}{S^{\mathrm{u}}(1) - S^{\mathrm{d}}(1)} \\ &= \frac{1}{A(0)(1+R)} \frac{H^{\mathrm{d}}(1)(1+U) - H^{\mathrm{u}}(1)(1+D)}{U - D}. \end{aligned}$$

As a consequence of the No-Arbitrage Principle, the price of the claim must be

$$H(0) = x(1)S(0) + y(1)A(0).$$

After substituting the above expressions for $x(1), y(1)$ and rearranging, we get

$$H(0) = \frac{1}{1+R}\left(H^{\mathrm{u}}(1)\frac{R-D}{U-D} + H^{\mathrm{d}}(1)\frac{U-R}{U-D}\right). \tag{6.1}$$

The coefficients $\frac{R-D}{U-D}$ and $\frac{U-R}{U-D}$ add up to one. Moreover, they are both greater than 0 and less than 1 if $D < R < U$, a condition for the lack of arbitrage in the binomial model; see Proposition 1.7. These coefficients can be regarded as probabilities and it is customary to use the notation

$$p_* = \frac{R-D}{U-D}.$$

Then

$$H(0) = \frac{1}{1+R}(H^{\mathrm{u}}(1)p_* + H^{\mathrm{d}}(1)(1-p_*))$$

$$= \frac{1}{1+R}\mathbb{E}_*[H(1)] = \frac{1}{1+R}\mathbb{E}_*[f(S(1))],$$

where the asterisk indicates that the probabilities $p_*, 1 - p_* \in (0,1)$ are used to compute the expectation.

Proposition 6.7

$$\mathbb{E}_*[K(1)] = R.$$

Proof

$$\mathbb{E}_*[K(1)] = Up_* + D(1-p_*) = U\frac{R-D}{U-D} + D\frac{U-R}{U-D} = R.$$

$\square$

Definition 6.8

A probability $(p_*, 1-p_*)$ such that $p_*, 1-p_* \in (0,1)$ under which the expected return $\mathbb{E}_*[K(1)]$ is equal to the risk-free rate R is called *risk-neutral*.

It is convenient to introduce the notation

$$\widetilde{S}(1) = \frac{S(1)}{1+R}, \quad \widetilde{V}(1) = \frac{V(1)}{1+R}, \quad \widetilde{H}(1) = \frac{H(1)}{1+R},$$

calling $\widetilde{S}(1), \widetilde{V}(1), \widetilde{H}(1)$ the *discounted values* of $S(1), V(1), H(1)$.

Proposition 6.9

Under the risk-neutral probability

$$\mathbb{E}_*[\widetilde{S}(1)] = S(0), \quad \mathbb{E}_*[\widetilde{V}(1)] = V(0), \quad \mathbb{E}_*[\widetilde{H}(1)] = H(0).$$

Proof

From Proposition 6.7 we have $\mathbb{E}_*[K(1)] = R$. It follows that

$$
\begin{aligned}
\mathbb{E}_*[\widetilde{S}(1)] &= \frac{1}{1+R}\mathbb{E}_*[S(1)] = \frac{S(0)}{1+R}\mathbb{E}_*[1 + K(1)] \\
&= \frac{S(0)}{1+R}(1 + \mathbb{E}_*[K(1)]) = S(0).
\end{aligned}
$$

Next, since $x(1)$ and $y(1)$ are deterministic (their values are decided at time 0),

$$
\begin{aligned}
\mathbb{E}_*[\widetilde{V}(1)] &= \mathbb{E}_*[x(1)\widetilde{S}(1) + y(1)A(0)] \\
&= x(1)\mathbb{E}_*[\widetilde{S}(1)] + y(1)A(0) \\
&= x(1)S(0) + y(1)A(0) = V(0).
\end{aligned}
$$

The prices of the claim H are just the values of the replicating portfolio so the result follows. □

Exercise 6.1

Show that the price of a call option grows with U, the other variables being kept constant. Analyse the impact of a change of D on the option price.

Exercise 6.2

Find a formula for the price $C_E(0)$ of a European call option if $R = 0$ and $S(0) = X = 1$. Compute the price for $U = 0.05$ and $D = -0.05$, and also for $U = 0.01$ and $D = -0.19$. Draw a conclusion about the relationship between the variance of the return on stock and that on the option.

Exercise 6.3

Find the initial value of the portfolio replicating a call option if proportional transaction costs are incurred whenever the underlying stock is sold. (No transaction costs apply when the stock is bought.) Compare this value with the case free of such costs. Assume that $S(0) = X = 100$, $U = 0.1$, $D = -0.1$ and $R = 0.05$, admitting transaction costs at $c = 2\%$ (the seller receiving 98% of the stock value).

Exercise 6.4

Let $S(0) = 75$ and let $U = 0.2$ and $D = -0.1$. Suppose that you can borrow money at 12%, but the rate for deposits is lower at 8%. Find the values of the replicating portfolios for a put and a call. Is the answer consistent with the put and call prices following from Proposition 6.9?

6.2.3 General Setting

Example 6.10 (Exotic Option)

Consider an interesting example of an option with exercise time $N = 2$ and payoff

$$C(2) = \max\left(0, \frac{S(1) + S(2)}{2} - X\right),$$

a European call option with the average stock price as the underlying security. It is an example of a *path-dependent* or *exotic* option. For $U = 10\%$, $D = -10\%$, $S(0) = 100$, $X = 90$,

$$C(2) = \begin{cases} C^{uu}(2) = 25.5, \\ C^{ud}(2) = 14.5, \\ C^{du}(2) = 4.5, \\ C^{dd}(2) = 0.0. \end{cases}$$

With $R = 5\%$ we find $p_* = 0.75$ and then

$$C^{u}(1) = \frac{1}{1+R}(p_* C^{uu}(2) + (1 - p_*)C^{ud}(2)) = 21.67,$$

$$C^{d}(1) = \frac{1}{1+R}(p_* C^{du}(2) + (1 - p_*)C^{dd}(2)) = 3.21.$$

This gives

$$C(0) = \frac{1}{1+R}(p_* C^{u}(1) + (1 - p_*)C^{d}(1)) = 6.24.$$

The single-step method is applied three times here, starting from the terminal values and going backwards in time.

Inspired by the example, we give a general definition.

Definition 6.11

A *path-dependent* European option with exercise time N and underlying asset S is a contingent claim with payoff of the form $H(N) = f(S(1), \dots, S(N))$ available to the holder only at time N, where f is a function of N variables. A *path-independent* European contingent claim has payoff of the form $H(N) = f(S(N))$ with function f of a single variable.

Particular cases are plain vanilla call and put options with exercise price X, given by $H(N) = f(S(N))$ where $f(x) = (x - X)^+$ for call and $f(x) = (X - x)^+$ for put.

European options can be priced in the binomial model by consecutive backward replication, or equivalently by computing the discounted expected payoff in each single-step subtree. The value obtained in each single-step phase must be the current option price, or otherwise we would see arbitrage opportunities. To formalise this statement we first discuss the No-Arbitrage Principle in the multi-step setting.

We extend the notion of a strategy $x(n), y(n)$ by allowing a third component $z(n)$, as indicated in Section 1.6, representing the position in the contingent claim we are going to price. The value of such a strategy is given by

$$V(0) = x(1)S(0) + y(1)A(0) + z(1)H(0),$$
$$V(n) = x(n)S(n) + y(n)A(n) + z(n)H(n) \quad \text{for } n > 0.$$

The strategy is called *admissible* if $V(n) \geq 0$ for all n.

Definition 6.12

A strategy is an *arbitrage opportunity* if it is admissible, self-financing, predictable, $V(0) = 0$, and such that $V^\omega(n) > 0$ for some n for at least one scenario $\omega \in \Omega$.

We repeat the fundamental assumption:

Assumption 6.13 (No-Arbitrage Principle)

There are no arbitrage opportunities.

Theorem 6.14

The initial value $V(0)$ of an (x, y)-strategy replicating a contingent claim $H(N)$ must be equal to the market price $H(0)$ of that claim.

Proof

Suppose $V(0) > H(0)$. Then we build an (x, y, z)-strategy $(-x(n), -y(n) + \frac{V(0)-H(0)}{A(0)}, 1)$, which has zero initial value, and is worth $A(N)\frac{V(0)-H(0)}{A(0)} > 0$ at time N since $x(N)S(N) + y(N)A(N) = H(N)$. This is a contradiction with the No-Arbitrage Principle. The case $V(0) < H(0)$ can be dealt with similarly. $\square$

6.2.4 Two Steps

We restrict our attention to European claims with path-independent payoff. For $N = 2$ we begin with

$$H(2) = f(S(2)),$$

which has three possible values. For each of the three subtrees in Figure 6.1 we use the one-step replication procedure as described above.

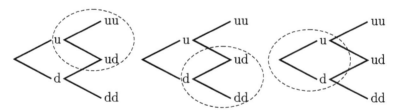

Figure 6.1 Subtrees in the two-step binomial tree

The claim price $H(1)$ has two values

$$H(1) = \begin{cases} \frac{1}{1+R}\left[p_* f(S^{uu}(2)) + (1 - p_*) f(S^{ud}(2))\right] & \text{on } B_u, \\ \frac{1}{1+R}\left[p_* f(S^{du}(2)) + (1 - p_*) f(S^{dd}(2))\right] & \text{on } B_d, \end{cases}$$

found by the one-step procedure applied to the two subtrees at nodes u and d. This gives

$$H(1) = \frac{1}{1+R}\left[p_* f(S(1)(1+U)) + (1 - p_*) f(S(1)(1+D))\right]$$
$$= g(S(1)),$$

where

$$g(x) = \frac{1}{1+R}\left[p_* f(x(1+U)) + (1 - p_*)f(x(1+D))\right].$$

As a result, $H(1)$ can be regarded as a derivative security expiring at time 1 with payoff defined by the function g. (Though it cannot be exercised at time 1, the derivative security can be sold for $H(1) = g(S(1))$.) This means that the one-step procedure can be applied once again to the single subtree at the root of the tree. We have, therefore,

$$H(0) = \frac{1}{1+R} \left[p_* g(S(0)(1+U)) + (1-p_*) g(S(0)(1+D)) \right].$$

It follows that

$$
\begin{aligned}
H(0) &= \frac{1}{1+R} \left[p_* g(S^{\mathrm{u}}(1)) + (1-p_*) g(S^{\mathrm{d}}(1)) \right] \\
&= \frac{1}{(1+R)^2} \left[p_*^2 f(S^{\mathrm{uu}}(2)) + 2p_* (1-p_*) f(S^{\mathrm{ud}}(2)) + (1-p_*)^2 f(S^{\mathrm{dd}}(2)) \right].
\end{aligned}
$$

The last expression in square brackets is the expectation of $f(S(2))$ under the risk-neutral probability P_* defined as

$$P_*(\mathrm{uu}) = p_*^2, \quad P_*(\mathrm{ud}) = P_*(\mathrm{du}) = p_*(1-p_*), \quad P_*(\mathrm{dd}) = (1-p_*)^2.$$

This shows the validity of the following result.

Theorem 6.15

The expectation of the discounted payoff computed with respect to the risk-neutral probability is equal to the present value of the European claim

$$H(0) = \mathbb{E}\left((1+R)^{-2} f(S(2)) \right).$$

Exercise 6.5

Let $S(0) = 120$ dollars and let $U = 0.2, D = -0.1, R = 0.1$. Consider a call option with strike price $X = 120$ dollars and exercise time $T = 2$. Find the option price and the replicating strategy.

Exercise 6.6

Using the data in the previous exercise, find the price of a call and the replicating strategy if a 15 dollar dividend is paid at time 1.

6.2.5 Several Steps

The extension of the results above to a multi-step model is straightforward. Beginning with the payoff at the final step, we proceed backwards, solving the one-step problem repeatedly. Here is the procedure for the three-step model:

$$H(3) = f(S(3)),$$

$$H(2) = \frac{1}{1+R} \left[p_* f(S(2)(1+U)) + (1-p_*)f(S(2)(1+D)) \right]$$
$$= g(S(2)),$$

$$H(1) = \frac{1}{1+R} \left[p_* g(S(1)(1+U)) + (1-p_*)g(S(1)(1+D)) \right]$$
$$= h(S(1)),$$

$$H(0) = \frac{1}{1+R} \left[p_* h(S(0)(1+U)) + (1-p_*)h(S(0)(1+D)) \right],$$

where

$$g(x) = \frac{1}{1+R} \left[p_* f(x(1+U)) + (1-p_*)f(x(1+D)) \right],$$

$$h(x) = \frac{1}{1+R} \left[p_* g(x(1+U)) + (1-p_*)g(x(1+D)) \right].$$

It follows that

$$H(0) = \frac{1}{1+R} \left[p_* h(S^u(1)) + (1-p_*)h(S^d(1)) \right]$$

$$= \frac{1}{(1+R)^2} \left[p_*^2 g(S^{uu}(2)) + 2p_*(1-p_*)g(S^{ud}(2)) + (1-p_*)^2 g(S^{dd}(2)) \right]$$

$$= \frac{1}{(1+R)^3} \left[p_*^3 f(S^{uuu}(3)) + 3p_*^2(1-p_*)f(S^{uud}(3)) \right.$$

$$\left. + 3p_*(1-p_*)^2 f(S^{udd}(3)) + (1-p_*)^3 f(S^{ddd}(3)) \right].$$

The emerging pattern is this: each term in the square bracket is characterised by the number k of upward stock price movements. This number determines the power of p_* and the choice of the payoff value. The power of $1-p_*$ is the number of downward price movements, equal to $3-k$ in the last expression, and $N-k$ in general, where N is the number of steps. The coefficients in front of each term give the number of scenarios (paths through the tree) that lead to the corresponding payoff, equal to $\binom{N}{k} = \frac{N!}{k!(N-k)!}$, the number of k-element combinations out of N elements. For example, there are three paths through the 3-step tree leading to the price $S^{udd}(3)$, namely udd, dud, ddu.

As a result, in the N-step model

$$H(0) = \frac{1}{(1+R)^N} \sum_{k=0}^{N} \binom{N}{k} p_*^k (1-p_*)^{N-k} f\left(S(0)(1+U)^k(1+D)^{N-k} \right). \quad (6.2)$$

The expectation of $f(S(N))$ under the risk-neutral probability

$$P_*(\omega) = p_*^k (1 - p_*)^{N-k},$$

where k is the number of upward stock price movements in the scenario $\omega \in \Omega$, can readily be recognised in this formula. The result can be summarised as follows.

Theorem 6.16

The value of a European path-independent contingent claim with payoff $f(S(N))$ in the N-step binomial model is the expectation of the discounted payoff under the risk-neutral probability:

$$H(0) = \mathbb{E}_* \Big((1 + R)^{-N} f(S(N)) \Big).$$

Remark 6.17

There is no need to know the actual probability p to compute $H(0)$. This remarkable property of the option price is important in practice, as the value of p may be difficult to estimate from market data. Instead, the formula for $H(0)$ features p_*, the risk-neutral probability, which may have nothing in common with p, but is easy to compute from the formula

$$p_* = \frac{R - D}{U - D}.$$

6.2.6 Cox–Ross–Rubinstein Formula

The payoff for a call option with strike price X satisfies $f(x) = 0$ for $x \le X$, which reduces the number of terms in (6.2). The summation starts with the least m such that

$$S(0)(1 + U)^m (1 + D)^{N-m} > X.$$

Hence

$$C_E(0) = (1 + R)^{-N} \sum_{k=m}^{N} \binom{N}{k} p_*^k (1 - p_*)^{N-k} \Big(S(0)(1 + U)^k (1 + D)^{N-k} - X \Big) q^k$$

$$= S(0) \sum_{k=m}^{N} \binom{N}{k} q^k (1 - q)^{N-k} - (1 + R)^{-N} X \sum_{k=m}^{N} \binom{N}{k} p_*^k (1 - p_*)^{N-k},$$

where

$$q = p_* \frac{1 + U}{1 + R}, \quad 1 - q = (1 - p_*) \frac{1 + D}{1 + R}.$$

A similar formula can be derived for put options, either directly or using put-call parity.

These important results are summarised in the following theorem, in which $\Phi(m, N, p)$ denotes the cumulative binomial distribution with N trials and probability p of success in each trial,

$$\Phi(m, N, p) = \sum_{k=0}^{m} \binom{N}{k} p^k (1-p)^{N-k}.$$

Theorem 6.18 (Cox–Ross–Rubinstein Formula)

In the binomial model, the price of a European call and put option with strike price X to be exercised after N time steps is given by

$$C_E(0) = S(0) \left[1 - \Phi(m-1, N, q)\right] - (1+R)^{-N} X \left[1 - \Phi(m-1, N, p_*)\right],$$
$$P_E(0) = -S(0)\Phi(m-1, N, q) + (1+R)^{-N} X\Phi(m-1, N, p_*).$$

Exercise 6.7

Let $S(0) = 50$, $R = 5\%$, $U = 30\%$ and $D = -10\%$. Find the price of a European call and put with strike price $X = 60$ to be exercised after $N = 3$ time steps.

Exercise 6.8

Let $S(0) = 50$, $R = 0.5\%$, $U = 1\%$ and $D = -1\%$. Find m and the price $C_E(0)$ of a European call option with strike $X = 60$ to be exercised after $N = 50$ time steps.

6.3 American Claims

An American option can be exercised at any time n such that $0 \le n \le N$, with payoff $f(S(n))$. The function f is the same for each n. Of course, it can be exercised only once. The price of an American claim at time n will be denoted by $H_A(n)$.

To begin with, we shall analyse an American claim expiring after 2 time steps. Unless it has already been exercised, at expiry it will be worth

$$H_A(2) = f(S(2)),$$

where we have three possible values depending on the values of $S(2)$. At time 1 the holder will have the choice to exercise immediately, with payoff $f(S(1))$, or to wait until time 2, when the value of the claim will become $f(S(2))$. The value of an option which is not exercised at time 1 can be computed by treating $f(S(2))$ as a one-step European contingent claim to be priced at time 1, which gives

$$\frac{1}{1+R}\left[p_* f(S(1)(1+U)) + (1-p_*)f(S(1)(1+D))\right].$$

In effect, the holder has the choice between this or the immediate payoff $f(S(1))$. The value of the American claim at time 1 will therefore be the higher of the two,

$$H_A(1) = \max\left\{f(S(1)), \frac{1}{1+R}\left[p_* f(S(1)(1+U)) + (1-p_*)f(S(1)(1+D))\right]\right\}$$

$$= f_1(S(1))$$

(a random variable with two values), where

$$f_1(x) = \max\left\{f(x), \frac{1}{1+R}\left[p_* f(x(1+U)) + (1-p_*)f(x(1+D))\right]\right\}.$$

A similar argument gives the value at time 0,

$$H_A(0) = \max\left\{f(S(0)), \frac{1}{1+R}\left[p_* f_1(S(0)(1+U)) + (1-p_*)f_1(S(0)(1+D))\right]\right\}.$$

Example 6.19

To illustrate the above procedure we consider an American put option with strike price $X = 80$ expiring at time 2 on a stock with initial price $S(0) = 80$ in a binomial model with $U = 0.1$, $D = -0.05$ and $R = 0.05$. (We consider a put, as we know that there is no difference between American and European call options, see Theorem 5.6.) The stock values are

n	0	1	2
			96.80
		88.00 $<$	
$S(n)$	80.00 $<$		83.60
		76.00 $<$	
			72.20

The price of the American put will be denoted by $P_A(n)$ for $n = 0, 1, 2$. At expiry the payoff will be positive only in the scenario with two downward stock price movements,

n	0		1		2
					0.00
			?	<	
$P_A(n)$	?	<			0.00
			?	<	
					7.80

At time 1 the option writer can choose between exercising the option immediately or waiting until time 2. In the up state at time 1 both the immediate
payoff and the value of the option if not exercised are zero. In the down state
the immediate payoff is 4, while the value of the option if not exercised is
$1.05^{-1} \times \frac{1}{3} \times 7.8 \cong 2.48$. The option holder will choose the higher value (exercising the option in the down state at time 1). This gives the time 1 values of
the American put,

n	0		1		2
					0.00
			0.00	<	
$P_A(n)$	?	<			0.00
			4.00	<	
					7.80

At time 0 the choice is, once again, between the payoff, which is zero, or the
value of the option if not exercised at that time, which is $1.05^{-1} \times \frac{1}{3} \times 4 \cong 1.27$
dollars. Taking the higher of the two completes the tree of option prices,

n	0		1		2
					0.00
			0.00	<	
$P_A(n)$	1.27	<			0.00
			4.00	<	
					7.80

For comparison, the price of a European put is $P_E(0) = 1.05^{-1} \times \frac{1}{3} \times 2.48 \cong 0.79$,
clearly less than the American put price $P_A(0) \cong 1.27$.

This can be generalised, leading to the following definition.

Definition 6.20

An *American derivative security* or *contingent claim* with payoff function f
expiring at time N is a sequence of random variables defined by backward

induction:

$$H_A(N) = f(S(N)),$$

$$H_A(N-1) = \max\left\{ f(S(N-1)), \frac{1}{1+R}[p_* f(S(N-1)(1+U)) \right.$$
$$\left. + (1-p_*)f(S(N-1)(1+D))]\right\} =: f_{N-1}(S(N-1)),$$

$$H_A(N-2) = \max\left\{ f(S(N-2)), \frac{1}{1+R}[p_* f_{N-1}(S(N-2)(1+U)) \right.$$
$$\left. + (1-p_*)f_{N-1}(S(N-2)(1+D))]\right\} =: f_{N-2}(S(N-2)),$$

$$\vdots$$

$$H_A(1) = \max\left\{ f(S(2)), \frac{1}{1+R}[p_* f_2(S(1)(1+U)) \right.$$
$$\left. + (1-p_*)f_2(S(1)(1+D))]\right\} =: f_1(S(1)),$$

$$H_A(0) = \max\left\{ f(S(0)), \frac{1}{1+R}[p_* f_1(S(0)(1+U)) \right.$$
$$\left. + (1-p_*)f_1(S(0)(1+D))]\right\}.$$

Exercise 6.9

Compute the value of an American put expiring at time 3 with strike price $X = 62$ on a stock with initial price $S(0) = 60$ in a binomial model with $U = 0.1$, $D = -0.05$ and $R = 0.03$.

Exercise 6.10

Compare the prices of an American call and a European call with strike price $X = 120$ expiring at time 2 on a stock with initial price $S(0) = 120$ in a binomial model with $U = 0.2$, $D = -0.1$ and $R = 0.1$.

Example 6.21

The last exercise can be modified to show that the equality of European and American call prices may not hold if a dividend is paid. Suppose that a dividend of 14 is paid at time 2. Otherwise, we shall use the same data as in Exercise 6.10.

The ex-dividend stock prices are

n	0	1	2
			158.80
		144.00 $<$	
$S(n)$ ex-div	120.00 $<$		115.60
		108.00 $<$	
			83.20

The corresponding European and American call values will be

n	0	1	2
			38.80
			38.80
		23.52	
		24.00 $<$	
$C_{\mathrm{E}}(n)$	14.25		0.00
$C_{\mathrm{A}}(n)$	14.55 $<$		0.00
		0.00	
		0.00 $<$	
			0.00
			0.00

The American call should be exercised early in the up state at time 1 with payoff 24 dollars (bold figures), which is more than the value of holding the option to expiry. As a result, the price of the American call is higher than that of the European call.

Exercise 6.11

Compute the prices of European and American puts with strike price $X = 14$ expiring at time 2 on a stock with $S(0) = 12$ in a binomial model with $U = 0.1$, $D = -0.05$ and $R = 0.02$, assuming that a dividend of 2 is paid at time 1.

6.4 Martingale Property

Looking into the future, we can see plenty of uncertainty and would like to have some guidance for our investment decisions. One possibility is to estimate the expected future price of the risky security. To this end we need a model: here we have at our disposal the binomial one, so we continue with the assumption that the prices follow this model.

Example 6.22

Our views will evolve with time as we gather some information and our perspective changes. To focus our attention consider a binomial model with $S(0) = 100$, $U = 10\%$, $D = -10\%$, $R = 5\%$ and $p = 0.8$.

Time step 0. Let us look just one step ahead into the future and compute the expected stock price at time 1, discounted back to time 0:

$$\mathbb{E}(\tilde{S}(1)) = \frac{pS^u(1) + (1-p)S^d(1)}{1+R} = 100.95.$$

We can see that the average growth in stock is above the risk-free rate.

Time step 1. We keep our time 0 perspective, working with discounted values and thinking in terms of 'what if' questions, covering all possible future developments. Two cases need to be analysed:

- The stock goes up to $S^u(1) = 110$, so the scenarios still remaining are in B_u. Working within this set, we evaluate the time 2 expected discounted price

$$\frac{pS^{uu}(2) + (1-p)S^{ud}(2)}{(1+R)^2} = 105.76.$$

- The stock goes down to $S^d(1) = 90$, the future lies in the set B_d, and the expected discounted price is

$$\frac{pS^{du}(2) + (1-p)S^{dd}(2)}{(1+R)^2} = 86.53.$$

We have obtained a random variable on Ω, constant on each element of the partition $\mathcal{P}_1$, and denoted by

$$\mathbb{E}(\tilde{S}(2)|\mathcal{P}_1) = \begin{cases} 105.75 & \text{on } B_u, \\ 85.53 & \text{on } B_d. \end{cases}$$

We call it the *conditional expectation* of $\tilde{S}(2)$ given $\mathcal{P}_1$.

Time step 2. Moving forward one more step in time, we encounter four possibilities:

$$\frac{pS^{uuu}(3) + (1-p)S^{uud}(3)}{(1+R)^3} = 110.80 \quad \text{on } B_{uu},$$
$$\frac{pS^{udu}(3) + (1-p)S^{udd}(3)}{(1+R)^3} = 90.65 \quad \text{on } B_{ud},$$
$$\frac{pS^{duu}(3) + (1-p)S^{dud}(3)}{(1+R)^3} = 90.65 \quad \text{on } B_{du},$$
$$\frac{pS^{ddu}(3) + (1-p)S^{ddd}(3)}{(1+R)^3} = 74.17 \quad \text{on } B_{dd}.$$

In fact there are just three distinct values to consider given that $S^{\mathrm{udu}}(3) = S^{\mathrm{duu}}(3)$ and $S^{\mathrm{udd}}(3) = S^{\mathrm{dud}}(3)$, so that the expected values on B_{ud} and B_{du} coincide. We have a random variable which is constant on each element of the partition $\mathcal{P}_2$, called the conditional expectation of $\widetilde{S}(3)$ given $\mathcal{P}_2$:

$$\mathbb{E}(\widetilde{S}(3)|\mathcal{P}_2) = \begin{cases} 110.80 & \text{on } B_{\mathrm{uu}}, \\ 90.65 & \text{on } B_{\mathrm{ud}} \cup B_{\mathrm{du}}, \\ 74.17 & \text{on } B_{\mathrm{dd}}. \end{cases}$$

Definition 6.23

If X is a random variable on Ω and $\mathcal{P}$ is a partition, then the *conditional expectation* $\mathbb{E}(X|\mathcal{P})$ is defined to be a random variable that is constant on each set $A \in \mathcal{P}$ with value $\mathbb{E}(X|A)$, the conditional expectation of X given event A (see Appendix 10.3).

Remark 6.24

Conditional expectation is often expressed in terms of a field rather than a partition. Namely, if $\mathcal{F}$ is a field extending a partition, then the conditional expectation of X given $\mathcal{F}$ is defined by

$$\mathbb{E}(X|\mathcal{F}) = \mathbb{E}(X|\mathcal{P}).$$

Exercise 6.12

Show that $\mathbb{E}_*(\widetilde{S}(2)|\mathcal{P}_1) = \widetilde{S}(1)$ and $\mathbb{E}_*(\widetilde{S}(3)|\mathcal{P}_2) = \widetilde{S}(2)$, where the expectation is computed under the risk-neutral probability P_*.

This is not a coincidence as we shall demonstrate next.

Theorem 6.25

For each $n = 0, \ldots, N-1$

$$\mathbb{E}_*(\widetilde{S}(n+1)|\mathcal{P}_n) = \widetilde{S}(n).$$

Proof

Since $S(n)$ is constant on any $A \in \mathcal{P}_n$ and $K(n+1)$ is independent of A,

$$
\begin{aligned}
\mathbb{E}_*(\widetilde{S}(n+1)|A) &= \frac{1}{(1+R)^{n+1}} \mathbb{E}_*(S(n+1)|A) \\
&= \frac{1}{(1+R)^{n+1}} S(n)\mathbb{E}_*(1+K(n+1)) \\
&= \frac{1}{(1+R)^{n+1}} S(n)\,(1 + p_*U + (1-p_*)D) \\
&= \frac{1}{(1+R)^{n+1}} S(n)(1+R) \\
&= \widetilde{S}(n).
\end{aligned}
$$

It follows that

$$
\mathbb{E}_*(\widetilde{S}(n+1)|\mathcal{P}_n) = \widetilde{S}(n).
$$

$\square$

Remark 6.26

A process $X(n)$ such that $\mathbb{E}(X(n+1)|\mathcal{P}_n) = X(n)$ for each n is called a *martingale*. According to Theorem 6.25 the discounted price process $\widetilde{S}(n)$ is a martingale under the probability P_*. Because of this, P_* is often referred to as *martingale probability*.

Theorem 6.27

The discounted value $\widetilde{V}(n)$ of a self-financing strategy $x(n), y(n)$ is a martingale under P_*.

Proof

The random variables $x(n+1), y(n+1)$ are constant on any $B \in \mathcal{P}_n$, so

$$
\begin{aligned}
\mathbb{E}_*(\widetilde{V}(n+1)|B) &= \frac{1}{(1+R)^{n+1}} \mathbb{E}_*(x(n+1)S(n+1) + y(n+1)A(n+1)|B) \\
&= x(n+1)\mathbb{E}_*(\widetilde{S}(n+1)|B) + y(n+1) \\
&= x(n+1)\widetilde{S}(n) + y(n+1) \\
&= \frac{1}{(1+R)^n}\,(x(n+1)S(n) + y(n+1)A(n)) \\
&= \widetilde{V}(n),
\end{aligned}
$$

implying that

$$\mathbb{E}_*(\widetilde{V}(n+1)|\mathcal{P}_n) = \widetilde{V}(n),$$

as required. □

Corollary 6.28

Since the value process $H(n)$ of a European contingent claim is equal to the value process $V(n)$ of the replicating strategy, it follows that discounted claim prices $\widetilde{H}(n)$ also form a martingale under P_*.

Remark 6.29

Applying the general property

$$\mathbb{E}(\mathbb{E}(Y|\mathcal{P})) = \mathbb{E}(Y),$$

valid for any random variable Y and partition $\mathcal{P}$, we can derive the following formula

$$H(0) = \mathbb{E}_*((1+R)^{-N}H(N)),$$

so we have an alternative proof of Theorem 6.16.

6.5 Hedging

6.5.1 European Options

Let us put ourselves in the position of an option writer who has sold a European call with strike price $X = 105$ and exercise time $N = 3$ for the no-arbitrage price resulting from the binomial model with $S(0) = 100$ dollars and $U = 20\%$, $D = -10\%, R = 10\%$. The option price can be computed using the CRR formula, which gives $C_E(0) = 23.3075$ dollars.

The writer of the option is exposed to risk. If the stock keeps going up during each of the three steps, reaching 172.80 dollars at time 3, it will be necessary to pay 67.80 dollars to the option holder. Of course, if the stock goes down, the writer may be able to keep the premium charged for the option. In order to make their position safe the writer can hedge the option by means of a replicating strategy.

It is possible, but not necessary, to construct the whole tree of replicating portfolios by working backwards from the payoff. It would not be hard to do it in the case in hand, but for large trees, with many time instants, it would be wasteful of computing resources. Instead, we can work a step at a time starting at time 0 and following the actual stock price changes without the need to analyse all scenarios.

The first portfolio to be constructed at time 0 requires finding the call prices at time 1, for which we again use the CRR formula,

$$C_E(1) = \begin{cases} 33.9394 & \text{on } B_u, \\ 9.0358 & \text{on } B_d, \end{cases}$$

and then computing the stock position

$$x(1) = \frac{C_E^u(1) - C_E^d(1)}{S^u(1) - S^d(1)} = 0.8301$$

and the money market position

$$y(1) = C_E(0) - x(1)S(0) = -59.7045.$$

In practice, the writer would charge a bit more for the option than $C_E(0) = 23.31$ dollars to be compensated for the service.

Next, we wait till time 1 and check the stock price. Suppose that it has gone up to $S^u(1)$. The stock position will need to be rebalanced to become

$$x^u(2) = \frac{C_E^{uu}(2) - C_E^{ud}(2)}{S^{uu}(2) - S^{ud}(2)} = 0.9343,$$

so we have to purchase more stock. The money market position will become

$$y^u(2) = y(1) + \frac{1}{1+R}(x(1) - x^u(2))S^u(1) = -71.0744.$$

We then wait till time 2 and check the stock price again. Suppose it has gone down to $S^{ud}(2)$, so we rebalance the portfolio to

$$x^{ud}(3) = \frac{C_E^{udu}(3) - C_E^{udd}(3)}{S^{udu}(3) - S^{udd}(3)} = 0.7593,$$

$$y^{ud}(3) = y^u(2) + \frac{1}{(1+R)^2}(x^u(2) - x^{ud}(3))S^{ud}(2) = -55.4470.$$

At the next time instant the option will be exercised. Suppose the stock has gone up again and reached $S^{udu}(3) = 129.60$. We are ready to close all positions:

- pay the option holder $-\left(S^{udu}(3) - X\right) = 24.60$ dollars;
- pay $-y^{ud}(3)(1 + R)^3 = 73.80$ dollars to close the money market position (with interest);
- sell $y^{ud}(3) = 0.7593$ of a share worth $S^{udu}(3) = 129.60$ dollars each, and receive $0.7593 \times 129.60 = 98.40$ dollars.

The balance of these three transactions is zero.

We have analysed the udu scenario in some detail, but it is not hard to convince oneself that a similar pattern can be followed no matter which scenario unfolds. As a writer of the option, we have been able to hedge all risk.

Exercise 6.13

Compute the hedging strategy along scenario duu for the same European call as above.

6.5.2 American Options

We are using the same data as in Section 6.5.1 and consider an American put with strike price $X = 100$ and expiry time $N = 2$. Building hedging positions is more difficult than in the European case since we cannot use the CRR formula and we have to compute the whole tree of option prices (on the left we give the stock prices so that we can see the cases where exercising the option prevails):

stock				option			
$S(0)$	$S(1)$	$S(2)$	$S(3)$	$P_A(0)$	$P_A(1)$	$P_A(2)$	$P_A(3)$
			172.8				0.0000
		144				0.0000	
	120		129.6		0.2571		0.0000
100		108		3.1861		0.8485	
	90		97.2		10.0000		2.8000
		81				19.0000	
			72.9				27.1000

Having written and sold one option for 3.1861 dollars, we compute the stock position

$$x(1) = \frac{P_A^u(1) - P_A^d(1)}{S^u(1) - S^d(1)} = -0.3248$$

in the hedging portfolio, which indicates the number of shares to be sold short. The money market position includes the premium received for the option and the short-selling proceeds,

$$y(1) = P_A(0) - x(1)S(0) = 35.6624.$$

We wait till time 1 holding this portfolio. Suppose that the stock has gone down to $S^d(1)$. The holder of the option has the choice to exercise it or not. If the option is exercised, the writer must pay $X - S^d(1) = 10$ dollars. Closing the positions in the hedging portfolio will provide $x(1)S^d(1) + y(1)(1+R) = 10$

dollars, so the writer will break even. If the holder decides not to exercise at this point, then the writer will need to rebalance the portfolio to

$$x^{\mathrm{d}}(2) = \frac{P_{\mathrm{A}}^{\mathrm{du}}(2) - P_{\mathrm{A}}^{\mathrm{dd}}(2)}{S^{\mathrm{du}}(2) - S^{\mathrm{dd}}(2)} = -0.6723,$$

$$y^{\mathrm{d}}(2) = y(1) + \frac{1}{1+R}(x(1) - x^{\mathrm{d}}(2))S^{\mathrm{d}}(1) = 64.0955,$$

which will be held till time 2. Suppose that the stock has gone down again and become $S^{\mathrm{dd}}(2)$ at time 2. If the holder exercises the option, the writer can liquidate the portfolio, receiving $x^{\mathrm{d}}(2)S^{\mathrm{dd}}(2) + y^{\mathrm{d}}(2)(1+R)^2 = 23.1010$ dollars, more than enough to cover the payoff of 19 dollars. If the option is not exercised, then the writer can rebalance the portfolio again as follows:

$$x^{\mathrm{dd}}(3) = \frac{P_{\mathrm{A}}^{\mathrm{ddu}}(3) - P_{\mathrm{A}}^{\mathrm{ddd}}(3)}{S^{\mathrm{ddu}}(3) - S^{\mathrm{ddd}}(3)} = -1.0000,$$

$$y^{\mathrm{dd}}(3) = y^{\mathrm{d}}(2) + \frac{1}{(1+R)^2}\left(x^{\mathrm{d}}(2) - x^{\mathrm{dd}}(3)\right)S^{\mathrm{dd}}(2) = 86.0339.$$

When time 3 is reached, suppose that the stock price turns out to be $S^{\mathrm{ddd}}(3)$. At this time the holder will exercise the option and the writer will have $x^{\mathrm{dd}}(3)S^{\mathrm{ddd}}(3) + y^{\mathrm{dd}}(3)(1+R)^3 = 41.6111$ dollars from liquidating the hedging portfolio, once again more than enough to cover the payoff of 27.10 dollars.

Observe that if the option holder does not exercise the option when the stock price turns out to be $S^{\mathrm{d}}(1)$ at time 1, then this hedging strategy provides more cash than necessary for the writer to cover the payoff when the option is exercised at a later time. The holder would do much better exercising when the stock is $S^{\mathrm{d}}(1)$ at time 1, receiving the payoff of 10 dollars. If the holder then still wants to have a put option that can be exercised at time 2 or 3, it would be possible to purchase one for just

$$\frac{1}{1+R}\left(p_*P_{\mathrm{A}}^{\mathrm{du}}(2) + (1-p_*)P_{\mathrm{A}}^{\mathrm{dd}}(2)\right) = 6.2718,$$

keeping the difference of 3.7282 dollars.

Though we have analysed only the ddd scenario, a similar pattern can be followed to construct a hedging portfolio along any other scenario.

Exercise 6.14

Compute the hedging strategy along scenario udd for the above American put. Should the holder exercise the option early along this scenario or keep it till expiry?

Case 6: Discussion

First we build a binomial tree of quarterly portfolio movements consistent with the expected annual return of 9% and standard deviation 8.5%. Taking three months (quarter of a year) as the time step, we need to find U, D such that $S(4) = S(0)(1 + K(1))(1 + K(2))(1 + K(3))(1 + K(4))$, where the $K(n)$ are independent $\{U, D\}$-valued random variables. To this end we formulate the conditions

$$\mathbb{E}(S(4)) = S(0)(1 + 9\%),$$

$$\text{Var}\left(\frac{S(4) - S(0)}{S(0)}\right) = (8.5\%)^2,$$

which represent a system of equations in three variables p, U, D. There is too much flexibility, and the simplest solution is to postulate $p = 0.5$. Then we get $U = 6.16\%$, $D = -1.80\%$ (all values here and in what follows are approximate). The quarterly risk-free return $R = 1.23\%$ can be computed from the annual interest rate of 5%. This gives the risk-neutral probability $p_* = 0.76$. We feed the binomial tree with $S(0) = 100$ dollars and have

time	0	1	2	3	4
unit price					$127.00
				$119.63	
			$112.69		$117.48
		$106.16		$110.66	
	$100		$104.24		$108.67
		$98.20		$102.37	
			$96.43		$100.52
				$94.69	
					$92.99

Consider the first phase of building up our portfolio (the first 40 years). Suppose that at the end of the first quarter we will be investing $1,000$ dollars (this sum of money will vary and we just consider an exemplary amount). We are going to analyse the various possibilities from the perspective of the beginning of the quarter, comparing the number of units we would purchase at the spot price, the forward price, and the prices resulting from using call options.

The forward price is $F(0,1) = 101.20$ dollars, so we can take the long forward position for $1,000/101.2 = 9.88$ units. The number of units we shall own at the end of the quarter is known in this case, their value being random, which will have an impact on the subsequent steps.

If we consider a call with strike price equal to the average price after the first quarter, equal to 102.18 dollars, the cost of buying a single option will be $C = 2.99$ dollars at the beginning of the period and 3.03 dollars at the end. We have to buy the option at the beginning of the quarter, but the investment will be performed at the end, so we have to borrow the money. We assume that we can borrow at the risk-free rate, an acceptable assumption since we are investing some of our money risk free anyway and would be 'borrowing' from that part of the fund.

We need to decide how many calls to buy (additionally, one could investigate other strike prices). Consider, for example, 5 options. Then in the 'up' case we would buy 5 units at the strike price and spend the remainder of the 1,000 dollars to buy risky asset units at the spot price. In the 'down' case all units will be purchased at the spot price. For instance, in the 'up' state our 1000 dollars will be converted into the following number of units:

$$5 + (1,000 - 15.14 - 5 \times 102.18)/106.2 = 9.46.$$

A summary of the results is given below:

	unhedged		forward		with 5 calls	
	units	worth	units	worth	units	worth
up	9.42	$1,000	9.88	$1,048.70	9.46	$1,004.75
down	10.18	$1,000	9.88	$970.08	10.03	$984.86

The strategy involving calls does not appear attractive.

Next, we discuss the use of derivatives to reduce the risk involved in the second phase of our plan, the final 20 years, when we will be gradually liquidating our position. We begin the second phase with a capital of 1,182,000 dollars accumulated over 40 years (see Case 2). Suppose we pay ourselves equal amounts at the end of each quarter (for simplicity, the growth assumption will be applied on an annual basis), so during the first year these payments are at 27,500 dollars per quarter, which gives an annual amount of 110,000 dollars (approximately 50% of the annual salary at the end of year 40).

We restrict out attention to a single year, so we consider a 4-step binomial tree. Suppose the initial unit price is again 100 dollars and the tree parameters are as above. Hypothetically, assume that all the money is invested in risky assets, so we have 11,820 units.

Consider a strategy where at each node of the tree we buy a certain number of puts with exercise times at the end of the subsequent quarter, with strike prices equal to the expected unit price at the end of the quarter, that is, 102.18, 104.40, 106.68, 109.00 dollars, respectively. The payoffs of these single-step puts are

time	1	2	3	4
put payoff				$0
			$0	
		$0		$0
	$0		$0	
		$0.1584		$0.3304
	$3.9795		$4.3102	
		$7.9739		$8.4777
			$11.9849	
				$16.0142

The corresponding put prices are

time	0	1	2	3
put price				$0
			$0	
		$0.0374		$0.0780
	$0.9391		$1.0171	
		$2.0010		$2.2490
			$6.0690	
				$10.1534

Along each scenario we compute the number of units left after some are sold to make the pension payments and put purchases, and some are bought for the payoffs of the options exercised. For instance, at node d at the end of the first quarter the number of units will be

$$11,820 - \frac{27,500}{98.20} - \frac{5,000 \times 0.9391}{100} + \frac{5,000 \times 3.9795}{98.20} = 11,696.$$

We continue in this fashion. The setting is path dependent: there will be 16 values after 4 steps. Multiplying the number of units by the final asset price, we find the terminal wealth. It turns out optimal to use $8,855$ put options if the worst scenario value is to be maximised.

no of puts	0	8,855
worst value	$992,029.50	$1,158,097.77
best value	$1,380,541.07	$1,369,584.85
average	$1,174,733.86	$1,212,799.96

An alternative would be to buy an American option expiring at the end of the fourth quarter. Here we should discuss various possibilities concerned with the strike price and the number of options. If the strike price is greater than 100.56 dollars, it is optimal to exercise the option at time 0, so this case is not interesting. If the strike is 100 dollars, the option will cost 0.42515 dollars, and will be exercised in the d scenario at time 1. The optimal number of American puts that maximises the worst case result turns out to be 51.253, in which case this result is $1,059,206.41$ dollars. It is better to buy puts with a lower strike price. For example, if the strike is 95 dollars, with $41,016$ puts the corresponding result becomes $1,074,404.40$ dollars (here the options are exercised at the end of step 4 in the dddd scenario), which is approximately the best we can get.

7
General Discrete Time Models

Case 7

Turning back again to our personal pension fund, we know that investing in risky securities has an advantage of high return, even though facing the risk is not desired. Risk can be managed, at least partially, by means of derivative securities. We have employed options written on one stock. However, portfolio theory taught us about the advantages of investing in many risky assets. This reduces risk as a result of diversification, but we would have to cope with many underlying assets. Here, the first problem is concerned with building an appropriate model. As a first step in this direction, we could tackle just two stocks.

7.1 Trinomial Tree Model

A natural generalisation of the binomial tree model extends the range of possible values of the one-step returns K to three. The idea is to allow the price not only to move up or down, but also to take an intermediate value at any given step. A natural probability space would be $\Omega = \{u, m, d\}$ with the probabilities $p, q, 1 - p - q$ of outcomes u, m, d determined by the choice of two numbers p, q such that $0 < p, q, p + q < 1$.

M. Capiński, T. Zastawniak, *Mathematics for Finance*,
Springer Undergraduate Mathematics Series,
© Springer-Verlag London Limited 2011

Condition 7.1

The return K is a random variable of the form

$$K^\omega = \begin{cases} U & \text{if } \omega = \mathrm{u}, \\ M & \text{if } \omega = \mathrm{m}, \\ D & \text{if } \omega = \mathrm{d}, \end{cases}$$

where $D < M < U$.

This means that U and D are the returns corresponding to the upward and downward price movements, as before, whereas M is the return on the intermediate price movement. As in the case of the binomial tree model, if there is more than one risky asset, the condition $D < M < U$ may need to be relaxed.

Condition 7.2

The one-step return R on a risk-free investment is the same at each time step and

$$D < R < U.$$

Since $S(1)/S(0) = 1 + K$, Condition 7.1 implies that $S(1)$ takes three different values,

$$S^\omega(1) = \begin{cases} S^{\mathrm{u}}(1) = S(0)(1+U) & \text{if } \omega = \mathrm{u}, \\ S^{\mathrm{m}}(1) = S(0)(1+M) & \text{if } \omega = \mathrm{m}, \\ S^{\mathrm{d}}(1) = S(0)(1+D) & \text{if } \omega = \mathrm{d}. \end{cases}$$

Exercise 7.1

Prove that the condition $D < R < U$ is equivalent to the absence of arbitrage.

7.1.1 Pricing

In the binomial tree model we successfully used the risk-neutral probability for contingent claim pricing. In the trinomial model the condition $\mathbb{E}_*(K) = R$ (cf. Definition 6.8) for a risk-neutral probability $(p_*, q_*, 1 - p_* - q_*)$ becomes

$$p_*(U - R) + q_*(M - R) + (1 - p_* - q_*)(D - R) = 0. \qquad (7.1)$$

This is one equation with two variables. In general, it can have none or infinitely many solutions such that $p_*, q_*, 1 - p_* - q_* \in (0, 1)$.

Exercise 7.2

Show that the condition $D < R < U$ guarantees that there are solutions to (7.1) such that all three probabilities $p_*, q_*, 1 - p_* - q_*$ belong to $(0, 1)$.

The triple $(p_*, q_*, 1 - p_* - q_*)$, regarded as a vector in $\mathbb{R}^3$, is orthogonal to the vector with coordinates $(U - R, M - R, D - R)$ representing the possible one-step gains (or losses) expressed by means of returns for an investor holding a single share of stock, the purchase of which was financed by a cash loan. This means that $(p_*, q_*, 1 - p_* - q_*)$ lies on the intersection of the triangle $\{(a, b, c) : a, b, c \geq 0, a + b + c = 1\}$ and the plane orthogonal to the gains vector $(U - R, M - R, D - R)$, as in Figure 7.1. Condition 7.2 guarantees that the intersection is non-empty, since the line containing the vector $(U - R, M - R, D - R)$ does not pass through the positive octant. In this case there are infinitely many risk-neutral probabilities, the intersection being a line segment.

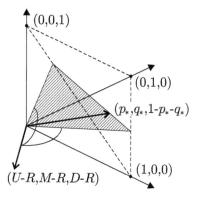

Figure 7.1 Geometric interpretation of risk-neutral probabilities p_*, q_*

Another interpretation of condition (7.1) for the risk-neutral probability is illustrated in Figure 7.2. If masses p_*, q_* and $1 - p_* - q_*$ are attached at the points with coordinates U, M and D on the real axis, then the centre of mass will be at R.

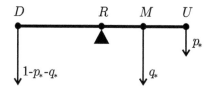

Figure 7.2 Barycentric interpretation of risk-neutral probabilities p_*, q_*

Exercise 7.3

Let $U = 0.2$, $M = 0$, $D = -0.1$ and $R = 0$. Find all risk-neutral probabilities.

In Chapter 6 we discussed the pricing of derivatives in the binomial model by two methods: replication or the discounted expected payoff under the risk-neutral probability. In the case of the trinomial model there are many different risk-neutral probabilities, which may produce different values for the corresponding discounted expected payoff. However, an option trading simultaneously at two different prices would create an instant arbitrage opportunity. The replication method, on the other hand, would also lead to difficulties. A contingent claim $H(1)$ on a probability space $\Omega = \{u, m, d\}$ takes three values $H^u(1)$, $H^m(1)$, $H^d(1)$. If we seek a portfolio (x, y) with value $V(1)$ equal to $H(1)$, this will yield three simultaneous equations

$$\begin{cases} xS^u(1) + yA(0)(1 + R) = H^u(1), \\ xS^m(1) + yA(0)(1 + R) = H^m(1), \\ xS^d(1) + yA(0)(1 + R) = H^d(1), \end{cases}$$

in two variables x, y, which, in general, may have no solution. Pricing by replication would fail in such cases.

7.1.2 Two Assets

As we have seen in Chapter 6, the binomial model is well suited for modelling the price movements of a single asset. However, it is insufficient if an additional risky asset is involved. Working with a single time step, denote by $S_1(n)$, $S_2(n)$, $n = 0, 1$, the prices of two stocks. If we insist on describing the price movements by $S_j(1) = S_j(0)(1 + K_j)$, $j = 1, 2$, with each K_j being non-constant and defined on the same probability space $\Omega = \{u, d\}$, then the random variables K_j must be perfectly correlated with correlation 1 or -1.

Exercise 7.4

Prove this fact.

Perfectly correlated returns are definitely not acceptable in a realistic model. Working in a trinomial model changes everything. Suppose that

$$\begin{cases} K_j^{\mathrm{u}} = U_j, \\ K_j^{\mathrm{m}} = M_j, \\ K_j^{\mathrm{d}} = D_j. \end{cases}$$

Then, provided some technical conditions are satisfied, we can solve the pricing problem. Replication of a contingent claim $H(1)$ is now feasible. For a portfolio (x_1, x_2, y) consisting of three assets, with x_j expressing the position in stock j, we obtain the system of three equations

$$\begin{cases} x_1 S_1^{\mathrm{u}}(1) + x_2 S_2^{\mathrm{u}}(1) + yA(0)(1 + R) = H^{\mathrm{u}}(1), \\ x_1 S_1^{\mathrm{m}}(1) + x_2 S_2^{\mathrm{m}}(1) + yA(0)(1 + R) = H^{\mathrm{m}}(1), \\ x_1 S_1^{\mathrm{d}}(1) + x_2 S_2^{\mathrm{d}}(1) + yA(0)(1 + R) = H^{\mathrm{d}}(1), \end{cases} \qquad (7.2)$$

in three unknowns x_1, x_2, y, so the system can have a single solution. The risk-neutral probability approach is also feasible. If we require that both returns should satisfy the condition $\mathbb{E}_*(K_j) = R$, we end up with a system of two equations for p^*, q^*:

$$\begin{cases} p_*(U_1 - R) + q_*(M_1 - R) + (1 - p_* - q_*)(D_1 - R) = 0, \\ p_*(U_2 - R) + q_*(M_2 - R) + (1 - p_* - q_*)(D_2 - R) = 0. \end{cases} \qquad (7.3)$$

Proposition 7.3

There is a unique solution to the system of equations (7.3) determining the risk-neutral probability if and only if the replication system (7.2) has a unique solution for each claim.

Exercise 7.5

Give a proof of Proposition 7.3.

Example 7.4

Assume $S_1(0) = S_2(0) = 30$, and $U_1 = 20\%$, $M_1 = 10\%$, $D_1 = -10\%$, $U_2 = -5\%$, $M_2 = 15\%$, $D_2 = 10\%$. Assume that $R = 5\%$. Consider a call option on S_1 with strike price 32 so that $C^{\mathrm{u}}(1) = 4$, $C^{\mathrm{m}}(1) = 1$, $C^{\mathrm{d}}(1) = 0$.

The replicating portfolio is $x_1 = 0.25926$, $x_2 = -0.37037$, $y = 4.97354$. The risk-neutral probabilities are $p_* = 0.16667$, $q_* = 0.38889$, $1 - p_* - q_* = 0.44444$.

We can compute the price of the option by taking the initial value of the replicating portfolio or the discounted expected payoff using the risk-neutral probabilities. Each method gives $C(0) = 1.6402$.

However, the existence of unique solutions to the above systems is insufficient as the following example shows.

Example 7.5

Take the same data as in the preceding example with different returns for stock
S_2, namely let $U_2 = 10\%$, $M_2 = 5\%$, $D_2 = -20\%$. Both systems have unique
solutions but the numbers $p_* = 2$, $q_* = -0.83333$, $1 - p_* - q_* = -0.166667$
do not represent probabilities. The replicating portfolio is now $x_1 = 1.55556$,
$x_2 = -1.11111$, $y = -14.5032$ and the option price given by each method is the
same but negative: $C(0) = -1.2698$. This of course violates the No-Arbitrage
Principle.

Later in this chapter we prove two theorems that give a proper foundation
for the pricing problem, consistent with the No-Arbitrage Principle.

Exercise 7.6

Suppose the returns are $U_1 = 15\%$, $M_1 = 5\%$, $D_1 = -10\%$, $U_2 = 20\%$,
$M_2 = 12\%$, $D_2 = -20\%$, $R = 5\%$. Find the risk-neutral probabilities.

7.2 General Model

This section is devoted to describing the so-called primary market consisting of
basic securities (also called underlying securities) such as stocks, in addition to
the money market account. This will then be extended to include contingent
claims (derivative securities). Stock prices are given exogenously and derivatives
prices will be obtained as endogenous quantities within the model of underlying
assets.

As in Chapter 6 we assume that time runs in steps of fixed length h. For
various time-dependent quantities we shall simplify the notation by writing n
in place of the time $t = nh$ of the nth step. For example, we shall write $A(n)$
in place of $A(nh)$.

Suppose that m risky assets are traded. These will be referred to as stocks.
Their prices at time $n = 0, 1, 2, \ldots$ are denoted by $S_1(n), \ldots, S_m(n)$. In addition, investors have at their disposal a risk-free asset. Unless stated otherwise,
we take the initial level of the risk-free investment to be one unit of the home
currency, $A(0) = 1$. Because the money market account can be manufactured
using bonds (see Chapter 2), we shall frequently refer to a risk-free investment

as a position in bonds, finding it convenient to think of $A(n)$ as the bond price at time n.

The positions in risky assets number $1, \ldots, m$ will be denoted by $x_1, \ldots, x_m$, respectively, and the risk-free position by y. The wealth of an investor holding a portfolio of such positions at time n will be

$$V(n) = \sum_{j=1}^{m} x_j S_j(n) + yA(n).$$

We introduce some natural assumptions.

Assumption 7.6 (Randomness)

The future stock prices $S_1(n), \ldots, S_m(n)$ are random variables for any $n = 1, 2, \ldots$. The future bond prices $A(n)$ for any $n = 1, 2, \ldots$ are known numbers.

Assumption 7.7 (Positivity of Prices)

All stock and bond prices are strictly positive,

$$S(n) > 0 \quad \text{and} \quad A(n) > 0 \quad \text{for } n = 0, 1, 2, \ldots .$$

Assumption 7.8 (Divisibility, Liquidity and Short Selling)

An investor may buy, sell and hold any number x_k of stock shares of each kind $k = 1, \ldots, m$ and take any risk-free position y (whether integer or fractional, negative, positive or zero) meaning that

$$x_1, \ldots, x_m, y \in \mathbb{R}.$$

Assumption 7.9 (Discrete Unit Prices)

For each $n = 0, 1, 2, \ldots$ the share prices $S_1(n), \ldots, S_m(n)$ are random variables taking only finitely many values. Mathematically this can be obtained by requiring that the probability space Ω be finite.

7.2.1 Investment Strategies

The positions held by an investor in the risky and risk-free assets can be altered at any time step by selling some assets and investing the proceeds in other assets. In real life, cash can be taken out of the portfolio for consumption or injected from other sources. Nevertheless, we shall assume that no consumption or injection of funds takes place in our model to keep things as simple as possible.

Decisions made by an investor on when to rebalance the portfolio and how many assets to buy or sell are based on the information currently available. We are going to exclude the unlikely possibility that investors could foresee the future, as well as the somewhat more likely (but illegal) one that they will act on insider information. However, all the historical information about the market, up to and including the time instant when a particular trading decision is executed, will be freely available.

Example 7.10

Let $m = 2$ and suppose that

$$
\begin{array}{llll}
S_1(0) = & 60, & S_1(1) = & 65, & S_1(2) = & 75, \\
S_2(0) = & 20, & S_2(1) = & 15, & S_2(2) = & 25, \\
A(0) = 100, & & A(1) = 110, & & A(2) = 121,
\end{array}
$$

in a certain market scenario. At time 0 an initial wealth of $V(0) = 3,000$ dollars is invested in a portfolio consisting of $x_1(1) = 20$ shares of stock number one, $x_2(1) = 65$ shares of stock number two, and $y(1) = 5$ bonds. Our notational convention will be to use 1 rather than 0 as the argument to denote the positions $x_1(1), x_2(1)$ and $y(1)$ created at time 0 so as to reflect the fact that this portfolio will be held over the first time step. At time 1 this portfolio will be worth $V(1) = 20 \times 65 + 65 \times 15 + 5 \times 110 = 2,825$ dollars. At that time the number of assets can be altered by buying or selling some of them, as long as the total value remains \$2,825. For example, we could form a new portfolio consisting of $x_1(2) = 15$ shares of stock one, $x_2(2) = 94$ shares of stock two, and $y(2) = 4$ bonds, which will be held during the second time step. The value of this portfolio will be $V(2) = 15 \times 75 + 94 \times 25 + 4 \times 121 = 3,959$ dollars at time 2, when the positions in stock and bonds can be adjusted once again, as long as the total value remains \$3,959, and so on. However, if no adjustments are made to the original portfolio, then it will be worth \$2,825 at time 1 and \$3,730 at time 2.

Definition 7.11

A *portfolio* is a vector $(x_1(n), \ldots, x_m(n), y(n))$ indicating the number of shares and bonds held by an investor between times $n - 1$ and n. A sequence of portfolios indexed by $n = 1, 2, \ldots$ is called an *investment strategy*. The *wealth* of an investor or the *value of the strategy* at time $n \geq 1$ is

$$V(n) = \sum_{j=1}^{m} x_j(n)S_j(n) + y(n)A(n).$$

At time $n = 0$ the *initial wealth* is given by

$$V(0) = \sum_{j=1}^{m} x_j(1)S_j(0) + y(1)A(0).$$

Strategies employing the underlying assets only will sometimes be referred to as (x, y)-strategies.

The contents of a portfolio can be adjusted by buying or selling some assets at any time step, as long as the current value of the portfolio remains unaltered.

Definition 7.12

An investment strategy is called *self-financing* if the portfolio constructed at time $n \geq 1$ to be held over the next time step $n + 1$ is financed entirely by the current wealth $V(n)$, that is,

$$\sum_{j=1}^{m} x_j(n+1)S_j(n) + y(n+1)A(n) = V(n). \qquad (7.4)$$

Example 7.13

Let the stock and bond prices be as in Example 7.10. Suppose that an initial wealth of $V(0) = 3,000$ dollars is invested by purchasing $x_1(1) = 18.22$ shares of the first stock, short selling $x_2(1) = -16.81$ shares of the second stock, and buying $y(1) = 22.43$ bonds. The time 1 value of this portfolio will be $V(1) = 18.22 \times 65 - 16.81 \times 15 + 22.43 \times 110 = 3,399.45$ dollars. The investor will benefit from the drop in the price of the shorted stock. This example illustrates the fact that portfolios containing fractional or negative numbers of assets are allowed.

As we have already assumed, we do not impose any restrictions on the numbers $x_1(n), \ldots, x_m(n), y(n)$. The fact that they can take non-integer values is referred to as *divisibility*. Negative $x_j(n)$ means that stock number j is *sold short* (in other words, a short position is taken in stock j), negative $y(n)$ corresponds to borrowing cash (taking a short position in the money market, for example, by issuing and selling a bond). The absence of any bounds on the size of these numbers means that the market is *liquid*, that is, any number of assets of each type can be purchased or sold at any time.

In practice some measures to control short selling may be implemented by stock exchanges. Typically, investors are required to deposit a certain percentage of the short sale as a collateral to cover possible losses. If their losses exceed the deposit, the position must be closed. The deposit creates a burden on the portfolio, particularly if it earns no interest for the investor. However, restrictions of this kind may not concern dealers who work for major financial institutions holding a large number of shares deposited by smaller investors. These shares may be borrowed internally in lieu of short selling.

Example 7.14

We continue assuming that stock prices follow the scenario in Example 7.10. Suppose that 20 shares of the first stock are sold short, $x_1(1) = -20$. The investor will receive $20 \times 60 = 1,200$ dollars in cash, but has to pay a security deposit of, say 50%, that is, \$600. One time step later she will suffer a loss of $20 \times 65 - 1,200 = 100$ dollars. This is subtracted from the deposit and the position can be closed by withdrawing the balance of $600 - 100 = 500$ dollars. On the other hand, if 60 shares of the second stock are shorted, that is, $x_2(1) = -60$, then the investor will make a profit of $1,200 - 60 \times 15 = 300$ dollars after one time step. The position can be closed with final wealth $600 + 300 = 900$ dollars. In both cases the final balance should be reduced by $600 \times 0.1 = 60$ dollars, the interest that would have been earned on the amount deposited, had it been invested in the money market.

An investor constructing a portfolio at time n has no knowledge of future stock prices. Investment decisions can be based on the information provided by the market to date and the resulting beliefs concerning the future. As in the case of the binomial model in Chapter 6, this can be formalised in terms of partitions. The Definition 6.1 of a partition extends, without any changes, to the case of an arbitrary finite probability space Ω.

The partition capturing the knowledge provided by the prices of several stocks up to and including time n can be constructed as follows. Take any $\omega_1 \in \Omega$ and consider the prices of all stocks at all times $0, 1, \ldots, n$ along scenario ω_1. Then consider the set B_1 of all scenarios $\omega \in \Omega$ such that the prices of all stocks at all times $0, 1, \ldots, n$ along scenario ω are the same as those along scenario ω_1. Next take any $\omega_2 \in \Omega \setminus B_1$ and repeat this procedure to obtain another subset B_2 of Ω. Repeat this again until all elements of Ω are exhausted. The resulting sets $B_1, B_2, \ldots$ form a partition of Ω, which will be denoted by $\mathcal{P}_n$.

We are now ready to give a formal definition of strategies that do not involve any knowledge of future stock prices.

Definition 7.15

An investment strategy is called *predictable* if for each $n = 0, 1, 2, \ldots$ the portfolio $(x_1(n+1), \ldots, x_m(n+1), y(n+1))$ constructed at time n is constant on each of the components of partition $\mathcal{P}_n$.

The next proposition shows that the position in the risk-free asset in a predictable self financing strategy is determined by the initial wealth and the positions in the risky assets.

Proposition 7.16

Given the initial wealth $V(0)$ and a predictable sequence $(x_1(n), \ldots, x_m(n))$, $n = 1, 2, \ldots$ of positions in risky assets, it is always possible to find a sequence $y(n)$ of risk-free positions such that $(x_1(n), \ldots, x_m(n), y(n))$ is a predictable self-financing investment strategy.

Proof

Put

$$y(1) = \frac{V(0) - x_1(1)S_1(0) - \cdots - x_m(1)S_m(0)}{A(0)}$$

and then compute

$$V(1) = x_1(1)S_1(1) + \cdots + x_m(1)S_m(1) + y(1)A(1).$$

Next, let

$$y(2) = \frac{V(1) - x_1(2)S_1(1) - \cdots - x_m(2)S_m(1)}{A(1)},$$

finding

$$V(2) = x_1(2)S_1(2) + \cdots + x_m(2)S_m(2) + y(2)A(2),$$

and so on. This clearly defines a self-financing strategy. The strategy is predictable because $y(n+1)$ can be expressed in terms of stock and bond prices up to time n:

$$y(n+1) = \frac{V(n) - x_1(n+1)S_1(n) - \cdots - x_m(n+1)S_m(n)}{A(n)},$$

which completes the proof. $\square$

Exercise 7.7

Find the number of bonds $y(1)$ and $y(2)$ held by an investor during the first and second steps of a predictable self-financing investment strategy with initial value $V(0) = 200$ dollars and risky asset positions

$$x_1(1) = 35.24, \quad x_1(2) = -40.50,$$
$$x_2(1) = 24.18, \quad x_2(2) = 10.13,$$

if the prices of assets follow the scenario in Example 7.10. Also find the time 1 value $V(1)$ and time 2 value $V(2)$ of this strategy.

Example 7.17

Once again, suppose that the stock and bond prices follow the scenario in Example 7.10. If an amount $V(0) = 100$ dollars were invested in a portfolio with $x_1(1) = -12$, $x_2(1) = 31$ and $y(1) = 2$, then it would lead to insolvency, since the time 1 value of this portfolio is negative, $V(1) = -12 \times 65 + 31 \times 15 + 2 \times 110 = -95$ dollars .

Such a portfolio might be impossible to construct in practice. No short position would be allowed unless it can be closed at any time and in any scenario (if necessary, by selling other assets in the portfolio to raise cash).

Definition 7.18

A strategy is called *admissible* if it is self-financing, predictable, and for each $n = 0, 1, 2, \ldots$

$$V(n) \geq 0$$

with probability 1.

Exercise 7.8

Consider a market consisting of one risk-free asset with $A(0) = 10$ and $A(1) = 11$ dollars, and one risky asset such that $S(0) = 10$ and $S(1) = 13$ or 9 dollars. On the x, y plane draw the set of all portfolios (x, y) such that the one-step strategy involving risky position x and risk-free position y is admissible.

7.2.2 The Principle of No Arbitrage

We are ready to formulate the fundamental principle underlying all mathematical models in finance. It generalises the simplified one-step version of the

No-Arbitrage Principle in Chapter 1 to models with several time steps and several risky assets. Whereas the notion of a portfolio is sufficient to state the one-step version, in the general setting we need to use a sequence of portfolios forming an admissible investment strategy. This is because investors can adjust their positions at each time step.

Assumption 7.19 (No-Arbitrage Principle)

There is no admissible strategy such that $V(0) = 0$ and $V(n) > 0$ with positive probability for some $n = 1, 2, \ldots$.

Exercise 7.9

Show that the No-Arbitrage Principle would be violated if there was a self-financing predictable strategy with initial value $V(0) = 0$ and final value $V(2) \geq 0$, such that $V(1) < 0$ with positive probability.

The strategy in Exercise 7.9 is not admissible since $V(1)$ may be negative. In fact, this assumption is not essential for the formulation of the No-Arbitrage Principle. An admissible strategy realising an arbitrage opportunity can be found whenever there is a predictable self-financing strategy such that $V(0) = 0$ and $0 \neq V(n) \geq 0$ for some $n > 0$.

Exercise 7.10

Consider a market with one risk-free asset and one risky asset that follows the binomial tree model. Suppose that whenever stock goes up, you know that it will go down at the next step. Find a self-financing (but not necessarily predictable) strategy with $V(0) = 0$, $V(1) \geq 0$ and $0 \neq V(2) \geq 0$.

This exercise indicates that predictability is an essential assumption in the No-Arbitrage Principle. An investor who could foresee the future behaviour of stock prices (here, if stock goes down at one step, you can predict what it will do at the next step) would always be able to find a suitable investment strategy to ensure a risk-free profit.

Exercise 7.11

Consider a market with a risk-free asset such that $A(0) = 100$, $A(1) =$

110, $A(2) = 121$ dollars and a risky asset, the price of which can follow
three possible scenarios,

Scenario	$S(0)$	$S(1)$	$S(2)$
ω_1	100	120	144
ω_2	100	120	96
ω_3	100	90	96

Is there an arbitrage opportunity if a) there are no restrictions on short
selling, and b) no short selling of the risky asset is allowed?

Exercise 7.12

Given the bond and stock prices in Exercise 7.11, is there an arbitrage
strategy if short selling of stock is allowed, but the number of units of
each asset in a portfolio must be an integer?

Exercise 7.13

Given the bond and stock prices in Exercise 7.11, is there an arbitrage
strategy if short selling of stock is allowed, but transaction costs of 5%
of the transaction volume apply whenever stock is traded.

7.3 Fundamental Theorems of Asset Pricing

We restrict ourselves to a simple version for a single-step model. The essential
ideas needed for the general case are covered by the proofs presented below.
These proofs are elementary but not easy. To some extent they go beyond the
average level of difficulty elsewhere in this book.

7.3.1 First Fundamental Theorem

The first fundamental theorem of mathematical finance, which we are now
going to formulate and prove, provides an equivalent condition for the absence
of arbitrage. This theorem generalises the results proved for the binomial model
in Chapter 6, where the lack of arbitrage was shown to be equivalent to $D <
R < U$, which in turn is equivalent to the risk-neutral probability $p_* = \frac{R-D}{U-D}$
satisfying $0 < p_* < 1$.

Here we assume that Ω has n elements $\omega_1, \ldots, \omega_n$ and a probability on

Ω is described by means of a vector $(p_1, \ldots, p_n)$ of positive numbers with $\sum_{j=1}^{n} p_j = 1$ such that p_i is the probability of scenario ω_i for $i = 1, \ldots, n$.

Definition 7.20

We say that probability $(p_1^*, \ldots, p_n^*)$ is risk-neutral if $0 < p_i < 1$ for each $i = 1, \ldots, n$ and

$$\mathbb{E}_*(S_i(1)) = S_i(0)(1 + R),$$

for all i, where $\mathbb{E}_*$ denotes expectation computed under probability $(p_1^*, \ldots, p_n^*)$.

Theorem 7.21

There is no arbitrage if and only if there exists a risk-neutral probability $(p_1^*, \ldots, p_n^*)$.

Proof

Assume first the existence of a risk-neutral probability $(p_1^*, \ldots, p_n^*)$ and consider any portfolio with $V(0) = 0$ and $V(1) \geq 0$. Compute

$$\mathbb{E}_*(V(1)) = \sum_{j=1}^{n} \left(\sum_{i=1}^{m} x_i S_i^{\omega_j}(1) \right) p_j^* = \sum_{i=1}^{m} x_i \left(\sum_{j=1}^{n} S_i^{\omega_j}(1) p_j^* \right)$$

$$= \sum_{i=1}^{m} x_i \mathbb{E}_*(S_i(1)) = (1 + R) \sum_{i=1}^{m} x_i S_i(0)$$

$$= (1 + R)V(0) = 0.$$

This implies that $V(1) = 0$ for all ω since for a non-negative random variable to have zero expectation the random variable must be identically zero (within a finite probability space context). This means that it would be impossible to find an arbitrage opportunity.

For the reverse implication, assume the absence of arbitrage. The set V consisting of all discounted portfolio gains

$$G = \sum_{i=1}^{n} x_i \left(\frac{S_i(1)}{1 + R} - S_i(0) \right), \quad \text{where } x_1, \ldots, x_n \in \mathbb{R},$$

is a vector space, which can be identified with a subspace of $\mathbb{R}^n$ by identifying G with the vector $(G^{\omega_1}, \ldots, G^{\omega_n})$ in $\mathbb{R}^n$.

We claim that each of the elements G of V has the following property: if $G^\omega \geq 0$ for all $\omega \in \Omega$, then $G^\omega = 0$ for all $\omega \in \Omega$. Indeed, suppose that

$$G^\omega = \sum_{i=1}^{n} x_i \left(\frac{S_i^\omega(1)}{1+R} - S_i(0) \right) \geq 0$$

for all $\omega \in \Omega$. Assume zero initial wealth $V(0) = 0$ for a portfolio consisting of positions $x_1, \ldots, x_n$ in the risky assets and a suitable money market position y. The discounted final value of this portfolio will be

$$\tilde{V}(1) = \frac{1}{1+R} \sum_{i=1}^{n} x_i S_i(1) + y$$

$$= \frac{1}{1+R} \sum_{i=1}^{n} x_i S_i(1) + y - V(0)$$

$$= \sum_{i=1}^{n} x_i \left(\frac{S_i(1)}{1+R} - S_i(0) \right) = G \geq 0.$$

Hence to avoid arbitrage, $V(1)$ must be identically equal to zero, and the same can be said about G, which proves the claim.

We consider the set

$$A = \{(q_1, \ldots, q_n) : \sum_{i=1}^{n} q_i = 1, q_i \geq 0\}$$

of all probabilities on Ω. We shall find a vector $(p_1^*, \ldots, p_n^*)$ in A with strictly positive coordinates, and orthogonal to V. The coordinates of such a vector will give the required probabilities p_i^*, since orthogonality means the inner product of $(p_1^*, \ldots, p_n^*)$ and $(G^{\omega_1}, \ldots, G^{\omega_n})$ will be zero, that is,

$$\mathbb{E}_*(G) = \sum_{i=1}^{n} p_i^* G^{\omega_i} = 0$$

for all G. In particular, if $x_i = 1$ with $x_k = 0$ for $k \neq i$, then $G = \frac{S_i(1)}{1+R} - S_i(0)$, so $\mathbb{E}_*(S_i(1)) = S(0)(1+R)$ for all $i = 1, \ldots, m$. The existence of such a vector $(p_1^*, \ldots, p_n^*)$ follows from Lemma 7.22 below. Indeed, A is a convex compact set in $\mathbb{R}^n$, $A \cap V = \emptyset$. According to the Lemma, there is a $z \in \mathbb{R}^n$ orthogonal to V and such that $\sum_{i=1}^{n} z_i a_i > 0$ for all $a \in A$. For any $j = 1, \ldots, n$ it follows that $z_j > 0$ by taking $a \in A$ such that $a_j = 1$ (and then $a_k = 0$ for all $k \neq j$). Putting

$$p_j^* = \frac{z_j}{\sum_{i=1}^{n} z_i}$$

completes the proof of the theorem.   □

Lemma 7.22

If $A \subset \mathbb{R}^n$ is convex and compact and V is a vector subspace of $\mathbb{R}^n$ disjoint with A, then there exists $z \in \mathbb{R}^n$ such that $\langle z, a \rangle = \sum_{i=1}^{n} z_i a_i > 0$ for all $a \in A$, and $\langle z, v \rangle = \sum_{i=1}^{n} z_i v_i = 0$ for all $v \in V$. (Here $\langle \cdot, \cdot \rangle$ is the inner product in $\mathbb{R}^n$).

Proof

Consider the set

$$A - V = \{a - v : a \in A, v \in V\}.$$

Since A and V are disjoint, $0 \notin A - V$.

Next, we can see that $A - V$ is convex and closed since both A and V are. In more detail, if $x, y \in A - V$, $x = a - v$, $y = b - w$, then $\alpha x + (1 - \alpha)y = \alpha a + (1 - \alpha)b - [\alpha v + (1 - \alpha)w] \in A - V$, which proves convexity. To show that $A - V$ is closed, take a sequence $x_n \in A - V$ convergent to some $x \in \mathbb{R}^n$. Then $x_n = a_n - v_n$, the sequence a_n has a convergent subsequence $a_{n_k} \to a$ since A is compact, so v_{n_k} also converges to some $v \in \mathbb{R}^n$. The limit of the sequence x_n must be the same as the limit of x_{n_k} so $x = a - v \in A - V$ as claimed.

Now take z to be the element of $A - V$ closest to 0. For any other element c of $A - V$ we have $|c| \geq |z| > 0$ (recall that $0 \notin A - V$). Existence of such an element z follows from the fact that the mapping $x \mapsto |x|$ is continuous, it is strictly positive on $A - V$, and since this set is closed, the minimum is attained and is different from 0. Consider the triangle with vertices $0, z, c$. We shall prove that $\langle z, c \rangle \geq \langle z, z \rangle = |z|^2 > 0$. The idea is best explained by Figure 7.3. We have

$$\langle z, c \rangle = |z| \, |c| \cos \alpha = |z| \, |c| \frac{|z|}{|y|} = |z|^2 \frac{|c|}{|y|},$$

where the triangle $0, z, y$ has a right angle at z, so it is sufficient to see that $|c| \geq |y|$. Suppose that $|c| < |y|$. The line segment joining c and z is contained in the convex set $A - V$. But the line intersects the interior of the ball with the centre at 0 and radius $|z|$. So this segment contains a point with distance to 0 strictly smaller than $|z|$, a contradiction with the definition of z.

This gives $\langle z, a - v \rangle = \langle z, a \rangle - \langle z, v \rangle \geq \langle z, z \rangle$ for each $v \in V$. However, V is a vector space, so $nv \in V$ for each n, which makes the inequality impossible unless $\langle z, v \rangle = 0$. This in turn yields $\langle z, a \rangle \geq \langle z, z \rangle > 0$, completing the proof. $\square$

7.3.2 Second Fundamental Theorem

In Section 7.1 we have seen that in the single-step trinomial model with one stock it may be impossible to find a portfolio replicating a given contingent

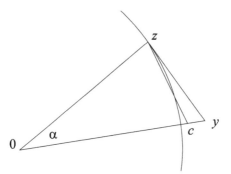

Figure 7.3 Segment z, c intersects the ball of radius $|z|$ and centre at 0

claim. The binomial model, on the other hand, allows for all claims to be replicated. The trinomial model for two stocks may also have this property.

Definition 7.23

A model of a market is called complete if for each European contingent claim there exists a replicating portfolio.

For pricing purposes, replication is a desired property. The pricing formulae derived in simple cases (of binomial trees) show that the price of a claim can be expressed by means of the expectation with respect to the risk-neutral probability. A unique risk-neutral probability gives a unique price, and it can be expected that this will be consistent with the possibility of replication. The following theorem closes this circle of ideas.

Theorem 7.24

A model is arbitrage-free and complete if and only if there exists a unique risk-neutral probability $(p_1^*, \ldots, p_n^*)$.

Proof

The absence of arbitrage implies the existence of a risk-neutral probability $(p_1^*, \ldots, p_n^*)$. We shall show that in a complete model such a vector is unique. We fix any $\omega \in \Omega$ and consider a claim with payoff $H(1) = \mathbb{I}_{\{\omega\}}$, the indicator function of the singleton $\{\omega\}$. By completeness, there exists a portfolio (x, y) such that $H(1) = xS(1) + y(1 + R)$. Take the expectation under $(p_1^*, \ldots, p_n^*)$

on both sides of this equality to get

$$p_j^* = \mathbb{E}_*(\mathbb{I}_{\{\omega\}}) = \mathbb{E}_*(xS(1) + y(1+R)) \tag{7.5}$$
$$= x\mathbb{E}_*(S(1)) + y(1+R) = xS(0)(1+R) + y(1+R),$$

since $\mathbb{E}_*(S(1)) = S(0)(1+R)$ for any risk-neutral probability. This shows that the risk-neutral probability is unique.

For the reverse implication, assume $(p_1^*, \ldots, p_n^*)$ is a unique risk-neutral probability and suppose there is a claim $H(1)$ that cannot be replicated. Our goal is to construct a risk-neutral probability $(q_1^*, \ldots, q_n^*)$ different from $(p_1^*, \ldots, p_n^*)$. Consider the set $V = \{V(1) : (x,y) \in R^{n+1}\}$ of values at time 1 of all portfolios, so that $H(1) \notin V$. These values form a vector subspace of $\mathbb{R}^n$ (with the random variable $V(1)$ identified with the vector $(V^{\omega_1}(1), \ldots, V^{\omega_n}(1)))$. This subspace contains the vector $\mathbf{1} = (1, \ldots, 1)$ because if $x = 0$ and $y = \frac{1}{1+R}$, then $V^\omega(1) = 1$ for all scenarios $\omega \in \Omega$. We equip $\mathbb{R}^n$ with the inner product $\langle z, w \rangle = \sum_{i=1}^n z_j w_j p_j^* = \mathbb{E}_*(zw)$ and use Lemma 7.22 with $A = \{C\}$ to get a $z \in \mathbb{R}^n$ such that $\langle z, C \rangle > 0$ and $\langle z, v \rangle = 0$ for all $v \in V$. The latter, in particular, implies, that $\langle z, \mathbf{1} \rangle = \mathbb{E}_* z = 0$, while the former condition guarantees that $z \neq 0$. Now we can define a risk-neutral probability different from $(p_1^*, \ldots, p_n^*)$:

$$q_j^* = \left(1 + \frac{z_j}{2a}\right) p_j^*$$

where $a = \max_{\omega_j \in \Omega} |z_j|$. This is a probability since

$$\sum_{j=1}^n q_j^* = \sum_{j=1}^n p_j^* + \sum_{j=1}^n \frac{z_j}{2a} p_j^*$$
$$= 1 + \frac{1}{2a}\mathbb{E}_*(z) = 1.$$

Clearly $-a \leq z_j$ so the number $1 + \frac{z_j}{2a}$ is always positive and for at least one value of j it is different from 1 since the vector z is non-zero, thus $q_j^* \neq p_j^*$ for at least one j and $0 < q_j^*$ for all j. Finally, we check the risk-neutral property of $(q_1^*, \ldots, q_n^*)$:

$$\sum_{j=1}^n S^{\omega_j}(1) q_j^* = \mathbb{E}_*\left(\left(1 + \frac{z}{2a}\right) S(1)\right)$$

$$= \mathbb{E}_*(S(1)) + \frac{1}{2a}\mathbb{E}_*(zS(1))$$
$$= S(0)(1+R)$$

since $S(1)$ belongs to V (take $x = 1$, $y = 0$) so $\mathbb{E}_*(zS(1)) = \langle z, S(1) \rangle = 0$, completing the proof. $\qquad\square$

7.4 Extended Models

The model of the primary market is based on trading m stocks and the money market account. Here we want to admit contingent claims among tradable assets, which requires a suitable extension of the model.

Suppose that, in addition to the primary assets, there are k contingent claims of European type with the same exercise date N. These are determined by payoffs $H_i(N)$ for $i = 1, \ldots, k$. The contingent claims can be traded, so we need to add them to our portfolios. The positions in contingent claims will be represented by $z_1, \ldots, z_k$. Other than that, we maintain a similar setup as for the primary market.

Assumption 7.25 (Divisibility, Liquidity and Short Selling)

An investor may buy, sell and hold any number of contingent claims (whether integer or fractional, negative, positive or zero). In general,

$$z_1, \ldots, z_k \in \mathbb{R}.$$

Definition 7.26

A *portfolio* is a vector

$$(x_1(n), \ldots, x_m(n), y(n), z_1(n), \ldots, z_k(n))$$

indicating the positions in the primary securities, the money market account and the derivative securities held by an investor between times $n - 1$ and n. A sequence of portfolios indexed by $n = 1, 2, \ldots$ is called an *extended investment strategy*. The *wealth* of an investor or the *value of the extended strategy* at time $n \geq 1$ is

$$V(n) = \sum_{j=1}^{m} x_j(n) S_j(n) + y(n) A(n) + \sum_{i=1}^{k} z_i(n) H_i(n).$$

At time $n = 0$ the *initial wealth* is given by

$$V(0) = \sum_{j=1}^{m} x_j(1) S_j(0) + y(1) A(0) + \sum_{i=1}^{k} z_i(1) H_i(0).$$

Definition 7.27

An extended investment strategy is called *self-financing* if the portfolio constructed at time $n \geq 1$ to be held over the next time step $n + 1$ is financed entirely by the current wealth $V(n)$, that is,

$$\sum_{j=1}^{m} x_j(n+1)S_j(n) + y(n+1)A(n) + \sum_{i=1}^{k} z_i(n+1)H_i(n) = V(n).$$

Definition 7.28

An extended investment strategy is called *predictable* if for each $n = 0, 1, 2, \ldots$ the portfolio

$$(x_1(n+1), \ldots, x_m(n+1), y(n+1), z_1(n+1), \ldots, z_k(n+1))$$

constructed at time n is constant on each component of partition $\mathcal{P}_n$ (see Section 7.2.1 for the construction of $\mathcal{P}_n$).

Definition 7.29

An extended strategy is called *admissible* if it is self-financing, predictable, and for each $n = 0, 1, 2, \ldots$

$$V(n) \geq 0$$

with probability 1.

The No-Arbitrage Principle extends without any major modifications.

Assumption 7.30 (No-Arbitrage Principle)

There is no admissible extended strategy such that $V(0) = 0$ and $V(n) > 0$ with positive probability for some n.

What we mean by this assumption is this: the processes of contingent claim values $H_i(n)$ have to be such that the extended market model is free of arbitrage opportunities.

Example 7.31

Consider the single-step binomial model of Section 1.3 ($S(0) = 100$ dollars, $U = 0.2$, $D = -0.2$, $R = 0.1$), which, as we know, is free of arbitrage. Extend it by a call with strike price $X = 100$ dollars, so $k = 1$, $H_1(T) = (S(T) - 100)^+$. With $H_1(0) = \mathbb{E}_*(H_1(T)) \cong 13.6364$ dollars the extended model is also free of arbitrage. However, adding $H_2(T) = (S(T) - 90)^+$, $H_2(0) = 19.4545$ creates arbitrage opportunities since the replication price is in fact $\mathbb{E}_*(H_2(T)) \cong 20.4545$ dollars.

When the primary market is free of arbitrage and complete, we can use the theory developed above to obtain unique option prices. For arbitrary contingent claim payoffs $H_i(N)$ we can find primary replicating strategies, and the values of such strategies give the contingent claim values $H_i(n)$. It is obvious that the extended market will then satisfy the No-Arbitrage Principle. In particular, we have the following pricing formulae: $H_i(0) = \mathbb{E}_*(H_i(N)/A(N))$.

On the other hand, for an incomplete primary model, there may be many extensions consistent with the No-Arbitrage Principle. In this case the derivative prices are, in general, non-unique.

Case 7: Discussion

In view of the fundamental theorems, we shall build a trinomial model for two stocks with a unique risk-neutral probability. Suppose that the current stock prices are $S_1(0) = 30$ and $S_2(0) = 50$. The data within the setup of portfolio theory is concerned with expected returns, their standard deviations and the correlation coefficient. Suppose that the expected returns are $\mu_1 = 9\%$, $\mu_2 = 6\%$ with risk-free rate $R = 5\%$, the standard deviations are $\sigma_1 = 11\%$, $\sigma_2 = 13.5\%$ and the correlation coefficient is $\rho_{12} = -0.36$. The parameters needed for the model are 6 returns and 2 probabilities, which leaves some room since the data give 5 equations. For simplicity we assume that the real-life probabilities are $\frac{1}{3}$ in each scenario, and we still have some flexibility left, so the choice proposed below is non-unique: we assume that

$$U_1 = 22\%, \quad M_1 = 10\%, \quad D_1 = -5\%,$$
$$U_2 = -5\%, \quad M_2 = 15\%, \quad D_1 = 8\%,$$

which gives the risk-neutral probabilities

$$p_* = 0.29948, \quad q_* = 0.12760, \quad 1 - p_* - q_* = 0.57292.$$

Suppose that portfolio theory suggests a simple choice of weights, namely: $w_1 = w_2 = 0.5$ (with obvious modifications of the subsequent analysis for any other values). These weights imply that for each share of stock S_2 we have to buy approximately 1.67 shares of stock S_1. The expected return on our portfolio is 7.5%. To insure its future values against unfavourable price movements we consider put options.

Step 1. First we analyse the idea of buying two put options, one for each stock. Suppose we have flexibility as far as the strike price is concerned, and let
$$X_1 = S_1(0)(1 + 7.5\%) = 32.25, \quad X_2 = S_2(0)(1 + 7.5\%) = 53.75$$

so as to guarantee the growth of each stock in line with the expected (required) overall return. It is easy to see that the put prices are

$$P_1 = 2.05, \quad P_2 = 1.78,$$

and the cost of hedging, bearing in mind the weights and the resulting need to cover 1.67 shares of stock S_1 for each share of S_2, gives the cost at $1.67 \times 2.05 + 1.78 = 5.19$. This has to be paid at the beginning of the period. Suppose we borrow the amount risk free (using for this the risk-free part of our investment fund), so that the cost at the end of the period will be 5% more, namely 5.45, approximately. Within the model, the payoff of the portfolio with the put options is

$$V_1^u = 1.67 \times 36.60 + 53.75 = 114.75,$$
$$V_1^m = 1.67 \times 33.00 + 57.50 = 112.50,$$
$$V_1^d = 1.67 \times 32.25 + 54.00 = 105.25,$$

(the put on stock S_1 is exercised in the 'down' scenario, and the put on stock S_2 in the 'up' scenario). The average payoff (using the real-world probabilities of $\frac{1}{3}$) is 110.83, so subtracting the cost of options we get the expected return of $\mu_{V_1} = 3.51\%$. This does not look appealing as it is lower than the risk-free rate.

Step 2. Consider a so-called basket option, whose payoff is a function of both stock prices. Here we take a put on the portfolio with strike 107.5 (the initial value is $30 \times 1.67 + 50 = 100$, so the strike is chosen in the same spirit as before). So we have the payoff of the option in the form

$$\max\{107.5 - (1.67 \times S_1(T) + S_2(T)), 0\}$$

with values 0, 0, 6 at the three scenarios.

The price of the option is 3.27. Increased by 5%, this gives the cost of hedging at the end of the period equal to 3.44. The portfolio with an added basket put has payoffs

$$V_2^u = 1.67 \times 36.60 + 47.50 = 108.50,$$
$$V_2^m = 1.67 \times 33.00 + 57.50 = 112.50,$$
$$V_2^d = 1.67 \times 28.50 + 54.00 + 6 = 107.5,$$

with expected value 109.5. After subtracting the cost of hedging, the expected return is 6.06%.

The latter is a better result than the previous one. This is a consequence of the low price of the basket option. (Note that in more sophisticated models, like the Black–Scholes model which we are about to present, the problem of pricing basket options is difficult with no closed-form solutions.)

The above result should be compared with the unhedged position, where the expected return of 7.5% is higher, but the possible final values of the portfolio are more risky, since in the worst case scenario d we would get only 101.5, with the same amounts in the other two. The choice between hedged and unhedged portfolios depends on the individual preferences of an investor.

8
Continuous Time Model

Case 8

A company manufactures goods in the UK for sale in the USA. The investment to start production is 5 million pounds. Additional funds can be raised by borrowing British pounds at 16% to finance a hedging strategy. The rate of return demanded by investors, bearing in mind the risk involved, is 25%. The sales are predicted to generate 8 million dollars at the end of the year. The costs are 3 million pounds per year. The interest rates are 8% for dollars and 11% for pounds (here we assume continuous compounding). The current rate of exchange is 1.6 dollars to a pound. The volatility of the logarithmic return on the rate of exchange is estimated at $\sigma = 15\%$. The company pays 20% tax on earnings.

8.1 Shortcomings of Discrete Models

Discrete models have considerable advantages, not least the simplicity of the mathematics involved. Very little beyond elementary probability and linear algebra is needed to express important ideas. However, discrete models also have some serious disadvantages as they restrict the range of asset price movements and the set of time instants at which these movements may occur.

As a possible improvement, we shall consider a model obtained as a con-

M. Capiński, T. Zastawniak, *Mathematics for Finance*,
Springer Undergraduate Mathematics Series,
© Springer-Verlag London Limited 2011

tinuous time limit from a sequence of suitably chosen binomial models. This
development parallels the introduction of derivatives and differential equations
at some point in the history of mathematics. Such tools give tremendous power
in addition to more accurate modelling and elegance.

A mathematically precise introduction to continuous models is beyond the
level of this book. While the presentation will be lacking full rigour, the facts
are solid, and we shall be walking largely along the lines of complete proofs.

8.2 Continuous Time Limit

Our aim will be to reach the celebrated continuous Black–Scholes market model
as a limit of a sequence of binomial models.

By $T > 0$ we denote the time horizon (measured in years). In the corresponding binomial model with N steps the duration of each time step will be
$h = \frac{T}{N}$. Time instants t between 0 and T will be represented as multiples of h,
so that $t = nh$, where $n = 0, \ldots, N$. For such t's the stock and risk-free asset prices in the N-step binomial model will be denoted by $S_N(t)$ and $A_N(t)$,
respectively.

8.2.1 Choice of N-Step Binomial Model

In Chapter 6 we expressed the changes of the risk-free asset and stock prices
in terms of single-step returns. Clearly, if the number of steps changes, so does
the length of a single step, and so do the returns, which should be reflected in
the notation. Using the new notational conventions, we shall write

$$A_N(t + h) = (1 + R_N)A_N(t),$$
$$S_N(t + h) = (1 + K_N(t))S_N(t),$$

for $t = nh$, $n = 0, 1, 2, \ldots$, where the returns $K_N(t)$ are independent identically
distributed random variables (see Appendix 10.3) such that

$$K_N(t) = \begin{cases} U_N & \text{if stock goes up in step } n, \\ D_N & \text{if stock goes down in step } n, \end{cases}$$

with $D_N < R_N < U_N$. In addition, it will be assumed that the probability of
up/down stock price movements is $\frac{1}{2}$ for each step, so that

$$P(K_N(t) = U_N) = P(K_N(t) = D_N) = \frac{1}{2}.$$

Because option prices in the binomial model do not depend on the real-world probability, this simplifying assumption can be adopted without loss of generality.

Remark 8.1

The probability space on which the random variables are defined is often ignored by practically oriented minds. What really matters in applications is the probability distribution. Nevertheless, it is sometimes convenient to work with a concrete probability space. This can be realised in many different ways. Here is one possibility, which has the advantage of being simple but sufficiently rich that it can conveniently accommodate infinite sequences of independent random variables. Take $\Omega = [0,1]$ and equip it with probability P such that $P([a,b]) = b - a$ for any $0 \le a \le b \le 1$ (the so-called uniform probability). Then $K_N(t) : \Omega \to \{U_N, D_N\}$ $(t = nh, \ h = \frac{T}{N})$ are defined for $\omega \in \Omega = [0,1]$ by

$$K_N^\omega(t) = \begin{cases} U_N & \text{if } \omega \in [\frac{k}{2^n}, \frac{k+1}{2^n}) \text{ for even } k, \\ D_N & \text{otherwise.} \end{cases} \tag{8.1}$$

Some simple (though tedious) work suffices to show that these random variables are independent and $P(K_N(t) = U_N) = P(K_N(t) = D_N) = \frac{1}{2}$.

When passing to the continuous time limit, it will become convenient to express the risk-free asset in terms of the continuously compounded interest rate r, so that

$$A(t) = e^{rt} A(0)$$

for any $t \ge 0$, where $e^{rh} = 1 + R_N$. We shall take $A(0) = 1$ for simplicity.
    We define the single-step *logarithmic returns* by

$$k_N(t) = \ln(1 + K_N(t)) = \ln \frac{S_N(t+h)}{S_N(t)}$$

for $t = nh$, $n = 0, 2, \ldots, N-1$. These are independent identically distributed random variables such that

$$k_N(t) = \begin{cases} \ln(1 + U_N) & \text{if stock goes up in step } n, \\ \ln(1 + D_N) & \text{if stock goes down in step } n. \end{cases}$$

    In general, the *logarithmic return* on stock between time instants $t < u$ is defined by

$$k_N(t, u) = \ln \frac{S_N(u)}{S_N(t)}.$$

We assume that the expectation and variance of the random variable $k_N(0, t)$ are of a special form:

$$\mathbb{E}(k_N(0, t)) = \mu t,$$
$$\text{Var}(k_N(0, t)) = \sigma^2 t$$

for some $\mu \in \mathbb{R}$, $\sigma > 0$ independent of N.

By the properties of the logarithm, for each $t = nh$

$$k_N(0, t) = k_N(h) + k_N(2h) + \cdots + k_N(nh),$$

and since the terms in this sum are independent identically distributed random variables,

$$\mathbb{E}(k_N(0, t)) = n\mathbb{E}(k_N(h)),$$
$$\text{Var}(k_N(0, t)) = n\text{Var}(k_N(h)),$$

so

$$\mu = \frac{1}{h}\mathbb{E}(k_N(h)),$$

is the expected logarithmic return per unit time and

$$\sigma^2 = \frac{1}{h}\text{Var}(k_N(h)),$$

is the variance of the logarithmic return per unit time. We call σ the *volatility* of the stock.

We can now express the single step returns by means of the parameters μ and σ:

$$1 + U_N = e^{\mu h + \sigma\sqrt{h}}, \quad 1 + D_N = e^{\mu h - \sigma\sqrt{h}}. \tag{8.2}$$

Exercise 8.1

Derive equalities (8.2).

Going back to stock prices, from (8.2) we get

$$S_N(h) = S(0)e^{\mu h + Y_1 \sigma\sqrt{h}},$$

where

$$Y_1 = \begin{cases} +1 & \text{with probability } \frac{1}{2}, \\ -1 & \text{with probability } \frac{1}{2}. \end{cases}$$

Extending this step by step, we arrive at

$$S_N(t) = S(0)e^{\mu t + \sigma W_N(t)} \tag{8.3}$$

for each $t = nh$, $n = 0, 1, \ldots, N$, where $h = \frac{T}{N}$ and

$$W_N(t) = \sqrt{h}(Y_1 + \cdots + Y_n), \qquad (8.4)$$

for an infinite sequence of independent identically distributed random variables $Y_1, Y_2, \ldots$ such that

$$Y_n = \begin{cases} +1 & \text{with probability } \frac{1}{2}, \\ -1 & \text{with probability } \frac{1}{2}. \end{cases}$$

In particular, $W_N(0) = 0$. We call $W_N(t)$ a *scaled random walk* (with time step h and jumps $\pm\sqrt{h}$).

Remark 8.2

The scaled random walk W_N can be constructed for each N using the same sequence of random variables $Y_1, Y_2, \ldots$ defined on the same probability space Ω. For example, this can be the probability space in Remark 8.1. To this end we put

$$Y_n = \frac{\ln\left(1 + K_N(nh - h)\right) - \mu h}{\sigma\sqrt{h}}$$

for $n = 1, 2, \ldots$ with $K_N(nh - h)$ defined by (8.1).

8.2.2 Wiener Process

We would like to consider the limit as $N \to 0$ (that is, as $h = \frac{T}{N} \to 0$) of the scaled random walk $W_N(t)$. From (8.4) we know that $W_N(T)$ can be written as

$$W_N(T) = \sqrt{h}\,(Y_1 + \cdots + Y_N).$$

The expectation of each of the independent identically distributed random variables $Y_1, Y_2, \ldots$ is 0 and the variance is 1. The expectation of $Y_1 + \cdots + Y_N$ is therefore 0 and the variance is N. According to the Central Limit Theorem (see Appendix 10.3), the distribution of

$$\frac{W_N(T)}{\sqrt{T}} = \frac{Y_1 + \cdots + Y_N}{\sqrt{N}}$$

tends as $N \to \infty$ to the *normal distribution* with mean 0 and variance 1, in the sense that for all $a \le b$

$$P\left(a \le \frac{W_N(T)}{\sqrt{T}} \le b\right) \to \int_a^b \frac{1}{\sqrt{2\pi}} e^{-\frac{x^2}{2}} \, dx \quad \text{as } N \to \infty.$$

As a consequence, the limit distribution of $W_N(T)$ is normal with mean 0 and variance T.

A similar argument shows that the limit distribution of

$$W_N(t) - W_N(s) = \sqrt{h} \sum_{s \leq nh < t} Y_n$$

is normal with mean 0 and variance $t - s$.

This motivates the next definition, where we introduce a process with such properties that it can be regarded as a limit (in a suitable sense) of a scaled random walk.

Definition 8.3

Wiener process (also known as *Brownian motion*) is a family of random variables $W(t)$ defined for $t \in [0, \infty)$ such that $W(0) = 0$, the increments $W(t) - W(s)$ have normal distribution with mean zero and variance $t - s$ for all $0 \leq s < t$, and $W(t_n) - W(t_{n-1}), \ldots, W(t_2) - W(t_1)$ are independent for all sequences $0 \leq t_1 < t_2 < \cdots < t_n$.

The existence of Wiener process, including the choice of a suitable probability space, is a matter of advanced considerations, beyond the scope of this text.

8.2.3 Black–Scholes Model

We have seen that in the N-step binomial model the stock price can be expressed as

$$S_N(t) = S(0)e^{\mu t + \sigma W_N(t)}, \tag{8.5}$$

where $W_N(t)$ is the scaled random walk. Replacing $W_N(t)$ by the Wiener process $W(t)$ (which corresponds to the limit as $N \to \infty$), we obtain a new model of stock prices, denoted by $S(t)$, defined for all real $t \geq 0$

$$S(t) = S(0)e^{\mu t + \sigma W(t)}. \tag{8.6}$$

Expression (8.6) is the essence of the famous *Black–Scholes model* of stock prices. It follows that

$$\ln S(t) = \ln S(0) + \mu t + \sigma W(t).$$

Since $W(t)$ has normal distribution $N(0, t)$, we can see that $\ln S(t)$ has normal distribution $N(\ln S(0) + \mu t, \sigma^2 t)$. For this reason the stock price $S(t)$ in the Black–Scholes model is said to have the *log normal distribution*.

The density of the distribution of $S(t)$ is shown in Figure 8.1 for $t = 10$, $S(0) = 1$, $\mu = 0$ and $\sigma = 0.1$. This can be compared with the discrete distribution in Figure 8.2.

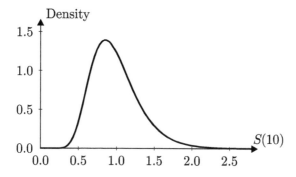

Figure 8.1 Density of the distribution of stock price $S(t)$ in the Black–Scholes model

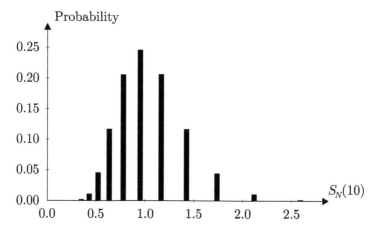

Figure 8.2 Distribution of stock price $S_N(t)$ in the binomial model with $N = 10$

Exercise 8.2

Show that

$$\mathbb{E}(S(t)) = S(0)e^{\left(\mu + \frac{1}{2}\sigma^2\right)t}.$$

Remark 8.4

According to Exercise 8.2,

$$\nu = \mu + \frac{1}{2}\sigma^2 \tag{8.7}$$

is the growth rate of the expected stock price. It is often used as a parameter instead of the expected logarithmic return μ when developing the Black–Scholes model, in which case (8.6) takes the form

$$S(t) = S(0)e^{(\nu-\frac{1}{2}\sigma^2)t+\sigma W(t)}.$$

8.3 Overview of Stochastic Calculus

This section provides an introduction to some of the mathematical tools used in continuous-time finance. We aim for an intuitive understanding based on properties established in the discrete setting and an informal treatment of the limit as the number of time steps goes to infinity, rather than precise constructions and proofs.

8.3.1 Itô Formula

Formulae (8.5) for the binomial approximation and (8.6) for the Black–Scholes model can be written as

$$S_N(t) = f(t, W_N(t)), \tag{8.8}$$
$$S(t) = f(t, W(t)), \tag{8.9}$$

where

$$f(t, x) = S(0)e^{\mu t + \sigma x}.$$

For any sufficiently regular function f, using the Taylor formula, we can write

$$\Delta f(t, W_N(t)) = f_t(t, W_N(t))h + f_x(t, W_N(t))\Delta W_N(t)$$
$$+ \frac{1}{2}f_{xx}(t, W_N(t))h + O(h^{3/2}) \tag{8.10}$$

for $t = nh$, $n = 0, 1, \ldots, N-1$, where the notation

$$f_t = \frac{\partial f}{\partial t}, \quad f_x = \frac{\partial f}{\partial x}, \quad f_{xx} = \frac{\partial^2 f}{\partial x^2}$$

is used for partial derivatives, and

$$\Delta f(t, W_N(t)) = f(t + h, W_N(t + h)) - f(t, W_N(t)),$$
$$\Delta W_N(t) = W_N(t + h) - W_N(t),$$

for increments, and where $O(h^{3/2})$ stands for terms of order $h^{3/2}$ or higher. We have used the fact that $\Delta W_N(t)$ has values $\pm\sqrt{h}$, so that $(\Delta W_N(t))^2 = h$ and $(\Delta W_N(t))^k = O(h^{3/2})$ for each $k \geq 3$.

Applying summation over time steps n in (8.10) such that $0 \leq nh < T$, we obtain

$$f(T, W_N(T)) = f(0, W_N(0)) + \sum_{0 \leq nh < T} f_t(nh, W_N(nh))h$$

$$+ \sum_{0 \leq nh < T} f_x(nh, W_N(nh))\Delta W_N(nh)$$

$$+ \frac{1}{2} \sum_{0 \leq nh < T} f_{xx}(nh, W_N(nh))h + O(h^{1/2}). \tag{8.11}$$

Observe that the sum of $N = \frac{T}{h}$ terms of order $O(h^{3/2})$ gives $O(h^{1/2})$.

The formula

$$f(T, W(T)) = f(0, W(0)) + \int_0^T f_t(t, W(t))dt$$

$$+ \int_0^T f_x(t, W(t))dW(t) + \frac{1}{2}\int_0^T f_{xx}(t, W(t))dt \tag{8.12}$$

corresponds to the limit as $N \to \infty$ of the expressions on both sides of the equality (8.11). The first and third integrals are the familiar Riemann integrals, while the second one is a special kind of integral called *stochastic* or *Itô integral*. Equality (8.12), called the *Itô formula*, is the cornerstone of *stochastic calculus*. For a precise definition of the stochastic integral and a proof of the Itô formula please refer to more advanced texts.

Remark 8.5

The term $\frac{1}{2}\int_0^T f_{xx}(t, W(t))dt$, called the *Itô correction*, is specific to stochastic calculus. Suppose that instead of Wiener process $W(t)$ we use a differentiable function $z(t)$. Then

$$f(T, z(T)) = f(0, z(0)) + \int_0^T f_t(t, z(t))dt + \int_0^T f_x(t, z(t))z'(t)dt.$$

No term corresponding to the Itô correction appears in this case.

It is convenient to write the Itô formula (8.12) in shorthand notation as

$$df(t, W(t)) = f_t(t, W(t))dt + f_x(t, W(t))dW(t) + \frac{1}{2}f_{xx}(t, W(t))dt, \quad (8.13)$$

omitting the integrals for brevity. In particular, the Itô formula for the composition of functions $h(t, W(t)) = g(t, f(t, W(t)))$ can be written as

$$dh(t, W(t)) = g_t(t, W(t))dt + g_x(t, W(t))df(t, W(t))$$
$$+ \frac{1}{2}g_{xx}(t, W(t))[f_x(t, W(t))]^2dt. \quad (8.14)$$

The last formula will be used repeatedly in the following sections.

Going back to (8.8) with $f(t, x) = S(0)e^{\mu t + \sigma x}$, we can write (8.10) as

$$\Delta S_N(t) = \left(\mu + \frac{1}{2}\sigma^2\right) S_N(t)h + \sigma S_N(t)\Delta W_N(t) + O(h^{3/2}). \quad (8.15)$$

This can be viewed as a difference equation for the stock price $S_N(t)$ in the binomial model. Moreover, applying the Itô formula (8.13) to (8.9) with $f(t, x) = S(0)e^{\mu t + \sigma x}$, we obtain

$$dS(t) = (\mu + \frac{1}{2}\sigma^2)S(t)dt + \sigma S(t)dW(t). \quad (8.16)$$

This is a *stochastic differential equation* satisfied by the stock price $S(t)$ in the Black–Scholes model.

Remark 8.6

The stochastic differential equation (8.16) with the parameter $\nu = \mu + \frac{1}{2}\sigma^2$ from (8.7) becomes

$$dS(t) = \nu S(t)dt + \sigma S(t)dW(t).$$

Written in this elegant form, the equation is often the starting point when constructing the Black–Scholes model.

8.3.2 Expected Value of Itô Integral

The Itô integral $\int_0^t f(u, W(u))dW(u)$ corresponds to the limit of

$$\sum_{0 \le nh < t} f(nh, W_N(nh))\Delta W_N(nh)$$

when $N \to \infty$. The precise meaning of such a limit of a sequence of random variables requires careful treatment, not covered here.

Because of the independence of increments of the random walk, it follows that $W_N(nh)$ and $\Delta W_N(nh)$ are independent, and so are $f(nh, W_N(nh))$ and $\Delta W_N(nh)$ for each n. As a result,

$$\mathbb{E}\left[\sum_{0 \le nh < t} f(nh, W_N(nh)) \Delta W_N(nh)\right]$$
$$= \sum_{0 \le nh < t} \mathbb{E}\left[f(nh, W_N(nh))\right]\mathbb{E}\left[\Delta W_N(nh)\right] = 0.$$

In the limit as $N \to \infty$ this corresponds to

$$\mathbb{E}\int_0^t f(u, W(u))\mathrm{d}W(u) = 0. \tag{8.17}$$

This is an informal argument, but the conclusion can be proved rigorously under suitable assumptions on f.

8.3.3 Girsanov Theorem

Our goal in this section will be to construct a probability in the Black–Scholes model that will play an analogous role to the martingale (risk-neutral) probability in the discrete case.

In the binomial model the discounted stock price is given by

$$\tilde{S}_N(t) = \frac{S_N(t)}{A_N(t)} = S(0)e^{(\mu-r)t + \sigma W_N(t)}$$

for $t = nh$, $n = 0, 1, \ldots, N$. The discounted stock price $\tilde{S}_N(t)$ is a martingale under the martingale probability on the binomial tree given by

$$p_* = \frac{R - D}{U - D} = \frac{e^{rh} - e^{\mu h - \sigma\sqrt{h}}}{e^{\mu h + \sigma\sqrt{h}} - e^{\mu h - \sigma\sqrt{h}}}.$$

A somewhat tedious computation using the Taylor formula, best performed with the aid of a symbolic algebra computer system, gives the following expansions in the powers of $\sqrt{h}$:

$$p_* = \frac{e^{rh} - e^{\mu h - \sigma\sqrt{h}}}{e^{\mu h + \sigma\sqrt{h}} - e^{\mu h - \sigma\sqrt{h}}}$$
$$= \frac{1}{2} - \frac{1}{2}b\sqrt{h} + O(h^{3/2}) = \frac{1}{2}e^{-b\sqrt{h} - \frac{1}{2}b^2 h + O(h^{3/2})}, \tag{8.18}$$

$$1 - p_* = \frac{e^{\mu h + \sigma\sqrt{h}} - e^{rh}}{e^{\mu h + \sigma\sqrt{h}} - e^{\mu h - \sigma\sqrt{h}}}$$
$$= \frac{1}{2} + \frac{1}{2}b\sqrt{h} + O(h^{3/2}) = \frac{1}{2}e^{+b\sqrt{h} - \frac{1}{2}b^2 h + O(h^{3/2})}, \tag{8.19}$$

where

$$b = \frac{\mu - r + \frac{1}{2}\sigma^2}{\sigma} \tag{8.20}$$

is substituted for brevity.

Denote by P_{W_N} and $P^*_{W_N}$ the true probability and, respectively, the martingale probability that the random walk follows a particular path of W_N. Then P_{W_N} is the product of factors p and $1 - p$, whereas $P^*_{W_N}$ is the product of factors p_* and $1 - p_*$ corresponding to the 'up' and 'down' movements depending on whether $\Delta W_N(t) = +\sqrt{h}$ or $-\sqrt{h}$ for $t = nh$, $n = 0, 1, \ldots, N - 1$. Because $p = 1 - p = \frac{1}{2}$ and using (8.18), (8.19) for p_* and $1 - p_*$, we get $P_{W_N} = 2^{-N}$ and

$$P^*_{W_N} = 2^{-N} e^{-bW_N(T) - \frac{1}{2}b^2 T + O(h^{1/2})}.$$

An informal argument that adding $N = \frac{T}{h}$ terms of order $O(h^{3/2})$ gives a term or order $O(h^{1/2})$ applies here just like in the derivation of the Itô formula. As a result,

$$P^*_{W_N} = P_{W_N} e^{-bW_N(T) - \frac{1}{2}b^2 T + O(h^{1/2})}. \tag{8.21}$$

From here, in the limit as $N \to \infty$ so that $h = \frac{T}{N} \to 0$, we can obtain a new probability P^* such that for any event A

$$P_*(A) = \mathbb{E}\left(e^{-bW(T) - \frac{1}{2}b^2 T}\mathbb{I}_A\right), \tag{8.22}$$

where $\mathbb{I}_A$ is the indicator function of A, equal to 1 on A and 0 otherwise, and where $\mathbb{E}$ is the expectation under probability P. Formula (8.22), even though reached informally, can serve as a precise definition of probability P_*.

Applying the Itô formula (8.13) with $f(t, x) = S(0)e^{(\mu-r)t + \sigma x}$, we can verify that the discounted stock price process

$$\tilde{S}(t) = \frac{S(t)}{A(t)} = S(0)e^{(\mu-r)t + \sigma W(t)}$$

in the Black–Scholes model satisfies

$$d\tilde{S}(t) = \left(\mu - r + \frac{1}{2}\sigma^2\right)\tilde{S}(t)dt + \sigma\tilde{S}(t)dW(t) = \sigma\tilde{S}(t)dW_*(t),$$

where

$$W_*(t) = bt + W(t) \tag{8.23}$$

with b given by (8.20). We know that $W(t)$ is a Wiener process under P. The following important and interesting proposition establishes a similar property of $W_*(t)$ under the new probability P_*.

Proposition 8.7 (special case of Girsanov theorem)

$W_*(t)$ defined by (8.23) is a Wiener process under the probability P_* defined by (8.22).

Remark 8.8

Proposition 8.7 is a special case of an important result known as the *Girsanov theorem*, which applies in a general setting when b is a stochastic process and not just a constant number. We shall not need this level of generality for our purposes.

Proof

Observe that if X is a random variable with normal distribution $N(0, t)$, then

$$
\mathbb{E}\left(e^{-bX - \frac{1}{2}b^2 t} \mathbb{I}_{\{X + bt \le a\}}\right) = \int_{\{x + bt \le a\}} e^{-bx - \frac{1}{2}b^2 t} \frac{1}{\sqrt{2\pi t}} e^{-\frac{x^2}{2t}} \, dx
$$

$$
= \int_{\{x + bt \le a\}} \frac{1}{\sqrt{2\pi t}} e^{-\frac{(x + bt)^2}{2t}} \, dx
$$

$$
= \int_{\{y \le a\}} \frac{1}{\sqrt{2\pi t}} e^{-\frac{y^2}{2t}} \, dy = P(X \le a). \quad (8.24)
$$

Let $0 = t_0 < t_1 < \cdots < t_n = T$ and $a_1, \ldots, a_n \in \mathbb{R}$. Since the increments $W(t_i) - W(t_{i-1})$ for $i = 1, \ldots, n$ have normal distribution $N(0, t_i - t_{i-1})$ and are independent under P,

$$
P_*\left(\bigcap_{i=1}^{n} \{W_*(t_i) - W_*(t_{i-1}) \le a_i\}\right)
$$

$$
= \mathbb{E}\left(e^{-bW(T) - \frac{1}{2}b^2 T} \mathbb{I}_{\bigcap_{i=1}^{n} \{W_*(t_i) - W_*(t_{i-1}) \le a_i\}}\right)
$$

$$
= \mathbb{E}\left(\prod_{i=1}^{n} e^{-b(W(t_i) - W(t_{i-1})) - \frac{1}{2}b^2(t_i - t_{i-1})} \mathbb{I}_{\{W(t_i) - W(t_{i-1}) + b(t_i - t_{i-1}) \le a_i\}}\right)
$$

$$
= \prod_{i=1}^{n} \mathbb{E}\left(e^{-b(W(t_i) - W(t_{i-1})) - \frac{1}{2}b^2(t_i - t_{i-1})} \mathbb{I}_{\{W(t_i) - W(t_{i-1}) + b(t_i - t_{i-1}) \le a_i\}}\right)
$$

$$
= \prod_{i=1}^{n} P(W(t_i) - W(t_{i-1}) \le a_i),
$$

where the last equality holds by (8.24). It follows, in particular, that for each i

$$
P_*(W_*(t_i) - W_*(t_{i-1}) \le a_i) = P(W(t_i) - W(t_{i-1}) \le a_i),
$$

so the increments $W_*(t_i) - W_*(t_{i-1})$ have normal distribution $N(0, t_i - t_{i-1})$. Moreover, it follows that

$$P_*\left(\bigcap_{i=1}^n \{W_*(t_i) - W_*(t_{i-1}) \le a_i\}\right) = \prod_{i=1}^n P(W(t_i) - W(t_{i-1}) \le a_i)$$

$$= \prod_{i=1}^n P_*(W_*(t_i) - W_*(t_{i-1}) \le a_i),$$

which means that the increments $W_*(t_i) - W_*(t_{i-1})$ for $i = 1, \ldots, n$ are independent under P_*. By Definition 8.3, $W_*(t)$ is therefore a Wiener process under P_*.   □

Exercise 8.3

Show that the Black–Scholes stock price (8.6) can be written as

$$S(t) = S(0)e^{(r-\frac{1}{2}\sigma^2)t + \sigma W_*(t)}. \tag{8.25}$$

Exercise 8.4

Show that

$$dS(t) = rS(t)dt + \sigma S(t)dW_*(t).$$

8.4 Options in Black–Scholes Model

We consider European options written on stock in the Black Scholes model. A connection will be made between the pricing and replication of such options, and the solution of a final value problem for a partial differential equation, the celebrated Black–Scholes equation. The option price will be expressed as the expectation under P_* of the discounted payoff. Explicit pricing formulae will be derived for European calls and puts in the Black–Scholes model.

8.4.1 Replicating Strategy

In an N-step binomial model consider a European option expiring at time T with payoff $g(S_N(T))$. A replicating strategy is a pair of processes $x_N(t), y_N(t)$, which represents the positions in stock and the risk-free asset, with the value of the strategy at time $t = nh$, $n = 0, \ldots, N$ given by

$$V_N(t) = x_N(t)S_N(t) + y_N(t)A_N(t),$$

such that:

- the strategy satisfies the self-financing condition

$$x_N(t)S_N(t) + y_N(t)A_N(t) = x_N(t-h)S_N(t) + y_N(t-h)A_N(t)$$

for $t = nh$, $n = 1, \ldots, N$, or equivalently

$$\Delta V_N(t) = x_N(t)\Delta S_N(t) + y_N(t)\Delta A_N(t)$$

for $t = nh$, $n = 0, \ldots, N-1$;
- the value of the strategy at time T matches the option payoff,

$$V_N(T) = g(S_N(T)).$$

Such replicating strategies in the binomial model are discussed in Chapter 6.

In the Black–Scholes model we consider a European option with expiry time T and payoff $g(S(T))$. By analogy with the binomial model, a *replicating strategy* will be a pair of processes $x(t), y(t)$ representing the positions in stock and the risk-free asset, with value at time $t \in [0, T]$ given by

$$V(t) = x(t)S(t) + y(t)A(t),$$

such that:

- the strategy satisfies the self-financing condition

$$dV(t) = x(t)dS(t) + y(t)dA(t) \tag{8.26}$$

for $t \in [0, T]$;
- the value of the strategy at time T matches the option payoff,

$$V(T) = g(S(T)). \tag{8.27}$$

8.4.2 Black–Scholes Equation

The discounted stock price in the Black–Scholes model is denoted by $\widetilde{S}(t) = e^{-rt}S(t)$. If there is a replicating strategy $x(t), y(t)$ for an option with expiry time T and payoff $g(S(T))$, the option is said to be *attainable*. The discounted value of the replicating strategy will be denoted by $\widetilde{V}(t) = e^{-rt}V(t)$.

If $x(t)$ and $y(t)$ can be expressed as sufficiently regular functions of t and $W(t)$, then so can $V(t)$ and $\widetilde{V}(t)$, and the Itô formula can be applied for any of these functions. Thus,

$$
\begin{aligned}
d\widetilde{V}(t) &= -re^{-rt}V(t)dt + e^{-rt}dV(t) \\
&= -re^{-rt}\left[x(t)S(t) + y(t)A(t)\right]dt + e^{-rt}\left[x(t)dS(t) + e^{-rt}y(t)dA(t)\right] \\
&= x(t)d\widetilde{S}(t) \\
&= \sigma x(t)\widetilde{S}(t)dW_*(t).
\end{aligned}
$$

We know that $W_*(t)$ is a Wiener process under P_*. Thus, using (8.17), we obtain

$$V(0) = \tilde{V}(0) = \mathbb{E}_* \left[\tilde{V}(0) + \int_0^T \sigma x(t) \tilde{S}(t) \mathrm{d}W_*(t) \right]$$
$$= \mathbb{E}_* \tilde{V}(T) = \mathrm{e}^{-rT} \mathbb{E}_* V(T) = \mathrm{e}^{-rT} \mathbb{E}_* g(S(T)),$$

where $\mathbb{E}_*$ is the expectation under P_*. A standard no-arbitrage argument shows that the price of the option must be equal to $V(0)$ and therefore to $\mathrm{e}^{-rT} \mathbb{E}_* g(S(T))$.

For this expression to be useful it is necessary to ensure that the option is attainable and that the positions $x(t), y(t)$ in the replicating strategy can be expressed as sufficiently regular functions of t and $W(t)$. This will be done for an arbitrary payoff $g(S(T))$ by representing the replicating strategy in terms of a solution to a final value problem for a partial differential equation, the Black–Scholes equation.

Let us take a sufficiently regular function $u(t, x)$ and put

$$f(t, x) = \mathrm{e}^{-rt} u(t, S(0) \mathrm{e}^{(r - \frac{1}{2}\sigma^2)t + \sigma x}),$$

so that, according to (8.25),

$$f(t, W_*(t)) = \mathrm{e}^{-rt} u(t, S(t)).$$

The Itô formula (8.13) applied to $f(t, W_*(t))$ gives

$$\mathrm{d}f(t, W_*(t)) = \left(f_t(t, W_*(t)) + \frac{1}{2} f_{xx}(t, W_*(t)) \right) \mathrm{d}t + f_x(t, W_*(t)) \mathrm{d}W_*(t).$$

If f satisfies the partial differential equation

$$f_t + \frac{1}{2} f_{xx} = 0, \tag{8.28}$$

then $\mathrm{d}f(t, W_*(t)) = f_x(t, W_*(t)) \mathrm{d}W_*(t)$. Because $W_*(t)$ is a Wiener process under P_*, it follows by (8.17) that

$$f(0, W_*(0)) = \mathbb{E}_* \left[f(0, W_*(0)) + \int_0^T f_x(t, W_*(t)) \mathrm{d}W_*(t) \right]$$
$$= \mathbb{E}_* f(T, W_*(T)).$$

Written in terms of $u(t, x)$, the partial differential equation (8.28) becomes

$$u_t + rx u_x + \frac{1}{2} \sigma^2 x^2 u_{xx} = ru. \tag{8.29}$$

This is the famous *Black–Scholes equation*. We can conclude that for any solution $u(t, x)$ to this equation

$$u(0, S(0)) = f(0, W_*(0)) = \mathbb{E}_* f(T, W_*(T)) = e^{-rT} \mathbb{E}_* u(T, S(T)).$$

In particular, the solution $u(t, x)$ to the Black–Scholes equation (8.29) that satisfies the final condition

$$u(T, x) = g(x) \tag{8.30}$$

plays an important role in pricing and replicating the option with payoff $g(S(T))$. The replicating strategy $x(t), y(t)$ can be expressed in terms of this solution. To this end, we apply the Itô formula (8.14) to $u(t, S(t))$, bearing in mind that $u(t, x)$ satisfies the Black–Scholes equation (8.29), $dS(t) = rS(t)dt + \sigma S(t)dW_*(t)$ and $dA(t) = rA(t)dt$:

$$\begin{aligned}
du(t, S(t)) &= \left[u_t(t, S(t)) + \frac{1}{2}\sigma^2 S(t)^2 u_{xx}(t, S(t)) \right] dt + u_x(t, S(t))dS(t) \\
&= r\left[u(t, S(t)) - S(t)u_x(t, S(t)) \right] dt + u_x(t, S(t))dS(t) \\
&= \frac{u(t, S(t)) - S(t)u_x(t, S(t))}{A(t)} dA(t) + u_x(t, S(t))dS(t).
\end{aligned}$$

It follows that the strategy

$$x(t) = u_x(t, S(t)), \quad y(t) = \frac{u(t, S(t)) - S(t)u_x(t, S(t))}{A(t)}$$

with value

$$V(t) = x(t)S(t) + y(t)A(t) = u(t, S(t))$$

for each $t \in [0, T]$ satisfies the self-financing condition (8.26) and the replication condition (8.27). This is a replicating strategy for the European option with expiry time T and payoff $g(S(T))$.

To complete the argument that the option is attainable, we need to know that there is a solution $u(t, x)$ to the Black–Scholes equation (8.29) with final condition (8.30). This can be obtained from general existence results for parabolic partial differential equations. Alternatively, it can be verified by direct computation that

$$u(t, x) = e^{r(T-t)} \int_{-\infty}^{+\infty} \frac{1}{\sqrt{2\pi(T-t)}} e^{-\frac{y^2}{2(T-t)}} g\left(xe^{(r-\frac{1}{2}\sigma^2)(T-t)+\sigma y} \right) dy$$

is such a solution.

These conclusions can be summarised in the following important theorem.

Theorem 8.9 (Black–Scholes Equation)

Let $g(x)$ be a sufficiently regular function, and let $u(t,x)$ be a solution to the final value problem for the Black–Scholes equation

$$u_t(t,x) + rxu_x(t,x) + \frac{1}{2}\sigma^2 x^2 u_{xx}(t,x) = ru(t,x),$$

$$u(T,x) = g(x).$$

Then the European option with exercise time T and payoff $g(S(T))$ is attainable with replicating strategy

$$x(t) = u_x(t, S(t)), \quad y(t) = \frac{u(t, S(t)) - S(t)u_x(t, S(t))}{A(t)},$$

and the no arbitrage price of the option, which is equal to the value of the replicating strategy, is

$$V(0) = u(0, S(0)) = e^{-rT}\mathbb{E}_* g(S(T)),$$

where $\mathbb{E}_*$ is the expectation under P_*.

Remark 8.10

The expected logarithmic return μ does not appear in the Black–Scholes equation (8.29). This means that the option price does not depend on μ, a seemingly paradoxical result analogous to the fact that the true probability does not affect the option price in the binomial model. It is an important result, because in practice it is difficult to get a reliable estimate of μ. On the other hand, the interest rate r is known precisely, and good estimates for the volatility σ can be obtained.

8.4.3 Black–Scholes Formula

It follows from Theorem 8.9 that the time 0 price of a European call option with strike price X in the Black–Scholes model is

$$C_E(0) = e^{-rT}\mathbb{E}_* \left(\max\{S(T) - X, 0\}\right).$$

Let us compute this expectation. From Exercise 8.3 we know that the Black–Scholes stock price can be written as

$$S(T) = S(0)e^{(r - \frac{1}{2}\sigma^2)T + \sigma W_*(T)}.$$

Since, by the Girsanov theorem, $W_*(T)$ has normal distribution under P_* with mean 0 and variance T,

$$
\begin{aligned}
C_{\mathrm{E}}(0) &= \mathbb{E}_* \left(\mathrm{e}^{-rT} \max\{S(T) - X, 0\} \right) \\
&= \mathbb{E}_* \left(\mathrm{e}^{-rT} \max\{S(0)\mathrm{e}^{(r - \frac{1}{2}\sigma^2)T + \sigma W_*(T)} - X, 0\} \right) \\
&= \int_{-d_-}^{\infty} \left(S(0)\mathrm{e}^{\sigma x - \frac{1}{2}\sigma^2 T} - X\mathrm{e}^{-rT} \right) \frac{1}{\sqrt{2\pi}} \mathrm{e}^{-\frac{x^2}{2}} \, dx \\
&= S(0) \int_{-d_+}^{\infty} \frac{1}{\sqrt{2\pi}} \mathrm{e}^{-\frac{x^2}{2}} \, dx - X\mathrm{e}^{-rT} \int_{-d_-}^{\infty} \frac{1}{\sqrt{2\pi}} \mathrm{e}^{-\frac{x^2}{2}} \, dx \\
&= S(0)N(d_+) - X\mathrm{e}^{-rT} N(d_-),
\end{aligned}
$$

where

$$
d_\pm = \frac{\ln \frac{S(0)}{X} + \left(r \pm \frac{1}{2}\sigma^2 \right) T}{\sigma\sqrt{T}}, \tag{8.31}
$$

and where

$$
N(x) = \int_{-\infty}^{x} \frac{1}{\sqrt{2\pi}} \mathrm{e}^{-\frac{y^2}{2}} \, dy = \int_{-x}^{\infty} \frac{1}{\sqrt{2\pi}} \mathrm{e}^{-\frac{y^2}{2}} \, dy \tag{8.32}
$$

is the normal cumulative distribution function.

Exercise 8.5

Show that

$$
N(d_-) = P_*(S(T) \geq X).
$$

The choice of time 0 to compute the price of the option is arbitrary. In general, the option price can be computed at any time $t < T$, in which case the time remaining before the option is exercised will be $T - t$. Substituting t for 0 and $T - t$ for T in the above formulae, we obtain the following result.

Theorem 8.11 (Black–Scholes Formula)

The time t price of a European call with strike price X and exercise time T, where $t < T$, is given by

$$
C_{\mathrm{E}}(t) = S(t)N(d_+(t)) - X\mathrm{e}^{-r(T-t)}N(d_-(t)) \tag{8.33}
$$

with

$$
d_\pm(t) = \frac{\ln \frac{S(t)}{X} + \left(r \pm \frac{1}{2}\sigma^2 \right)(T - t)}{\sigma\sqrt{T - t}}. \tag{8.34}
$$

Exercise 8.6

Derive the Black–Scholes formula

$$P_E(t) = Xe^{-r(T-t)}N(-d_-(t)) - S(t)N(-d_+(t)),$$

with $d_+(t)$ and $d_-(t)$ given by (8.34), for the price of a European put
with strike price X and exercise time T.

Remark 8.12

Observe that the Black–Scholes formula does not contain μ. It is a property
analogous to that in Remarks 6.17 and 8.10, and of similar practical significance: there is no need to know μ to work out the price of a European call or
put option in the Black–Scholes model.

It is interesting to compare numerically the Black–Scholes formula for the
price of a European call with the Cox–Ross–Rubinstein formula. Figure 8.3
shows the price of a European call with strike $X = 100$ on a stock with $S(0) =$
100, $\sigma = 0.3$ and $\mu = 20\%$. (Though μ is irrelevant for the Black–Scholes
formula, it still features in the discrete time approximation (8.3).) The interest
rate under continuous compounding is taken to be $r = 8\%$. The option price is
computed in two ways as a function of the time T remaining before the option
is exercised:

a) (*solid line*) from the Black–Scholes formula for T between 0 and 1;

b) (*dots*) using the Cox–Ross–Rubinstein formula with T increasing from 0
   to 1 over $N = 10$ steps of duration $h = 0.1$ each; the discrete growth rates
   for each step are computed using formulae (8.2).

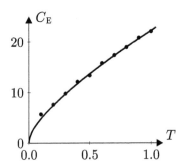

Figure 8.3 Call price in continuous and discrete time models as a function
of time T remaining before the option is exercised

Even with as few as 10 steps there is remarkably good agreement between the discrete and continuous time formulae.

Exercise 8.7

Show that the replicating strategy for a European call is

$$x(t) = N(d_+(t)), \quad y(t) = -Xe^{-r(T-t)}N(d_-(t))$$

with $d_+(t)$ and $d_-(t)$ given by (8.34).

8.5 Risk Management

The writer of an option is exposed to risk. The premium due to the writer at time 0 is the expectation of discounted future payoffs under the risk-neutral probability. In the case of some unfavourable scenarios, the amounts due to the holder of the option can be substantial and indeed unlimited, for example, in the case of a call option. This risk to the writer needs to be addressed in a satisfactory manner.

8.5.1 Delta Hedging

The value of a European call or put option as given by the Black–Scholes formula clearly depends on the price of the underlying asset. This can be seen in a slightly broader context.

Consider a portfolio whose value depends on the current stock price $S = S(0)$ and is hence denoted by $V(S)$. Its dependence on S can be measured by the derivative $\frac{d}{dS}V(S)$, called the *delta* of the portfolio. For small price variations from S to $S + \Delta S$ the value of the portfolio will change by

$$\Delta V(S) \cong \frac{d}{dS}V(S) \times \Delta S.$$

The principle of *delta hedging* is based on embedding derivative securities in a portfolio, so that the extended portfolio value does not alter much when S varies. This can be achieved by ensuring that the delta of the portfolio is equal to zero. Such a portfolio is called *delta neutral*.

We take a portfolio composed of stock, money market account, and the hedged derivative security, its value given by

$$V(S) = xS + y + zH(S),$$

where the derivative security price is denoted by $H(S)$ and a bond with current value 1 is used in the risk-free position. Specifically, suppose that a single derivative security has been written, that is, $z = -1$. Then

$$\frac{\mathrm{d}}{\mathrm{d}S}V(S) = x - \frac{\mathrm{d}}{\mathrm{d}S}H(S).$$

The last term $\frac{\mathrm{d}}{\mathrm{d}S}H(S)$, which is the *delta* of the derivative security, can readily be computed in the Black–Scholes model of stock prices, so that an explicit formula for $H(S)$ is available.

Proposition 8.13

Denote the European call option price in the Black–Scholes model by $C_E(S)$. The delta of the option is given by

$$\frac{\mathrm{d}}{\mathrm{d}S}C_E(S) = N(d_+),$$

where $N(x)$ is the standard normal distribution function and d_+ is defined by (8.31).

Proof

The price $S = S(0)$ appears in three places in the Black–Scholes formula, see Theorem 8.11, so the differentiation requires a bit of work, with plenty of nice cancellations in due course, and is left to the reader, see Exercise 8.7. Bear in mind that the derivative $\frac{\mathrm{d}}{\mathrm{d}S}C_E(S)$ is computed at time $t = 0$.   ☐

Exercise 8.8

Find a similar expression for the delta $\frac{\mathrm{d}}{\mathrm{d}S}P_E(S)$ of a European put option in the Black–Scholes model.

For the remainder of this section we shall consider a European call option within the Black–Scholes model. By Proposition 8.13 the portfolio $(x, y, z) = (N(d_+), y, -1)$, where the position in stock $N(d_+)$ is computed for the initial stock price $S = S(0)$, has delta equal to zero for any y. Consequently, its value

$$V(S) = N(d_+)S + y - C_E(S)$$

does not vary much under small changes of the stock price about the initial value. It is convenient to choose y so that the initial value of the portfolio is equal to zero. By the Black–Scholes formula for $C_E(S)$ this gives

$$y = -Xe^{-Tr}N(d_-),$$

with d_- given by (8.31).

Let us analyse the following example, which will subsequently be expanded and modified. Suppose that the risk-free rate is 8% and consider a 90-day call option with strike price $X = 60$ dollars written on a stock with current price $S = 60$ dollars. Assume that the stock volatility is $\sigma = 30\%$. The Black–Scholes formula gives the option price $C_E = 4.14452$ dollars, the delta of the option being equal to 0.581957.

Suppose that we write and sell $1,000$ call options, cashing the premium of $\$4,144.52$. To construct the hedge we buy 581.96 shares for $\$34,917.39$, borrowing $\$30,772.88$. Our portfolio will be (x, y, z) with $x = 581.96$, $y = -30,772.88$, $z = -1,000$ and with total value zero. (While it might be more natural mathematically to consider a single option with $z = -1$, in practice options are traded in batches.)

We shall analyse the value of the portfolio after one day by considering some possible scenarios. The time to expiry will then be 89 days. Suppose that the stock volatility and the risk-free rate do not vary, and consider the following three scenarios of stock price movements:

1. The stock price remains unaltered, $S(\frac{1}{365}) = 60$ dollars. A single option is now worth $\$4.11833$, so our liability due to the short position in options is reduced. Our debt on the money market is increased by the interest due. The position in stock is worth the same as initially. The balance on day one is

stock	34,917.39
money	−30,779.62
options	−4,118.33
TOTAL	19.45

   Without hedging ($x = 0$, $y = 4,118.33$, $z = -1,000$) our wealth would have been $\$27.10$, that is, we would have benefited from the reduced value of the option and the interest due on the premium invested without risk.

2. The stock price goes up to $S(\frac{1}{365}) = 61$ dollars. A single option is now worth $\$4.72150$, which is more than initially. The unhedged (naked) position would have suffered a loss of $\$576.07$. On the other hand, for a holder of a delta neutral portfolio the loss on the options is almost completely balanced out

by the increase in stock value:

stock	35,499.35
money	−30,779.62
options	−4,721.50
TOTAL	**−1.77**

3. The stock price goes down to $S(\frac{1}{365}) = 59$ dollars. The value of the written options decreases, a single option now being worth \$3.55908. The value of the stock held decreases too. The portfolio brings a small loss:

stock	34,335.44
money	−30,779.62
options	−3,559.08
TOTAL	**−3.26**

In this scenario it would have been much better not to have hedged at all, since then we would have gained \$586.35.

It may come as a surprise that the hedging portfolio brings a profit when the stock price remains unchanged. As we shall see later in Exercise 8.12, a general rule is at work here.

Exercise 8.9

Find the stock price on day one for which the hedging portfolio attains its maximum value.

Exercise 8.10

Suppose that 50,000 puts with exercise date in 90 days and strike price $X = 1.80$ dollars are written on a stock with current price $S(0) = 1.82$ dollars and volatility $\sigma = 14\%$. The risk-free rate is $r = 5\%$. Construct a delta neutral portfolio and compute its value after one day if the stock price drops to $S(\frac{1}{365}) = 1.81$ dollars.

Going back to our example, let us collect the values V of the delta neutral portfolio for various stock prices after one day as compared to the values U of the unhedged position:

S	V	U
58.00	-71.35	$1,100.22$
58.50	-31.56	849.03
59.00	-3.26	586.35
59.50	13.69	312.32
60.00	19.45	27.10
60.50	14.22	-269.11
61.00	-1.77	-576.07
61.50	-28.24	-893.53
62.00	-64.93	$-1,221.19$

Now, let us see what happens if the stock price changes are considerable:

S	V	U
50	$-2,233.19$	$3,594.03$
55	-554.65	$2,362.79$
60	19.45	27.10
65	-481.60	$-3,383.73$
70	$-1,765.15$	$-7,577.06$

If we fear that such large changes might happen, the above hedge is not a satisfactory solution. If we do not hedge, at least we have a gamble with a positive outcome whenever the stock price goes down. Meanwhile, no matter whether the stock price goes up or down, the delta neutral portfolio may bring losses, though considerably smaller than the naked position.

Let us see what can happen if some other variables, in addition to the stock price, change after one day:

1. Suppose that the interest rate increases to 9% with volatility as before. Some loss will result from an increase in the option value. The interest on the cash loan due on day one is not affected because the new rate will only have an effect on the interest payable on the second day or later. The values of the hedging portfolio are given in the second column in the table below.

2. Now suppose that σ grows to 32%, with the interest rate staying at the original level of 8%. The option price will increase considerably, which is not compensated by the stock position even if the stock price goes up. The

results are given in the third column in the following table:

S	V	
	$r = 9\%, \sigma = 30\%$	$r = 8\%, \sigma = 32\%$
58.00	-133.72	-299.83
58.50	-97.22	-261.87
59.00	-72.19	-234.69
59.50	-58.50	-218.14
60.00	-55.96	-212.08
60.50	-64.38	-216.33
61.00	-83.51	-230.68
61.50	-113.07	-254.90
62.00	-152.78	-288.74

As we can see, in some circumstances delta hedging may be far from satisfactory. We need to improve the stability of hedging when the underlying asset
price changes considerably and/or some other variables change simultaneously.
In what follows, after introducing some theoretical tools, we shall return again
to the current example.

Exercise 8.11

Find the value of the delta neutral portfolio in Exercise 8.10 if the risk-
free rate of interest decreases to 3% on day one.

8.5.2 Greek Parameters

We shall define so-called *Greek parameters* describing the sensitivity of a portfolio with respect to the various variables determining the option price. The
strike price X and expiry date T are fixed once the option is written, so we
have to analyse the four remaining variables S, t, r, σ.

Let us write the value of a general portfolio containing stock and some
contingent claims based on this stock as a function $V(S, t, \sigma, r)$ of these variables

and denote

$$\text{delta}_V = \frac{\partial V}{\partial S},$$

$$\text{gamma}_V = \frac{\partial^2 V}{\partial S^2},$$

$$\text{theta}_V = \frac{\partial V}{\partial t},$$

$$\text{vega}_V = \frac{\partial V}{\partial \sigma},$$

$$\text{rho}_V = \frac{\partial V}{\partial r}.$$

For small changes ΔS, Δt, $\Delta \sigma$, Δr of the variables we have the following approximate equality (by the Taylor formula):

$$\Delta V \cong \text{delta}_V \times \Delta S + \text{theta}_V \times \Delta t + \text{vega}_V \times \Delta \sigma + \text{rho}_V \times \Delta r$$
$$+ \frac{1}{2}\text{gamma}_V \times (\Delta S)^2.$$

Hence, a way to immunise a portfolio against small changes of a particular variable is to ensure that the corresponding Greek parameter is equal to zero. For instance, to hedge against volatility movements we should construct a *vega neutral* portfolio, with vega equal to zero. To retain the benefits of delta hedging, we should design a portfolio with both delta and vega equal to zero (*delta-vega neutral*). A *delta-gamma neutral* portfolio will be immune against larger changes of the stock price. Examples of such hedging portfolios will be examined below.

The Black–Scholes formula allows us to compute the derivatives explicitly for a single option. For a European call we have

$$\text{delta}_{C_E} = N(d_+),$$

$$\text{gamma}_{C_E} = \frac{1}{S\sigma\sqrt{2\pi T}}e^{-\frac{d_+^2}{2}},$$

$$\text{theta}_{C_E} = -\frac{S\sigma}{2\sqrt{2\pi T}}e^{-\frac{d_+^2}{2}} - rXe^{-rT}N(d_-),$$

$$\text{vega}_{C_E} = \frac{S\sqrt{T}}{\sqrt{2\pi}}e^{-\frac{d_+^2}{2}},$$

$$\text{rho}_{C_E} = TXe^{-rT}N(d_-).$$

(The Greek parameters are computed at time $t = 0$.)

Remark 8.14

It is easy to see from the above that

$$\text{theta}_{C_E} + rS\,\text{delta}_{C_E} + \frac{1}{2}\sigma^2 S^2 \text{gamma}_{C_E} = rC_E.$$

This implies that the function u expressing the dependence of call price on t and S, that is, $C_E = u(t, S)$ satisfies the Black–Scholes equation:

$$\frac{\partial u}{\partial t} + rS\frac{\partial u}{\partial S} + \frac{1}{2}\sigma^2 S^2 \frac{\partial^2 u}{\partial S^2} = ru.$$

Exercise 8.12

Show that a delta neutral portfolio with initial value zero hedging a single call option will gain in value with time if the stock price, volatility and risk-free rate remain unchanged.

Exercise 8.13

Derive formulae for the Greek parameters of a put option.

To show some possibilities offered by Greek parameters we consider hedging the position of a writer of European call options.

Delta-Gamma Hedging. The construction is based on making both delta and gamma zero. A portfolio of the form (x, y, z) is insufficient for this. Given the position in options, say $z = -1,000$, there remains only one parameter that can be adjusted, namely the position x in the underlying. This allows us to make the delta of the portfolio zero. To make the gamma also equal to zero an additional degree of freedom is needed. To this end we consider another option on the same underlying stock, for example, a call expiring after 60 days, $\widehat{T} = 60/365$, with strike price $\widehat{X} = 65$, and construct a portfolio $(x, y, z, \widehat{z})$, where $\widehat{z}$ is a position in the additional option. The other variables are as in the previous examples: $r = 8\%$, $\sigma = 30\%$, $S(0) = 60$.

Let us sum up all the information about the prices and selected Greek parameters (we also include vega, which will be used later):

option	time to expiry	strike price	option price	delta	gamma	vega
original	90/365	60	4.14452	0.581957	0.043688	11.634305
additional	60/365	65	1.37826	0.312373	0.048502	8.610681

We choose x and $\hat{z}$ so that the delta and gamma of the portfolio are zero,

$$\text{delta}_V = x - 1,000\,\text{delta}_{C_E} + \hat{z}\,\text{delta}_{\hat{C}_E} = 0,$$
$$\text{gamma}_V = -1,000\,\text{gamma}_{C_E} + \hat{z}\,\text{gamma}_{\hat{C}_E} = 0,$$

and the money position y so that the value of the portfolio is zero,

$$V(S) = xS + y - 1,000 C_E(S) + \hat{z}\hat{C}_E(S) = 0.$$

This gives the following system of equations:

$$x - 581.957 + 0.312373\,\hat{z} = 0,$$
$$-43.688 + 0.048502\,\hat{z} = 0,$$

with solution $x \cong 300.58$, $\hat{z} \cong 900.76$. It follows that $y \cong -15,131.77$. That is, we take long positions in stock and the additional option, and a short cash position. (We already have a short position $z = -1,000$ in the original option.)

After one day, if stock goes up, the original option will become more expensive, increasing our liability, which will be set off by increases in the value of stock and the additional options held. The reverse happens if the stock price declines. Our money debt increases in either case by the interest due after one day. The values of the portfolio are given below (for comparison we also recall the values of the delta neutral portfolio):

$S(\frac{1}{365})$	delta-gamma	delta
58.00	−2.04	−71.35
58.50	0.30	−31.56
59.00	1.07	−3.26
59.50	0.81	13.69
60.00	0.02	19.45
60.50	−0.79	14.22
61.00	−1.11	−1.77
61.50	−0.49	−28.24
62.00	1.52	−64.93

We can see that we are practically safe within the given range of stock prices. For larger changes we are also in a better position as compared with delta hedging:

$S(\frac{1}{365})$	delta-gamma	delta
50	−614.08	−2,233.19
55	−78.22	−554.65
60	0.02	19.45
65	63.13	−481.60
70	440.81	−1,765.15

As predicted, a delta-gamma neutral portfolio offers better protection against stock price changes than a delta neutral one.

Delta-Vega Hedging. Next we shall hedge against an increase in volatility, while retaining cover against small changes in the stock price. This will be achieved by constructing a delta-vega neutral portfolio containing, as before, an additional option. The conditions imposed are

$$\text{delta}_V = x - 1,000\,\text{delta}_{C_{\mathrm{E}}} + \widehat{z}\,\text{delta}_{\widehat{C}_{\mathrm{E}}} = 0,$$
$$\text{vega}_V = -1,000\,\text{vega}_{C_{\mathrm{E}}} + \widehat{z}\,\text{vega}_{\widehat{C}_{\mathrm{E}}} = 0.$$

They lead to the system of equations

$$x - 581.957 + 0.312373\,\widehat{z} = 0,$$
$$-1,1634.305 + 8.610681\,\widehat{z} = 0,$$

with an approximate solution $x \cong 159.89$, $\widehat{z} \cong 1,351.15$. The corresponding money position is $y \cong -7,311.12$.

Suppose that the volatility increases to $\sigma = 32\%$ on day one. Let us compare the results for delta-vega and delta hedging:

$S(1/365)$	delta-vega	delta
58.00	−5.90	−299.83
58.50	−12.81	−261.87
59.00	−16.05	−234.69
59.50	−14.99	−218.14
60.00	−9.06	−212.08
60.50	2.27	−216.33
61.00	19.52	−230.68
61.50	43.17	−254.90
62.00	73.62	−288.74

Exercise 8.14

Using the data in our ongoing example (stock price $60, volatility 30%, interest rate 8%), construct a delta-rho neutral portfolio to hedge a short position of 1,000 call options expiring after 90 days with strike price $60, taking as an additional component a call option expiring after 120 days with strike price $65. Analyse the sensitivity of the portfolio value to stock price variations if the interest rate goes up to 9% after one day, comparing with the previous results. What will happen if the interest rate jumps to 15%?

The examples above illustrate the variety of possible hedging strategies. The choice between them depends on individual aims and preferences. We have not touched upon questions related to transaction costs or long term hedging. Nor have we discussed the optimality of the choice of an additional derivative instrument. Portfolios based on three Greek parameters would require yet another derivative security as a component. They could provide comprehensive cover, though their performance might deteriorate if the variables remain unchanged. In addition, they might prove expensive if transaction costs were included.

8.5.3 Value-at-Risk

For the purpose of measuring the risk involved in running a business it is common to adopt the intuitive understanding of risk as the size and likelihood of a possible loss.

Let us present the basic idea using a simple example. We buy a share of stock for $S(0) = 100$ dollars to sell it after one year. The selling price $S(1)$ is random. We shall suffer a loss if $S(1) < 100e^r$, where r is the risk-free rate under continuous compounding. (The purchase can either be financed by a loan, or, if the initial sum is already at our disposal, we take into account the foregone opportunity of a risk-free investment.) What is the probability of the loss being less than a given amount, for example,

$$P(100e^r - S(1) < 20) = ?$$

Let us reverse the question and fix the probability, 95% say. Now we seek an amount such that the probability of a loss not exceeding this amount is 95%. This is referred to as *Value-at-Risk* at 95% confidence level and denoted by VaR. (Other confidence levels can also be used.) So, VaR is an amount such that

$$P(100e^r - S(1) < \text{VaR}) = 95\%.$$

It should be noted that the majority of textbooks neglect the time value of money in this context, stating the definition of VaR only for $r = 0$.

Example 8.15

Suppose that the distribution of the stock price is log normal, the logarithmic return $k = \ln(S(1)/S(0))$ having normal distribution with mean $\mu = 12\%$ and standard deviation $\sigma = 30\%$. (Note that we use the actual mean return, not the risk-free one since we are working under the true probability rather than the martingale one.) With probability 95% the return will satisfy $k > \mu + x\sigma \cong -37.50\%$, where $N(x) \cong 5\%$, so $x \cong -1.645$. (Here $N(x)$ is the normal distribution function with mean 0 and variance 1.) Hence, with probability 95% the future price $S(1)$ will satisfy

$$S(1) > S(0)e^{\mu+x\sigma} \cong 68.83 \text{ dollars},$$

and so, given that $r = 8\%$,

$$\text{VaR} = S(0)e^{r} - S(0)e^{\mu+x\sigma} \cong 39.50 \text{ dollars}.$$

Exercise 8.15

Evaluate VaR at 95% confidence level for a one-year investment of $1,000 into euros if the interest rate for risk-free investments in euros is $r_{\text{EUR}} = 4\%$ and the exchange rate from euros into US dollars follows the log normal distribution with $\mu = 1\%$ and $\sigma = 15\%$. Take into account the foregone opportunity of investing dollars without risk, given that the risk-free interest rate for dollars is $r_{\text{USD}} = 5\%$.

Exercise 8.16

Suppose that $1,000 is invested in European call options on a stock with current price $S(0) = 60$ dollars. The options expire after 6 months with strike price $X = 40$ dollars. Assume that $\sigma = 30\%$, $r = 8\%$, and the expected logarithmic return on stock is 12%. Compute VaR after 6 months at 95% confidence level. Find the final wealth if the stock price grows at the expected rate. Find the stock price level that will be exceeded with 5% probability and compute the corresponding final payoff.

Case 8: Discussion

We discuss the problem from the point of view of controlling Value-at-Risk. This approach is often applied by banks to assess companies when they apply for loans.

First note that to satisfy the expectations of their investors the company should be able to achieve a profit of 1.25 million pounds a year to pay the dividend. A lower profit would mean a dividend lower than required, with any shortfall regarded as a loss from the point of view of the investors. The profit depends on the rate of exchange d at the end of the year, hence some risk emerges. (We assume that the other values will be as predicted.)

To begin with, suppose that no action is taken to manage the risk.

1. **Unhedged Position.** If the exchange rate d turns out to be 1.6 dollars to a pound at the end of the year, then the net earnings will be 1.6 million pounds, as shown in the following statement (all amounts in pounds):

sales	5,000,000
costs	−3,000,000
earnings before tax	2,000,000
tax	−400,000
earnings after tax	1,600,000
dividend	−1,250,000
result	350,000

The surplus income will be 0.35 million pounds.

However, if the exchange rate d becomes 2 dollars to a pound, the dividend will in fact have to be reduced and the investors will end up with a loss of 0.45 million pounds (taking the full dividend as a hypothetical cost):

sales	4,000,000
costs	−3,000,000
earnings before tax	1,000,000
tax	−200,000
earnings after tax	800,000
dividend	−1,250,000
result	−450,000

Let us compute VaR. We assume that the rate of exchange has log normal distribution of the form

$$d(t) = d(0) \exp\left((r_{\mathrm{USD}} - r_{\mathrm{GBP}} - \frac{1}{2}\sigma^2)t + \sigma\sqrt{t}N \right),$$

where N has the standard normal distribution.[1] With probability 95%, N will not exceed 1.6449, which corresponds to an exchange rate $d = d(1) \cong$ 1.9650 dollars to a pound. The statement for this borderline rate of exchange will be as follows (all amounts rounded to the nearest pound):

sales	4,071,290
costs	−3,000,000
earnings before tax	1,071,290
tax	−214,258
earnings after tax	857,032
dividend	−1,250,000
result	−392,968

---

[1] See Etheridge (2002), Section 5.3.

As a result, VaR $\cong 392,968$ dollars. The final result as a function of the exchange rate d is

$$b(d) = 80\% \times \left(\frac{8,000,000}{d} - 3,000,000 \right) - 1,250,000$$

$$= \frac{6,400,000}{d} - 3,650,000.$$

The break even exchange rate, which solves $b(d) = 0$, is approximately equal to 1.7534 dollars to a pound. In an optimistic scenario in which the pound weakens, for example, down to 1.5 dollars, the final balance will be about £616, 667.

The question is how to manage this risk exposure.

2. **Forward Contract.** The easiest solution would be to fix the exchange rate in advance by entering into a long forward contract. The forward rate is $d(0)e^{r_{\text{USD}} - r_{\text{GBP}}} = 1.6 \times e^{-3\%} \cong 1.5527$ dollars to a pound. As a result, the company can obtain the following statement with guaranteed surplus, but no possibility of further gains should the exchange rate become more favourable:

sales	5, 152, 273
costs	−3, 000, 000
earnings before tax	2, 152, 273
tax	−430, 455
net income	1, 721, 818
dividend	−1, 250, 000
result	471, 818

3. **Full Hedge with Options.** Options can be used to ensure that the rate of exchange is capped at a certain level, whilst the benefits associated with favourable exchange rate movements are retained. However, this may be costly because of the premium paid for options.

The company can buy call options on the pound. A European call to buy one pound with strike price 1.6 dollars to a pound will cost $0.0669.[2] Suppose that the company buys 5 million of such options, paying £209, 069, an amount they have to borrow at 16%. The interest is tax deductible,

---

[2] For options on currencies the Black–Scholes formula applies with $r = r_{\text{USD}}$, $t = 1$, $S(0) = d(0)\exp(-r_{\text{GBP}})$.

making the loan less costly. Nevertheless, the final result is disappointing:

sales	5,000,000
costs	−3,000,000
earnings before interest and tax	2,000,000
interest	−33,451
earnings before tax	1,966,549
tax	−393,310
net income	1,573,239
loan repaid	−209,069
dividend	−1,250,000
result	114,170

If the exchange rate drops to 1.5 dollars to a pound, the options will not be exercised and the sum obtained from sales will reach £5,333,333, with a positive final result of £380,837. This strategy leads to a better result than the hedge involving a forward contract only if the rate of exchange drops below 1.4687 dollars to a pound.

4. **Partial Hedge with Options.** To reduce the cost of options the company can hedge partially by buying call options to cover only a fraction of the dollar amount from sales. Suppose that the company buys 2,500,000 units of the same call option as above, paying half of the previous premium. Half of the revenue is then exposed to risk. To find VaR at 95% confidence level we assume that this sum is exchanged at 1.9650 dollars to a pound, as in the case of an unhedged position, the other half being exchanged at the exercise price:

sales	4,535,645
costs	−3,000,000
earnings before interest and tax	1,535,645
interest	−16,726
earnings before tax	1,518,919
tax	−303,784
net income	1,215,136
loan repaid	−104,534
dividend	−1,250,000
result	−139,399

If the exchange rate drops to 1.5 dollars to a pound, the investors will have a surplus of £498,752.

5. **Combination of Options and Forward Contracts.** Finally, let us investigate what happens if the company hedges with both kinds of derivatives.
   Half of their position will be hedged with options. In the worst case scenario they will buy pounds for half of their dollar revenue at the rate of
   1.6 dollars to a pound, the remaining half being exchanged at the forward
   rate of 1.5527 dollars to a pound. The outcome is shown below, where we
   summarise the resulting VaR for all strategies considered (the result below
   is equal to minus VaR):

strategy	1	2	3	4	5
result	$-392,968$	$471,818$	$114,170$	$-139,399$	$292,994$

   These values are computed at 95% confidence level, corresponding to the
   exchange rate of 1.9887 dollars to a pound.

   Clearly, VaR provides only partial information about possible outcomes of
various strategies. Figure 8.4 shows the graphs of the final result as a function
of the exchange rate d for each of the above strategies. The graphs are labelled
by the strategy number as above. The strategy using a forward contract (strat-

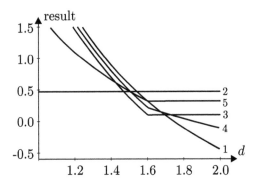

Figure 8.4 Comparison of various strategies

egy 2) appears to be the safest one. An adventurous investor who strongly
believes that the pound will weaken considerably may prefer to remain uncov-
ered (strategy 1). A variety of middle-of-the-road strategies are also available.
The probability distribution of the exchange rate d should also be taken into
account when examining the graphs.

9
Interest Rates

Case 9

This is the year 2050. You have reached retirement age with a substantial sum of money saved in your personal pension fund. It is time to supplement the state pension and use the money for monthly withdrawals. A source of worry is that risk-free rates may change and this will affect your income. Your bank manager offers an annuity consisting of 12 monthly payments of $85.08 for each $1000 invested at the beginning of the year. You want to find out if you might be better off to manufacture an annuity by investing in interest rate products.

9.1 Tools

This chapter begins with a model in which the interest rates implied by bonds do not depend on maturity. If the rates are deterministic, then it turns out that they must be constant, and the model would become too simple to describe any real-life situation. In an extension allowing random changes of interest rates the problem of risk management will be dealt with by introducing a mathematical tool called the *duration* of bond investments. Then we shall show that maturity-dependent rates cannot be deterministic either, preparing the motivation and notation for the final sections, in which a simple model of stochastic rates will be explored.

M. Capiński, T. Zastawniak, *Mathematics for Finance*,
Springer Undergraduate Mathematics Series,
© Springer-Verlag London Limited 2011

As in Chapter 2, $B(t, T)$ will denote the price at time t (the running time) of a zero-coupon bond paying one unit of currency (the face value) at time T (the maturity time). The dependence on two time variables gives rise to some difficulties in mathematical models of bond prices. These prices are exactly what is needed to describe the time value of money. In Chapter 2 we saw how bond prices imply the interest rate, under the assumption that the rate is constant. We relax this restriction, allowing variable interest rates.

In this chapter, time will be discrete, though some parts of the theory can easily be extended to continuous time. We fix a time step h, writing $t = nh$ for the running time and $T = Nh$ for the maturity time. In the majority of examples we take either $h = \frac{1}{12}$ or $h = 1$. The notation $B(n, N)$ will be employed instead of $B(t, T)$ for the price of a zero-coupon unit bond. We use continuous compounding, bearing in mind that it simplifies notation and makes it possible to handle time steps of any length consistently.

9.2 Maturity-Independent Yields

The present value of a zero-coupon unit bond determines an interest rate called the *yield* and denoted by $y(0)$ to emphasise the fact that it is computed at time 0:

$$B(0, N) = e^{-Nhy(0)}.$$

For a different running time instant $n < N$ the implied yield may in general be different from $y(0)$. For each such n we thus have a rate $y(n)$ satisfying

$$B(n, N) = e^{-(N-n)hy(n)}.$$

Generally (and in most real cases), bonds with different maturities will imply different yields. Nevertheless, in this section we consider the simplified situation when $y(n)$ is independent of N, that is, bonds with different maturities generate the same yield. Independence of maturity will be relaxed later in Section 9.3.

Proposition 9.1

If the yield $y(n)$ for some $n > 0$ were known at time 0, then $y(0) = y(n)$ or else an arbitrage strategy could be found.

Proof

Suppose that $y(0) < y(n)$. (We need to know not only $y(0)$ but also $y(n)$ at time 0 to see whether or not this inequality holds.)

- Borrow a dollar for the period between 0 and $n+1$ and deposit it for the period between 0 and n, both at the rate $y(0)$. (The yield can be regarded as the interest rate for deposits and loans.)
- At time n withdraw the deposit with interest, $e^{nhy(0)}$ in total, and invest this sum for a single time step at the rate $y(n)$. At time $n+1$ this brings $e^{nhy(0)+hy(n)}$. The initial loan requires repayment of $e^{(n+1)hy(0)}$, leaving a positive balance $e^{nhy(0)}(e^{hy(n)} - e^{hy(0)})$, which is the arbitrage profit.

The reverse inequality $y(0) > y(n)$ can be dealt with in a similar manner. □

Exercise 9.1

Let $h = \frac{1}{12}$. Find arbitrage if the yields are independent of maturity, and unit bonds maturing at time 6 (half a year) are traded at $B(0,6) = 0.9320$ dollars and $B(3,6) = 0.9665$ dollars, both prices being known at time 0.

As a consequence of Proposition 9.1, if the yield is independent of maturity and deterministic (that is, $y(n)$ is known in advance for any $n \geq 0$), then it must be constant, $y(n) = y$ for all n. This is the situation in Chapter 2, where all bond prices were determined by a single interest rate. The yield $y(n) = y$, independent of n, is then equal to the constant risk-free interest rate denoted previously by r.

Historical bond prices show a different picture: the yields implied by the bond prices recorded in the past clearly vary with time. In an arbitrage-free model, to admit yields varying with time but independent of maturity we should allow them to be random, so it is impossible to predict in advance whether $y(n)$ will be higher or lower than $y(0)$.

We assume, therefore, that at each time instant the yield $y(n)$ is a positive random number independent of the maturity of the underlying bond.

Our goal is to analyse the return on a bond investment and the imminent risk arising from random changes of interest rates. Suppose that we intend to invest a certain sum of money P for a fixed period of N time steps. If the yield y remains constant, then, as observed in Chapter 2, our terminal wealth will be Pe^{Nhy}. This will be our benchmark for designing strategies hedged against unpredictable interest rate movements.

9.2.1 Investment in Single Bonds

If we invest in zero-coupon bonds and keep them to maturity, the rate of return is guaranteed, since the final payment is fixed in advance and is not affected by any future changes of interest rates. However, if we choose to close our investment prior to maturity by selling the bonds, we face the risk that the interest rates may change in the meantime with an adverse effect on the final value of the investment.

Example 9.2

Suppose we invest in bonds for a period of six months. Let $h = \frac{1}{12}$. We buy a number of unit bonds that will mature after one year, paying $B(0, 12) = 0.9300$ for each. This price implies a rate $y(0) \cong 7.26\%$. Since we are going to sell the bonds at time $n = 6$, we are concerned with the price $B(6, 12)$ or, equivalently, with the corresponding rate $y(6)$. Let us discuss some possible scenarios:

1. The rate is stable, $y(6) = 7.26\%$. The bond price is $B(6, 12) \cong 0.9644$ and the logarithmic return on the investment is 3.63%, a half of the interest rate, in line with the additivity of logarithmic returns.

2. The rate decreases to $y(6) = 6.26\%$, say. (The convention is that 0.01% is one *basis point*, so here the rate drops by 100 basis points.) Then $B(6, 12) \cong 0.9692$, which is more than in scenario 1. As a result, we are going to earn more, achieving a logarithmic return of 4.13%.

3. The rate increases to $y(6) = 8.26\%$. In this case the logarithmic return on our investment will be 3.13%, which is lower than in scenario 1, the bond price being $B(6, 12) \cong 0.9596$.

We can see a pattern here: one is better off if the rate drops and worse off if the rate increases. A general formula for the return on this kind of investment is easy to find.

Suppose that the initial yield changes randomly to become $y(n)$ at time n. Hence

$$B(0, N) = e^{-y(0)hN}, \quad B(n, N) = e^{-y(n)h(N-n)},$$

and the return on an investment closed at time n will be

$$k(0, n) = \ln \frac{B(n,N)}{B(0,N)} = \ln e^{y(0)hN - y(n)h(N-n)} = y(0)hN - y(n)h(N - n).$$

We can see that the return decreases as the rate $y(n)$ increases. The impact of a rate change on the return depends on the timing. For example, if $h = \frac{1}{12}$, $N = 12$ and $n = 6$, then a rate increase of 120 basis points will reduce the return by 0.6% as compared to the case when the rate remains unchanged.

Exercise 9.2

Let $h = \frac{1}{12}$. Invest \$100 in 6-month zero-coupon bonds trading at $B(0, 6) = 0.9400$ dollars. After 6 months reinvest the proceeds in bonds of the same kind, now trading at $B(6, 12) = 0.9368$ dollars. Find the implied interest rates and compute the number of bonds held at each time. Compute the logarithmic return on the investment over one year.

Exercise 9.3

Suppose that $B(0, 12) = 0.8700$ dollars. What is the interest rate after 6 months if an investment for 6 months in zero-coupon bonds gives a logarithmic return of 14% ?

Exercise 9.4

In this exercise we take a finer time scale with $h = \frac{1}{360}$. (A year is assumed to have 360 days here.) Suppose that $B(0, 360) = 0.9200$ dollars, the rate remains unchanged for the first six months, goes up by 200 basis points on day 180, and remains at this level until the end of the year. If a bond is bought at the beginning of the year, on which day should it be sold to produce a logarithmic return of 4.88% or more?

An investment in coupon bonds is more complicated. Even if the bond is kept to maturity, the coupons are paid in the meantime and can be reinvested. The return on such an investment depends on the interest rates prevailing at the times when the coupons are due. First consider the relatively simple case of an investment terminated as soon as the first coupon is paid.

Example 9.3

Let us invest the sum of $1,000 in 4-year bonds with face value $100 and $10 annual coupons. A coupon bond of this kind can be regarded as a collection of four zero-coupon bonds maturing after 1, 2, 3 and 4 years with face value $10, $10, $10 and $110, respectively. Suppose that such coupon bonds trade at $91.78, which can be expressed as the sum of the prices of the four zero-coupon bonds,

$$91.78 = 10e^{-y(0)} + 10e^{-2y(0)} + 10e^{-3y(0)} + 110e^{-4y(0)}.$$

(The length of a time step is $h = 1$.) This equation can be solved to find the yield, $y(0) \cong 12\%$. We can afford to buy 10.896 coupon bonds. After one year we cash the coupons, collecting $108.96, and sell the bonds, which are now 3-year coupon bonds. Consider three scenarios:

1. After one year the interest rate remains unchanged, $y(1) = 12\%$, the coupon

bonds being valued at

$$10e^{-0.12} + 10e^{-2\times0.12} + 110e^{-3\times0.12} \cong 93.48$$

dollars, and we shall receive $108.96 + 1,018.52 \cong 1,127.48$ dollars in total.

2. The rate drops to 10%. As a result, the coupon bonds will be worth

$$10e^{-0.1} + 10e^{-2\times0.1} + 110e^{-3\times0.1} \cong 98.73$$

dollars each. We shall end up with $1,184.63.

3. The rate goes up to 14%, the coupon bonds trading at $88.53. The final value of our investment will be $1,073.51.

Exercise 9.5

Find the rate $y(1)$ such that the logarithmic return on the investment in Example 9.3 will be a) 12%, b) 10%, c) 14%.

If the lifetime of our investment exceeds one year, we will be facing the problem of reinvesting coupons. In the following example we assume that the coupons are used to purchase the same bond.

Example 9.4

We begin as in Example 9.3, but our intention is to terminate the investment after three years. After one year we reinvest the coupons obtained in the same, now a 3-year, coupon bond. Consider the following scenarios after one year:

1. The rate remains the same for the period of our investment, $y(0) = y(1) = y(2) = y(3) = 12\%$. The bond price is $93.48, so for the $108.96 received from coupons we can buy 1.17 additional bonds, increasing the number of bonds held to 12.06. We can monitor the value of our investment by simply multiplying the number of bonds held by the current bond price. We repeat this in the following year. After three years we cash the coupons and sell the bonds, the final value of the investment being $1,433.33. This number will be used as a benchmark for other scenarios. Observe that

$$1,433.33 \cong 1,000e^{3\times12\%},$$

the same as the value after 3 years of $1,000 invested in zero-coupon bonds.

The building blocks of our investment are summarised in the table below.

Year	0	1	2	3
Rate	12%	12%	12%	12%
PV of coupon 1	$8.87	$10.00		
PV of coupon 2	$7.87	$8.87	$10.00	
PV of coupon 3	$6.98	$7.87	$8.87	$10.00
PV of coupon 4	$6.19	$6.98	$7.87	$8.87
PV of face value	$61.88	$69.77	$78.66	$88.69
Bond price	$91.78	$93.48	$95.40	$97.56
Cashed coupons		$108.96	$120.60	$133.26
Additional bonds		1.17	1.26	
Number of bonds	10.90	12.06	13.33	
Value of investment	$1,000.00	$1,127.50	$1,271.25	$1,433.33

2. Suppose that the rate goes down by 2% after one year and then remains
   at the new level. The drop in the rate results in an increase of all bond
   prices. The number of additional bonds that can be bought for the coupons
   is lower than in scenario 1. Nevertheless, the final value of the investment is
   higher because so is the price at which we sell the bonds after three years.

Year	0	1	2	3
Rate	12%	10%	10%	10%
PV of coupon 1	$8.87	$10.00		
PV of coupon 2	$7.87	$9.05	$10.00	
PV of coupon 3	$6.98	$8.19	$9.05	$10.00
PV of coupon 4	$6.19	$7.41	$8.19	$9.05
PV of face value	$61.88	$74.08	$81.87	$90.48
Bond price	$91.78	$98.73	$99.11	$99.53
Cashed coupons		$108.96	$119.99	$132.10
Additional bonds		1.10	1.21	
Number of bonds	10.90	12.00	13.21	
Value of investment	$1,000.00	$1,184.65	$1,309.25	$1,446.94

3. If the rate increases to 14% and stays there, the bonds will be cheaper than

in scenario 1. The final value of the investment will be disappointing.

Year	0	1	2	3
Rate	12%	14%	14%	14%
PV of coupon 1	$8.87	$10.00		
PV of coupon 2	$7.87	$8.69	$10.00	
PV of coupon 3	$6.98	$7.56	$8.69	$10.00
PV of coupon 4	$6.19	$6.57	$7.56	$8.69
PV of face value	$61.88	$65.70	$75.58	$86.94
Bond price	$91.78	$88.53	$91.83	$95.63
Cashed coupons		$108.96	$121.26	$134.46
Additional bonds		1.23	1.32	
Number of bonds	10.90	12.13	13.45	
Value of investment	$1,000.00	$1,073.53	$1,234.85	$1,420.41

As a motivation for certain theoretical notions, consider the above investment, with the same possible scenarios, but involving a specially designed security, a coupon bond with annual coupons paying $32, all other parameters remaining unchanged. The results are as follows:

Scenario	Value after 3 years
12%, 12%, 12%, 12%	$1,433.33
12%, 10%, 10%, 10%	$1,433.68
12%, 14%, 14%, 14%	$1,433.78

It is remarkable that any change in interest rates improves the result of our investment. We do not lose if the rates change unfavourably. On the other hand, we do not gain in other circumstances. This is explained by the fact that a certain parameter of the bond, called duration and defined below, is exactly equal to the lifetime of our investment. In some sense, the bond behaves approximately like a zero-coupon bond with prescribed maturity.

Exercise 9.6

Check the numbers given in the above tables.

Exercise 9.7

Compute the value after three years of $1,000 invested in a 4-year bond with $32 annual coupons and $100 face value if the rates in consecutive years are as follows:

Scenario 1: 12%, 11%, 12%, 12%;

Scenario 2: $12\%, 13\%, 12\%, 12\%$.

Design a spreadsheet and experiment with various interest rates.

9.2.2 Duration

We have seen that variable rates lead to uncertainty as to the future value of an investment in bonds. This may be undesirable, or even unacceptable, for example, for a pension fund manager. We shall introduce a tool which makes it possible to immunise such an investment, at least in the special situation of the maturity-independent rates considered in this section.

For notational simplicity we denote the current yield $y(0)$ by y. Consider a coupon bond with coupons $C_1, C_2, \ldots, C_N$ payable at times $0 < hn_1 < hn_2 < \ldots < hn_N$ and face value F, maturing at time hn_N. Its current price is given by

$$P(y) = C_1 e^{-hn_1 y} + C_2 e^{-hn_2 y} + \cdots + (C_N + F)e^{-hn_N y}. \qquad (9.1)$$

The *duration* of the coupon bond is defined to be

$$D(y) = \frac{hn_1 C_1 e^{-hn_1 y} + hn_2 C_2 e^{-hn_2 y} + \cdots + hn_N(C_N + F)e^{-hn_N y}}{P(y)}. \qquad (9.2)$$

The numbers $C_1 e^{-hn_1 y}/P(y), C_2 e^{-hn_2 y}/P(y), \ldots, (C_N + F)e^{-hn_N y}/P(y)$ are non-negative and add up to one, so they may be regarded as weights or probabilities. It can be said that the duration is a weighted average of future payment times. The duration of any future cash flow can be defined in a similar manner.

Duration measures the sensitivity of the bond price to changes in the interest rate. To see this we compute the derivative of the bond price with respect to y,

$$\frac{d}{dy}P(y) = -hn_1 C_1 e^{-hn_1 y} - hn_2 C_2 e^{-hn_2 y} - \cdots - hn_N(C_N + F)e^{-hn_N y},$$

which gives

$$\frac{d}{dy}P(y) = -D(y)P(y).$$

The last formula is sometimes taken as the definition of duration.

Example 9.5

A 6-year bond with \$10 annual coupons, \$100 face value and yield of 6% has duration of 4.898 years. A 6-year bond with the same coupons and yield but with \$500 face value will have duration of 5.671 years. The duration of any zero-coupon bond is equal to its lifetime.

Exercise 9.8

A 2-year bond with $100 face value pays a $6 coupon each quarter and has 11% yield. Compute the duration.

Exercise 9.9

What should be the face value of a 5-year bond with 10% yield, paying $10 annual coupons to have duration 4? Find the range of duration that can be obtained by altering the face value, as long as a coupon cannot exceed the face value. If the face value is fixed, say $100, find the level of coupons for the duration to be 4. What durations can be manufactured in this way?

Exercise 9.10

Show that P is a convex function of y.

If we invest in a bond with the intention to close the investment at time t, then the future value of the money invested in a single bond will be $P(y(0))e^{ty(0)}$, provided that the interest rate remains unchanged (being equal to the initial yield $y(0)$). To see how sensitive this amount is to interest rate changes compute the derivative

$$\frac{d}{dy}(P(y)e^{ty}) = \left(\frac{d}{dy}P(y)\right)e^{ty} + tP(y)e^{ty} = (t - D(y))P(y)e^{ty}.$$

If the duration of the bond is exactly t, then

$$\left.\frac{d}{dy}(P(y)e^{ty})\right|_{y=y(0)} = 0.$$

If the derivative is zero at some point, then the graph of the function is 'flat' near this point. This means that small changes in the rate will have little effect on the future value of the investment.

9.2.3 Portfolios of Bonds

If a bond of desirable duration is not available, it may be possible to create a synthetic one by investing in a suitable portfolio of bonds of different durations.

Example 9.6

If the initial interest rate is 14%, then a 4-year bond with annual coupons $C = 10$ and face value $F = 100$ has duration 3.44 years. A zero-coupon bond with $F = 100$ and $N = 1$ has duration 1. A portfolio consisting of two bonds, one of each kind, can be regarded as a single bond with coupons $C_1 = 110$, $C_2 = C_3 = C_4 = 10$, $F = 100$. Its duration can be computed using the general formula (9.2), which gives 2.21 years.

We shall derive a formula for the duration of a portfolio in terms of the durations of its components. Denote by $P_A(y)$ and $P_B(y)$ the values of two bonds A and B with durations $D_A(y)$ and $D_B(y)$. Take a portfolio consisting of a bonds A and b bonds B, its value being $aP_A(y) + bP_B(y)$. The task of finding the duration of the portfolio will be divided into two steps:

Step 1. Find the duration of a portfolio consisting of a bonds of type A. We shall write aA to denote such a portfolio. Its price is obviously $aP_A(y)$. Since

$$\frac{d}{dy}(aP_A(y)) = -D_A(y)(aP_A(y)),$$

it follows that

$$D_{aA}(y) = D_A(y).$$

This is clear if we examine the cash flow of aA. Each coupon and the face value are multiplied by a, which cancels out in the computation of duration directly from (9.2).

Step 2. Find the duration of a portfolio consisting of one bond A and one bond B, which will be denoted by $A + B$. The price of this portfolio is $P_A(y) + P_B(y)$. Differentiating the last expression, we obtain

$$\frac{d}{dy}(P_A(y) + P_B(y)) = \frac{d}{dy}P_A(y) + \frac{d}{dy}P_B(y)$$
$$= -D_A(y)P_A(y) - D_B(y)P_B(y).$$

The last term can be written as $-D_{A+B}(y)(P_A(y) + P_B(y))$ if we put

$$D_{A+B}(y) = D_A(y)\frac{P_A(y)}{P_A(y) + P_B(y)} + D_B(y)\frac{P_B(y)}{P_A(y) + P_B(y)}.$$

This means that $D_{A+B}(y)$ is a linear combination of $D_A(y)$ and $D_B(y)$, the coefficients being the percentage weights of each bond in the portfolio.

From the above considerations we obtain the general formula

$$D_{aA+bB}(y) = D_A(y)w_A + D_B(y)w_B,$$

where
$$w_A = \frac{aP_A(y)}{aP_A(y) + bP_B(y)}, \quad w_B = \frac{bP_B(y)}{aP_A(y) + bP_B(y)},$$
are the percentage weights of individual bonds.

If we allow negative values of a or b (which corresponds to writing a bond instead of purchasing it, in other words, to borrowing money instead of investing), then, given two durations $D_A \neq D_B$, the duration D of the portfolio can take any value because $w_B = 1 - w_A$ and

$$D = D_A w_A + D_B(1 - w_A) = D_B + w_A(D_A - D_B).$$

The value of D can even be negative, which may happen in the case of a cash flow where some payments are negative, that is, they are to be made rather than received.

Example 9.7

Let $D_A = 1$ and $D_B = 3$. We wish to invest $\$1,000$ for 6 months. For the duration to match the lifetime of the investment we need $0.5 = w_A + 3w_B$. Since $w_A + w_B = 1$, it follows that $w_B = -0.25$ and $w_A = 1.25$. With $P_A = 0.92$ dollars and $P_B = 1.01$ dollars, we invest $\$1,250$ in $\frac{1250}{0.92} \cong 1,358.70$ bonds A and issue $\frac{250}{1.01} \cong 247.52$ bonds B.

Exercise 9.11

Find the number of bonds of type A and B to be bought if $D_A = 2$, $D_B = 3.4$, $P_A = 0.98$, $P_B = 1.02$ and you need a portfolio worth $\$5,000$ with duration 6.

Exercise 9.12

Invest $\$1,000$ in a portfolio of bonds with duration 2 using 1-year zero-coupon bonds with $\$100$ face value and 4-year bonds with $\$15$ annual coupons and $\$100$ face value that trade at $\$102$.

A portfolio with duration matching the investment lifetime is insensitive to small changes of interest rates. However in practice we shall have to modify the portfolio if, for example, the investment is for 3 years and one of the bonds is a zero-coupon bond expiring after one year. In addition, the duration may, as we shall see below, go off the target. As a result, it will become necessary to update the portfolio during the lifetime of the investment. This is the subject of the next subsection.

9.2.4 Dynamic Hedging

Even if a portfolio is selected with duration matching the desired investment lifetime, this will only be valid at the initial instant, since duration changes with time as well as with the interest rate.

Example 9.8

Take a 5-year bond with $10 annual coupons and $100 face value. If $y = 10\%$, then the duration will be about 4.16 years. Before the first coupon is paid the duration decreases in line with time: after 6 months it will be 3.66, and after 9 months $4.16 - 0.75 = 3.31$. If the duration matches our investment's lifetime and the interest rates do not change, no action will be necessary until a coupon becomes payable. As soon as the first coupon is paid after one year, the bond will become a 4-year one with duration 3.48, no longer consistent with the investment lifetime.

Exercise 9.13

Assuming that the interest rate does not change, show that before the first coupon is paid the duration after time t will $D - t$, where D is the duration computed at time 0.

The next example shows the impact of the interest rate on duration.

Example 9.9

The bond in Example 9.8 will have duration 4.23 if $y = 6\%$, and 4.08 if $y = 14\%$.

Exercise 9.14

Show that the duration of a 2-year bond with annual coupons decreases as the yield increases.

Duration will now be applied to design an investment strategy immune to interest rate changes. This will be done by monitoring the position at the end of each year, or more frequently if needed. For clarity of exposition we restrict ourselves to an example.

Set the lifetime of the investment to be 3 years and the target value to be $100,000. Suppose that the interest rate is 12\% initially. We invest $69,767.63, which would be the present value of $100,000$ if the interest rate remained constant.

We restrict our attention to two instruments, a 5-year bond A with $10 annual coupons and $100 face value, and a 1-year zero-coupon bond B with the same face value. We assume that a new bond of type B is always available. In subsequent years we shall combine it with bond A.

At time 0 the bond prices are $90.27 and $88.69, respectively. We find $D_A \cong 4.12$ and the weights $w_A \cong 0.6405$, $w_B \cong 0.3595$ which give a portfolio with duration 3. We split the initial sum according to the weights, spending $44,687.93 to buy $a \cong 495.05$ bonds A and $25,079.70 to buy $b \cong 282.77$ bonds B. Consider some possible scenarios of future interest rate changes.

1. After one year the rate increases to 14%. The value of our portfolio is the sum of:

   - the first coupons of bonds A: $4,950.51$;
   - the face value of cashed bonds B: $28,277.29$;
   - the market value of bonds A held, which are now 4-year bonds selling at $85.65: $42,403.53$.

   This gives $75,631.32 altogether. The duration of bonds A is now 3.44. The desired duration is 2, so we find $w_A \cong 0.4094$ and $w_B \cong 0.5906$ and arrive at the number of bonds to be held in the portfolio: 361.53 bonds A and 513.76 bonds B. (This means that we have to sell 133.52 bonds A and buy 513.76 new bonds B.)

   a) After two years the rate drops to 9%. To compute our wealth we add:

      - the coupons of A: $3,615.30$;
      - the face values of B: $51,376.39$;
      - the market value of A, selling at $101.46: $36,682.22$.

      The result is $91,673.92. We invest all the money in bonds B, since the required duration is now 1. (The payoff of these bonds is guaranteed next year.) We can afford to buy 1,003.07 bonds B selling at $91.39. The terminal value of the investment will be about **$100,307**.

   b) After two years the rate goes up to 16%. We cash the same amount as above for the coupons and zero-coupon bonds, but bonds A are now cheaper, selling at $83.85, so we have less money in total: $85,305.68. However, the zero-coupon bonds are now cheap as well, selling at $85.21, and we can afford to buy 1,001.07 of them, ending up with **$100,107**.

2. After one year the rate drops to 9%. In a similar way as before, we arrive at the current value of the investment by adding the coupons of A, the face value of B and the market value of bonds A held, obtaining $83,658.73. Then we find the weights $w_A \cong 0.4013$, $w_B \cong 0.5987$, determining our new

portfolio of 329.56 bonds A and 548.04 bonds B. (We have to sell 165.50 bonds A and buy 548.04 new bonds B.)

a) After two years the rate goes up to 14%. We cash $3,295.55$ from the coupons of A, which together with the $54,803.77$ obtained from B and the market value of $29,174.39$ of bonds A gives $87,273.72$ in total. We buy 1003.89 new zero-coupon bonds B, ending up with **$100,389** after 3 years.

b) After two years the rate drops to 6%. Our wealth will then be $94,405.29$, we can afford to buy $1,002.43$ bonds B, and the final value of our investment will be **$100,243**.

As we can see, we end up with more than $100,000$ in each scenario. As a result of convexity, the future value at time t of a bond investment with duration equal to t has a minimum if the rate y remains unchanged. This means that in a model with yields independent of maturity, rate jumps can lead to arbitrage.

Exercise 9.15

Design an investment of $20,000$ in a portfolio of duration 2 years consisting of two kinds of coupon bonds maturing after 2 years, with annual coupons, bond A with $20 coupons and $100 face value, and bond B with $5 coupons and $500 face value, given that the initial rate is 8%. How much will this investment be worth after 2 years?

9.3 General Term Structure

We shall discuss a model of bond prices without the condition that the yield should be independent of maturity.

The prices $B(n, N)$ of zero-coupon unit bonds with various maturities determine a family of yields $y(n, N)$ by

$$B(n, N) = e^{-(N-n)hy(n,N)}.$$

Note that the yields have to be positive, since $B(n, N)$ has to be less than 1 for $n < N$. The function $y(n, N)$ of two variables $n < N$ is called the *term structure of interest rates*. The yields $y(0, N)$ dictated by the current prices are called the *spot rates*.

The *initial term structure* $y(0, N)$ formed by the spot rates is a function of one variable N. Typically, it is an increasing function, but other shapes have also been observed in financial markets. In particular, the initial term structure may be *flat*, that is, the yields may be independent of N, which is the case considered in the previous section.

Exercise 9.16

If $B(0,6) = 0.96$ dollars, find $B(0,3)$ and $B(0,9)$ such that the initial term structure is flat.

The price of a coupon bond as the present value of future payments can be written using the spot rates in the following way:

$$P = C_1 e^{-hn_1 y(0,n_1)} + C_2 e^{-hn_2 y(0,n_2)} + \cdots + (C_N + F)e^{-hn_N y(0,n_N)} \quad (9.3)$$

for a bond with coupons $C_1, C_2, \ldots, C_N$ due at times $0 < hn_1 < hn_2 < \cdots < hn_N$ and with face value F, maturing at time hn_N.

Despite the fact that for a coupon bond we cannot use a single rate for discounting future payments, such a rate can be introduced just as an artificial quantity. It is called the *yield to maturity*, and is defined to be the rate y solving the equation

$$P = C_1 e^{-hn_1 y} + C_2 e^{-hn_2 y} + \cdots + (F + C_N)e^{-hn_N y}.$$

Yield to maturity provides a convenient simple description of coupon bonds and is quoted in the financial press. Of course, if the interest rates are independent of maturity, then this formula is the same as (9.1).

Remark 9.10

To determine the initial term structure we need the prices of zero-coupon bonds. However, for longer maturities (typically over one year) only coupon bonds may be traded, making it necessary to decompose coupon bonds into zero-coupon bonds with various maturities. This can be done by applying formula (9.3) repeatedly to find the yield with the longest maturity, given the bond price and all the yields with shorter maturities. This procedure was recognised by the U.S. Treasury, who in 1985 introduced a programme called STRIPS (Separate Trading of Registered Interest and Principal Securities), allowing an investor to keep the required cash payments (for certain bonds) by selling the rest (the 'stripped' bond) back to the Treasury.

Example 9.11

Suppose that a one-year zero-coupon bond with face value $100 is trading at $91.80 and a two-year bond with $10 annual coupons and face value $100 is trading at $103.95. This gives the following equations for the yields

$$91.80 = 100e^{-y(0,1)},$$
$$103.95 = 10e^{-y(0,1)} + 110e^{-2y(0,2)}.$$

From the first equation we obtain $y(0,1) \cong 8.56\%$. On substituting this into the second equation, we find $y(0,2) \cong 7.45\%$. As a result, the price of the 'stripped' two-year bond, a zero-coupon bond maturing in two years with face value $100, will be $100e^{-2y(0,2)} \cong 86.16$ dollars. Given the price of a three-year coupon bond, we could then evaluate $y(0,3)$, and so on.

Going back to our general study of bonds, let us consider a deterministic term structure (thus known in advance with certainty). The next proposition indicates that this is unrealistic.

Proposition 9.12

If the term structure is deterministic, then the No-Arbitrage Principle implies that

$$B(0,N) = B(0,n)B(n,N). \tag{9.4}$$

Proof

If $B(0,N) < B(0,n)B(n,N)$, then:

- Buy a bond maturing at time N and write a fraction $B(n,N)$ of a bond maturing at n. (Here we use the assumption that the future bond prices are known today.) This gives $B(0,n)B(n,N) - B(0,N)$ dollars now.
- At time n settle the written bonds, raising the required sum of $B(n,N)$ by issuing a single unit bond maturing at N.
- At time N close the position, retaining the initial profit.

The reverse inequality $B(0,N) > B(0,n)B(n,N)$ can be dealt with in a similar manner, by adopting the opposite strategy.   □

Employing the representation of bond prices in terms of yields, we have

$$B(n,N) = \frac{B(0,N)}{B(0,n)} = e^{hny(0,n)-hNy(0,N)}.$$

This would mean that all bond prices (and so the whole term structure) are determined by the initial term structure. However, it is clear that one cannot expect this to hold in real bond markets. In particular, this relation is not supported by historical data.

This shows that assuming deterministic bond prices would go too far in reducing the complexity of the model. We have to allow the future term structure
to be random. In what follows, future bond prices will be random, as will be
the quantities determined by them.

9.3.1 Forward Rates

We begin with an example showing how to secure in advance the interest rate
for a deposit to be made or a loan to be taken at some future time.

Example 9.13

Suppose that the business plan of your company will require taking a loan of
$100,000 one year from now in order to purchase new equipment. You expect to
have the means to repay the loan after another year. You would like to arrange
the loan today at a fixed interest rate, rather than to gamble on future rates.
Suppose that the spot rates are $y(0,1) = 8\%$ and $y(0,2) = 9\%$ (with $h = 1$).
You buy $1,000$ one-year bonds with $100 face value, paying $100,000e^{-8\%} \cong$
$92,311.63$ dollars. This sum is borrowed for 2 years at 9%. After one year
you will receive the $100,000 from the bonds, and after two years you can
settle the loan with interest, the total amount to pay being $92,311.63e^{2\times9\%} \cong$
$110,517.09$ dollars. Thus, the interest rate on the constructed future loan will
be $\ln(110,517.09/100,000) \cong 10\%$. Financial intermediaries may simplify your
task by offering a so-called Forward Rate Agreement and perform the above
construction of the loan on your behalf.

Exercise 9.17

Explain how a deposit of $50,000 for six months can be arranged to start
in six months, and find the rate if $y(0,6) = 6\%$ and $y(0,12) = 7\%$, where
$h = \frac{1}{12}$.

In general, the *initial forward rate* $f(0, M, N)$ is an interest rate such that

$$B(0, N) = B(0, M)e^{-(N-M)hf(0,M,N)},$$

so

$$f(0, M, N) = -\frac{1}{h(N-M)} \ln \frac{B(0, N)}{B(0, M)} = -\frac{\ln B(0, N) - \ln B(0, M)}{h(N-M)}.$$

Note that this rate is deterministic since it is worked out using the present bond
prices. It can be conveniently expressed in terms of the initial term structure.

Insert into the above expression the bond prices as determined by the yields, $B(0, N) = e^{-hNy(0,N)}$ and $B(0, M) = e^{-hMy(0,M)}$, to get

$$f(0, M, N) = \frac{Ny(0, N) - My(0, M)}{N - M}. \tag{9.5}$$

Exercise 9.18

Suppose that the following spot rates are provided by central London banks (LIBOR, the London Interbank Offered Rate, is the rate at which money can be deposited; LIBID, the London Interbank Bid Rate, is the rate at which money can be borrowed):

Rate	LIBOR	LIBID
1 month	3.46%	3.62%
2 months	4.06%	4.24%
3 months	4.19%	4.33%
6 months	4.35%	4.54%

These rates use the rule of simple interest. As a bank manager acting for a customer who wishes to arrange a loan of $100,000 in a month's time for a period of 5 months, what rate could you offer and how would you construct the loan? Suppose that another institution offers the possibility of making a deposit for 4 months, starting 2 months from now, at a rate of 6.91%, simple interest. Does this present an arbitrage opportunity?

As time passes, the bond prices will change and, consequently, so will the forward rates. The *forward rate over the interval* $[M, N]$ determined at time $n < M < N$ is defined by

$$B(n, N) = B(n, M)e^{-(N-M)hf(n,M,N)},$$

that is,

$$f(n, M, N) = -\frac{\ln B(n, N) - \ln B(n, M)}{(N - M)h}.$$

The *instantaneous forward rates* $f(n, N) = f(n, N, N + 1)$ are the forward rates over a one-step interval. Typically, when h is one day, the instantaneous forward rates correspond to overnight deposits or loans. The formula

$$f(n, N) = -\frac{\ln B(n, N + 1) - \ln B(n, N)}{h} \tag{9.6}$$

will enable us to reconstruct the bond prices, given the forward rates at a particular time n.

Example 9.14

Let $h = \frac{1}{12}$, $n = 0$, $N = 0, 1, 2, 3$, and suppose that the bond prices are

$$B(0, 1) = 0.9901,$$
$$B(0, 2) = 0.9828,$$
$$B(0, 3) = 0.9726.$$

Then we have the implied yields

$$y(0, 1) \cong 11.94\%,$$
$$y(0, 2) \cong 10.41\%,$$
$$y(0, 3) \cong 11.11\%,$$

and forward rates

$$f(0, 0) \cong 11.94\%,$$
$$f(0, 1) \cong \ \ 8.88\%,$$
$$f(0, 2) \cong 12.52\%.$$

Observe that, using the formula for the forward rates, we get

$$\exp(-(0.1194 + 0.0888 + 0.1252)/12) \cong 0.9726 = B(0, 3),$$

which illustrates the next proposition.

Proposition 9.15

The bond price is given by

$$B(n, N) = \exp\{-h(f(n, n) + f(n, n + 1) + \cdots + f(n, N - 1))\}.$$

Proof

For this purpose note that

$$f(n, n) = -\frac{\ln B(n, n + 1)}{h},$$

since $B(n, n) = 1$, so

$$B(n, n + 1) = \exp\{-hf(n, n)\}.$$

Next,

$$f(n, n + 1) = -\frac{\ln B(n, n + 2) - \ln B(n, n + 1)}{h}$$

and, after inserting the expression for $B(n, n + 1)$,

$$B(n, n + 2) = \exp\{-h(f(n, n) + f(n, n + 1))\}.$$

Repeating this a number of times, we arrive at the required general formula. $\square$

We have a simple representation of the forward rates in terms of the yields:

$$f(n, N) = (N + 1 - n)y(n, N + 1) - (N - n)y(n, N). \qquad (9.7)$$

In particular,

$$f(n, n) = y(n, n + 1)$$

and, as a direct consequence of the above proposition, we have the intuitive formula

$$y(n, N) = \frac{f(n, n) + f(n, n + 1) + \cdots + f(n, N - 1)}{N - n}.$$

Example 9.16

We can see from the above formulae that if the term structure is flat, that is, $y(n, N)$ is independent of N, then $f(n, N) = y(n, N)$. Now consider an example of $f(n, N)$ increasing with N for a fixed n, and compute the corresponding yields:

$$f(0, 0) = 8.01\%, \quad y(0, 1) = 8.01\%,$$
$$f(0, 1) = 8.03\%, \quad y(0, 2) = 8.02\%,$$
$$f(0, 2) = 8.08\%, \quad y(0, 3) = 8.04\%.$$

We can see that the yields also increase. (See Exercise 9.20 below for a generalisation of this.)

However, the forward rates do not have to increase with maturity even if the yields do:

$$f(0, 0) = 9.20\%, \quad y(0, 1) = 9.20\%,$$
$$f(0, 1) = 9.80\%, \quad y(0, 2) = 9.50\%,$$
$$f(0, 2) = 9.56\%, \quad y(0, 3) \cong 9.52\%.$$

Exercise 9.19

Can a forward rate be negative?

Exercise 9.20

Prove that if $f(n, N)$ increases with N, then the same is true for $y(n, N)$.

9.3.2 Money Market Account

The *short rate* is defined by $r(n) = f(n, n)$. An alternative expression is $r(n) = y(n, n + 1)$, so this is a rate valid for one step starting at time n. The short rates are unknown in advance, except for the current one, $r(0)$. It is important to distinguish between $r(n)$ and $f(0, n)$. Both rates apply to a single step from time n to $n + 1$, but the former is random, whereas the latter is known at the present moment and determined by the initial term structure.

The money market account, denoted by $A(n)$, $n \geq 1$, is defined by

$$A(n) = \exp\{h(r(0) + r(1) + \cdots + r(n - 1))\}$$

with $A(0) = 1$, and represents the value at time n of one dollar invested in an account attracting interest at the short rate under continuous compounding. For example, if $h = \frac{1}{365}$, then the interest is given by the overnight rate.

The money market account defined in Chapter 2 was a deterministic sequence independent of the particular way the initial dollar is invested. Here $A(n)$ is random and, as will be seen below, in general different from $\exp\{hny(0, n)\}$, the latter being deterministic and constructed by using zero-coupon bonds maturing at time n.

Example 9.17

In the setting introduced in Example 9.14, suppose that the bond prices change as follows:

$$\begin{aligned}
B(0, 1) &= 0.9901, \\
B(0, 2) &= 0.9828, \quad B(1, 2) = 0.9947, \\
B(0, 3) &= 0.9726, \quad B(1, 3) = 0.9848, \quad B(2, 3) = 0.9905.
\end{aligned}$$

The corresponding yields are

$$\begin{aligned}
y(0, 1) &\cong 11.94\%, \\
y(0, 2) &\cong 10.41\%, \quad y(1, 2) \cong 6.38\%, \\
y(0, 3) &\cong 11.11\%, \quad y(1, 3) \cong 9.19\%, \quad y(2, 3) \cong 11.45\%.
\end{aligned}$$

The forward rates are

$$\begin{aligned}
f(0, 0) &\cong 11.94\%, \\
f(0, 1) &\cong \ \ 8.88\%, \quad f(1, 1) \cong \ \ 6.38\%, \\
f(0, 2) &\cong 12.52\%, \quad f(1, 2) \cong 12.00\%, \quad f(2, 2) \cong 11.45\%.
\end{aligned}$$

We can read off the short rates and compute the values of the money market account

$$\begin{aligned}
&& A(0) &= 1, \\
r(0) = f(0, 0) &\cong 11.94\%, & A(1) &\cong 1.0100, \\
r(1) = f(1, 1) &\cong \ \ 6.38\%, & A(2) &\cong 1.0154, \\
r(2) = f(2, 2) &\cong 11.45\%, & A(3) &\cong 1.0251.
\end{aligned}$$

Exercise 9.21

Which bond prices in Example 9.17 can be altered so that the values of the money market remain unchanged?

Exercise 9.22

Using the data in Example 9.17, compare the logarithmic return on an investment in the following securities over the period from 0 to 3: a) zero-coupon bonds maturing at time 3; b) single-period zero-coupon bonds; c) the money market account.

9.4 Binomial Model of Stochastic Interest Rates

This section is devoted to modelling the time evolution of random interest rates adopting an approach similar to the binomial model of stock prices in Chapter 6. Modelling the evolution of interest rates can be reduced to modelling the evolution of the bond prices since the latter determine the former. We begin with some properties that a model of bond prices should satisfy, emphasising the differences between bonds and stock.

First, let us recall that the evolution of interest rates or bond prices is described by functions of two variables, the running time and the maturity time, whereas stock prices are functions of just one variable, the running time.

Second, there are many ways of describing the term structure: bond prices, implied yields, forward rates, short rates. Bond prices and yields are clearly equivalent, being linked by a simple formula. Bond prices and forward rates are also equivalent. The short rates are different, easier to handle, but the problem of reconstructing the term structure emerges. This may be non-trivial, since short rates usually carry less information.

Third, the model needs to match the initial data. For a stock this is just the current price. In the case of bonds the whole initial term structure is given, imposing more restrictions on the model, which has to be consistent with currently available market information.

Fourth, bonds become non-random at maturity. This is in sharp contrast with stock prices. The fact that a bond gives a sure dollar at maturity has to be included in the model.

Finally, the dependence of yields on maturity must be quite special. Bonds with similar maturities will typically behave in a similar manner. In statistical terms this means that they are strongly positively correlated.

We shall develop a binomial tree model of bond prices. The shape of the tree will be similar to that in Chapter 6. However, to facilitate the necessary level of sophistication of the model, it has to be more complex. Namely, the probabilities and returns will depend on the position in the tree. We need suitable notation to distinguish between different positions. This added level of complexity is necessary. In particular, note that the same return at each step would lead to unlimited growth of the asset value along some scenarios, contradicting the fact that the bond price cannot exceed 1.

By a *state* we mean a finite sequence of consecutive up or down movements. The state depends, first of all, on time or, in other words, on the number of steps. We shall use sequences of letters u and d of various lengths, the length corresponding to the time elapsed (the number of steps from the root of the tree). At time 1 we have just two states $s_1 = $ u or d, at time 2 four states $s_2 = $ ud, dd, du, or uu. We shall write $s_2 = s_1$u or s_1d, meaning that we go up or, respectively, down at time 2, having been at s_1 at time 1. In general, $s_{n+1} = s_n$u or s_nd.

The probabilities will be allowed to depend on particular states. We write p^{s_n} to denote the probability of going up at time $n + 1$, having started at state s_n at time n. At the first step the probability of going up will be denoted by p without an argument. In Figure 9.1 we have $p = 0.3$, $p^u = 0.1$, $p^d = 0.4$, $p^{uu} = 0.4$, $p^{ud} = 0.2$, $p^{du} = 0.5$, $p^{dd} = 0.4$.

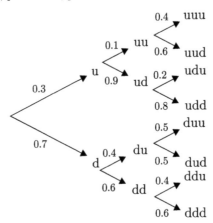

Figure 9.1 States and probabilities

Let us fix a natural number N as the time horizon. It will be the upper bound of the maturities of all the bonds considered. The states s_N at time N represent the complete scenarios of bond price movements.

Next, we shall describe the evolution of bond prices. At time 0 we are given the initial bond prices for all maturities up to N, that is, a sequence of N

numbers

$$B(0,1), B(0,2), B(0,3), \ldots, B(0, N-1), B(0, N).$$

At time 1 one of the prices becomes redundant, namely the first bond matures and only the remaining $N-1$ bonds are still being traded. We introduce randomness by allowing two possibilities distinguished by the states u and d, so we have two sequences

$$B^u(1,2), B^u(1,3), \ldots, B^u(1, N-1), B^u(1, N),$$
$$B^d(1,2), B^d(1,3), \ldots, B^d(1, N-1), B^d(1, N).$$

At time 2 we have four states and four sequences of length $N-2$:

$$B^{uu}(2,3), \ldots, B^{uu}(2, N-1), B^{uu}(2, N),$$
$$B^{ud}(2,3), \ldots, B^{ud}(2, N-1), B^{ud}(2, N),$$
$$B^{du}(2,3), \ldots, B^{du}(2, N-1), B^{du}(2, N),$$
$$B^{dd}(2,3), \ldots, B^{dd}(2, N-1), B^{dd}(2, N).$$

We do not require that the ud and du prices coincide, which was the case for stock price movements.

This process continues in the same manner. At each step the length of the sequence decreases by one and the number of sequences doubles. At time $N-1$ we have just single numbers, 2^{N-1} of them,

$$B^{s_{N-1}}(N-1, N)$$

indexed by all possible states s_{N-1}. The tree structure breaks down here because the last movement is certain: the last bond matures, becoming a sure dollar at time N, $B^{s_N}(N, N) = 1$ for all states.

Example 9.18

A particular evolution of bond prices for $N = 3$, with monthly steps ($h = \frac{1}{12}$) is given in Figure 9.2. The prices of three bonds with maturities 1, 2, and 3 are shown.

The evolution of bond prices can be described be means of returns. Suppose we have reached state s_{n-1} and the bond price $B^{s_{n-1}}(n-1, N)$ becomes known. Then we can write

$$B^{s_{n-1}u}(n, N) = B^{s_{n-1}}(n-1, N)\exp\{k^{s_{n-1}u}(n, N)\},$$
$$B^{s_{n-1}d}(n, N) = B^{s_{n-1}}(n-1, N)\exp\{k^{s_{n-1}d}(n, N)\},$$

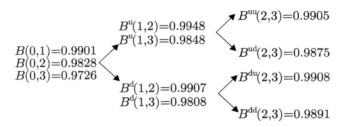

Figure 9.2 Evolution of bond prices in Example 9.18

implicitly defining the logarithmic returns

$$k^{s_{n-1}u}(n, N) = \ln \frac{B^{s_{n-1}u}(n, N)}{B^{s_{n-1}}(n-1, N)},$$

$$k^{s_{n-1}d}(n, N) = \ln \frac{B^{s_{n-1}d}(n, N)}{B^{s_{n-1}}(n-1, N)}.$$

We assume here that $k^{s_{n-1}u}(n, N) \geq k^{s_{n-1}d}(n, N)$.

Remark 9.19

Note that there are some places in the tree where the returns are non-random given the state s_{n-1} is known. Namely,

$$k^{s_{n-1}u}(n, n) = k^{s_{n-1}d}(n, n) = \ln \frac{1}{B^{s_{n-1}}(n-1, n)},$$

since $B^{s_n}(n, n) = 1$ for all s_n.

Example 9.20

From the data in Example 9.18 we extract the prices of bonds with maturity 3, completing the picture with the final value 1. The tree shown in Figure 9.3 describes the random evolution of a single bond purchased at time 0 for 0.9726. The returns are easy to compute, for instance

$$k^{ud}(2, 3) = \ln \frac{B^{ud}(2, 3)}{B^u(1, 3)} \cong 0.27\%.$$

The results are gathered in Figure 9.4. (Recall that the length of each step is one month.)

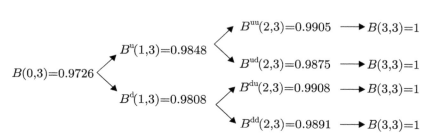

Figure 9.3 Prices of the bond maturing at time 3 in Example 9.20

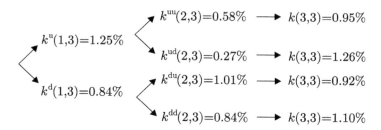

Figure 9.4 Returns on the bond maturing at time 3 in Example 9.20

Exercise 9.23

For the tree of returns shown in Figure 9.5 construct the tree of bond prices and fill in the missing returns.

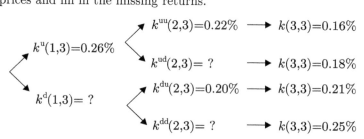

Figure 9.5 Returns in Exercise 9.23

The evolution of bond prices is in perfect correspondence with the evolution of implied yields to maturity. Namely,

$$y^{s_n}(n,m) = \frac{1}{h(m-n)} \ln \frac{1}{B^{s_n}(n,m)}$$

with the same tree structure as for bond prices. In particular, the final yields are non-random given that the state s_{n-1} at the penultimate step is known. Note that the words 'up' and 'down' lose their meaning here because the yield goes down as the bond price goes up. Nevertheless, we keep the original indicators u and d.

Example 9.21

We continue Example 9.18 and find the yields, bearing in mind that $h = \frac{1}{12}$. The results are collected in Figure 9.6.

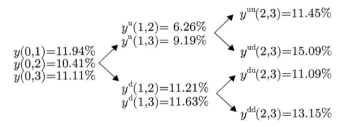

$y^{uu}(2,3)=11.45\%$

$y^u(1,2)= 6.26\%$
$y^u(1,3)= 9.19\%$

$y(0,1)=11.94\%$
$y(0,2)=10.41\%$
$y(0,3)=11.11\%$

$y^{ud}(2,3)=15.09\%$

$y^{du}(2,3)=11.09\%$

$y^d(1,2)=11.21\%$
$y^d(1,3)=11.63\%$

$y^{dd}(2,3)=13.15\%$

Figure 9.6   Yields in Example 9.21

Exercise 9.24

Take the returns in Exercise 9.23 and find the yield $y(0, 3)$. What is the general relationship between the returns and yields to maturity? Can you complete the missing returns without computing the bond prices?

Now consider the instantaneous forward rates. At the initial time 0 there are N forward rates

$$f(0,0), f(0,1), f(0,2), \ldots, f(0, N-1)$$

generated by the initial bond prices. Note that the first number is the short rate $r(0) = f(0,0)$.

For all subsequent steps the current bond prices imply the forward rates. Formula (9.6) applied to random bond prices allows us to find the random evolution of forward rates:

$$f^{s_n}(n, N) = -\frac{\ln B^{s_n}(n, N+1) - \ln B^{s_n}(n, N)}{h}. \tag{9.8}$$

At time 1 we have two possible sequences of $N-1$ forward rates obtained from two sequences of bond prices

$$f^u(1,1), f^u(1,2), \ldots, f^u(1, N-1),$$
$$f^d(1,1), f^d(1,2), \ldots, f^d(1, N-1).$$

At time 2 we have four sequences of $N-2$ forward rates, and so on. At time $N-1$ we have 2^{N-1} single numbers $f^{s_{N-1}}(N-1, N-1)$.

Example 9.22

Using (9.8), we can evaluate the forward rates for the data in Example 9.18. For instance,

$$f^u(1,2) = -\frac{\ln B^u(1,3) - \ln B^u(1,2)}{h}.$$

Alternatively, we can use the yields found in Example 9.21 along with formula (9.7):

$$f^u(1,2) = 2y^u(1,3) - y^u(1,2).$$

The results are gathered in Figure 9.7.

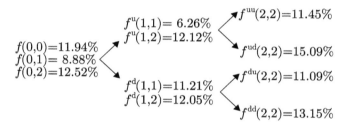

Figure 9.7 Forward rates in Example 9.22

The information contained in forward rates is sufficient to reconstruct the bond prices, as was shown in Proposition 9.15.

Exercise 9.25

Suppose a tree of forward rates is given as in Figure 9.8. Find the corresponding bond prices (using one-month steps).

The short rates are just special cases of forward rates,

$$r^{s_n}(n) = f^{s_n}(n, n)$$

for $n \geq 1$, with deterministic $r(0) = f(0,0)$. The short rates are also given by $r^{s_n}(n) = y^{s_n}(n, n+1)$, $n \geq 1$, and $r(0) = y(0,1)$, that is, by the rates of return on a bond maturing at the next step. This is obvious from the relationship between the forward rates and yields.

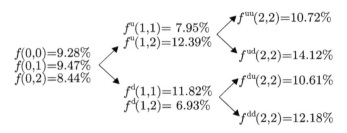

Figure 9.8 Forward rates in Exercise 9.25

We are now ready to describe the money market account. It starts with $A(0) = 1$. The next value

$$A(1) = \exp(hr(0))$$

is still deterministic. It becomes random at subsequent steps. At time 2 there are two values depending on the state at time 1:

$$A^{\mathrm{u}}(2) = \exp(h(r(0) + r^{\mathrm{u}}(1))) = A(1)\exp(hr^{\mathrm{u}}(1)),$$
$$A^{\mathrm{d}}(2) = \exp(h(r(0) + r^{\mathrm{d}}(1))) = A(1)\exp(hr^{\mathrm{d}}(1)).$$

Next, for example,

$$A^{\mathrm{ud}}(3) = \exp(h(r(0) + r^{\mathrm{u}}(1) + r^{\mathrm{ud}}(2))) = A^{\mathrm{u}}(2)\exp(hr^{\mathrm{ud}}(2)).$$

In general,

$$A^{s_{n-1}\mathrm{u}}(n+1) = A^{s_{n-1}}(n)\exp(hr^{s_{n-1}\mathrm{u}}(n)),$$
$$A^{s_{n-1}\mathrm{d}}(n+1) = A^{s_{n-1}}(n)\exp(hr^{s_{n-1}\mathrm{d}}(n)).$$

Exercise 9.26

Find the evolution of the money market account if the forward rates are the same as in Exercise 9.25.

For bond investments the money market account plays a similar role to the risk-free component of investment strategies on the stock market in earlier chapters.

9.5 Arbitrage Pricing of Bonds

Suppose that we are given a binomial tree of bond prices $B^{s_n}(n, N)$ for a bond maturing at time horizon N. In addition, we are given the money market process $A^{s_{n-1}}(n)$. As was mentioned in the introduction to this chapter, the prices of other bonds cannot be completely arbitrary. We shall show that the prices $B^{s_n}(n, M)$ for $M < N$ can be replicated by means of bonds with maturity N and the money market account. As a consequence of the No-Arbitrage Principle, the price $B^{s_n}(n, M)$ will have to be equal to the time n value of the corresponding replicating strategy.

Example 9.23

Consider the data in Example 9.18. At the first step the short rate is deterministic, being implied by the price $B(0, 1)$. The first two values of the money market account are $A(0) = 1$ and $A(1) = 1.01$. As the underlying instrument we take the bond maturing at time 3. The prices of this bond at time 0 and 1 are given in Figure 9.9, along with the prices of the bond maturing at time 2. We can find a portfolio (x, y), with x being the number of bonds of maturity 3

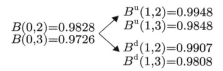

$$B(0,2)=0.9828$$
$$B(0,3)=0.9726$$

$$B^u(1,2)=0.9948$$
$$B^u(1,3)=0.9848$$
$$B^d(1,2)=0.9907$$
$$B^d(1,3)=0.9808$$

Figure 9.9 Bond prices in Example 9.18

and y the position in the money market, such that the value of this portfolio matches the time 1 prices of the bond maturing at time 2. To this end we solve the system of equations

$$0.9848x + 1.01y = 0.9948,$$
$$0.9808x + 1.01y = 0.9907,$$

obtaining $x = 1$ and $y \cong 0.0098$. The value of this portfolio at time 0 is $1 \times B(0, 3) + 0.0098 \times A(0) \cong 0.9824$, which is *not* equal to $B(0, 2)$. The prices in Figure 9.9 provide an arbitrage opportunity:

- sell a bond maturing at time 2 for $0.9828 and buy the portfolio constructed above for $0.9824;
- whatever happens at time 1, the value of the portfolio will be sufficient to buy the bond back, the initial balance $0.0004 being the arbitrage profit.

The model in Example 9.18 turns out to be *inconsistent with the No-Arbitrage Principle* and needs to be rectified. We can only adjust some of the future prices since the present prices of all bonds are dictated by the market. By taking $B^u(1, 2) = 0.9958$ with $B^d(1, 2)$ unchanged, or letting $B^d(1, 2) = 0.9913$ with $B^u(1, 2)$ unchanged, we can eliminate the arbitrage opportunity. Of course, there are many other ways of repairing the model by a simultaneous change of both values $B^u(1,2)$ and $B^d(1,2)$. Let us put $B^d(1,2) = 0.9913$ and leave $B^u(1,2)$ unchanged. The rectified tree of bond prices is shown in Figure 9.10 and the corresponding yields in Figure 9.11.

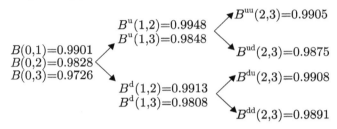

Figure 9.10   Rectified tree of bond prices in Example 9.23

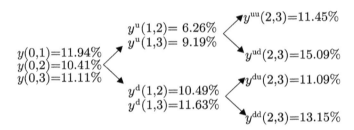

Figure 9.11   Rectified tree of yields in Example 9.23

Remark 9.24

The process of rectifying bond prices in Example 9.23 bears resemblance to the pricing of derivative securities described in Chapter 6. The role of the derivative security is played by the bond maturing at time 2. The bond maturing at time 3 serves as the underlying security. The difference is that the present price of the bond with maturity 2 is fixed and we can only adjust the future prices in the model to eliminate arbitrage. At this stage we are concerned only with building consistent models rather than with pricing securities.

Remark 9.25

If the initial model has been built in line with certain assumptions, for instance, parameters like the standard deviation being consistent with historical data, then altering the model to remove arbitrage would most likely destroy this feature. A way around this would be to allow more states by adopting a trinomial tree or one with even more branches. We would gain some flexibility, which gives a broader scope for arbitrage-free models. Models of this kind are called multi-factor models.

Exercise 9.27

Evaluate the prices of a bond maturing at time 2 given a tree of prices of a bond maturing at time 3 and short rates as shown in Figure 9.12, with $h = 1/12$.

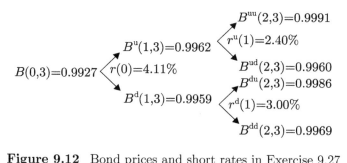

Figure 9.12 Bond prices and short rates in Exercise 9.27

We can readily generalise Example 9.23. The underlying bond matures at time N and we can find the prices of any bond maturing at $M < N$. Replication proceeds backwards step-by-step starting from time M, for which $B^{s_M}(M, M) = 1$ in each state s_M. The first step is easy: for each state s_{M-1} we take a portfolio with $x = 0$ and $y = 1/A^{s_{M-1}}(M)$ since the bond becomes risk free one step prior to maturity.

Next, consider time $M - 2$. For any state s_{M-2} we find $x = x^{s_{M-2}}(M-1)$, the number of bonds maturing at N, and $y = y^{s_{M-2}}(M-1)$, the money market account position, by solving the system

$$xB^{s_{M-2}u}(M-1, N) + yA^{s_{M-2}}(M-1) = B^{s_{M-2}u}(M-1, M),$$
$$xB^{s_{M-2}d}(M-1, N) + yA^{s_{M-2}}(M-1) = B^{s_{M-2}d}(M-1, M).$$

In this way we can find the prices at time $M - 2$ of the bond maturing at time M,

$$B^{s_{M-3}u}(M-2, M) = xB^{s_{M-3}u}(M-2, N) + yA^{s_{M-3}}(M-2),$$
$$B^{s_{M-3}d}(M-2, M) = xB^{s_{M-3}d}(M-2, N) + yA^{s_{M-3}}(M-2).$$

We can iterate the replication process moving backwards through the tree.

Remark 9.26

Replication is possible if a no-arbitrage condition analogous to the condition $D < R < U$ of Chapter 6 is satisfied for the binomial tree of bond prices. This condition has the form

$$k^{s_{n-1}\mathrm{u}}(n, N) > hr^{s_{n-1}}(n-1) > k^{s_{n-1}\mathrm{d}}(n, N). \qquad (9.9)$$

Any future cash flow can be replicated in a similar fashion. Consider, for example, a coupon bond with fixed coupons.

Example 9.27

Take a coupon bond maturing at time 2 with face value $F = 100$, paying coupons $C = 10$ at times 1 and 2. We price the future cash flow by using the zero-coupon bond maturing at time 3 as the underlying security. The coupon bond price P at a particular time will not include the coupon due (the so-called ex-coupon price). Assume that the structure of bond prices is as in Figure 9.10.

Consider time 1. In state u the short rate is determined by the price $B^{\mathrm{u}}(1,2) = 0.9948$, so we have $r^{\mathrm{u}}(1) \cong 6.26\%$. Hence $P^{\mathrm{u}}(1) \cong 109.43$. In state d we use $B^{\mathrm{d}}(1,2) = 0.9913$ to find $r^{\mathrm{d}}(1) \cong 10.49\%$ and $P^{\mathrm{d}}(1) \cong 109.04$.

Consider time 0. The cash flow at time 1, which we are to replicate includes the coupon due, so it is given by $P^{\mathrm{u}}(1) + 10 \cong 119.43$ and $P^{\mathrm{d}}(1) + 10 \cong 119.04$. The short rate $r(0) \cong 11.94\%$ determines the money market account as in Example 9.23, $A(1) = 1.01$, and we find $x \cong 96.25$, $y \cong 24.40$. Hence $P(0) \cong 118.01$ is the present price of the coupon bond.

An alternative is to use the spot yields: $y(0,1) \cong 11.94\%$ and $y(0,2) \cong 10.41\%$ to discount the future payments with the same result: $118.01 \cong 10 \times \exp(-\frac{1}{12} \times 11.94\%) + 110 \times \exp(-\frac{2}{12} \times 10.41\%)$.

In general,

$$P(0) = C_1 \exp\{-hy(0,1)\} + C_2 \exp\{-2hy(0,2)\}$$
$$+ \cdots + (C_N + F) \exp\{-Nhy(0, N)\}. \qquad (9.10)$$

(For simplicity we include all time steps, so $C_k = 0$ at each time step k when no coupon is paid.) At each time k when a coupon is paid, the cash flow is the sum of the (deterministic) coupon and the (stochastic) price of the remaining bond:

$$C_k + P^{s_k}(k) = C_k + C_{k+1} \exp\{-hy^{s_k}(k, k+1)\}$$
$$+ \cdots + (C_n + F) \exp\{-h(n-k)y^{s_k}(k, n)\}.$$

Quite often the coupons depend on other quantities. In this way a coupon bond may become a derivative security. An important benchmark case is described below, where the coupons are computed as fractions of the face value. These fractions, defining the *coupon rate*, are obtained by converting the short rate to an equivalent discrete compounding rate. In practice, when h is one day, the coupon rate will be the overnight LIBOR rate.

Proposition 9.28

A coupon bond maturing at time N with random coupons

$$C_k^{s_{k-1}} = (\exp\{hr^{s_{k-1}}(k-1)\} - 1)F \tag{9.11}$$

for $0 < k \leq N$ is trading *at par*. (That is, the price $P(0)$ is equal to the face value F.)

Proof

Fix time $N-1$ and a state s_{N-1}. In this state the value $P^{s_{N-1}}(N-1)$ of the bond is $F + C_N^{s_{N-1}}$ discounted at the short rate, which gives $P^{s_{N-1}}(N-1) = F$ if the coupon is expressed by (9.11). Proceeding backwards through the tree and applying the same argument for each state, we finally arrive at $P(0) = F$. □

Exercise 9.28

Find the coupons of a bond trading at par and maturing at time 2, given the yields as in Example 9.23, see Figure 9.11.

9.5.1 Risk-Neutral Probabilities

In Chapter 6 we learnt that the stock price $S(n)$ at time n is equal to the expectation under the risk-neutral probability of the stock price $S(n+1)$ at time $n+1$ discounted to time n. The situation is similar in the binomial model of interest rates.

The discount factors are determined by the money market account, or, in other words, by the short rates. In general, they are random, being of the form $\exp\{-hr^{s_n}(n)\}$.

Suppose that state s_n has occurred at time n. The short rate determining the time value of money for the next step is now known with certainty. Consider a bond maturing at time N with $n < N - 1$. We are given the bond price $B^{s_n}(n, N)$ and two possible values at the next step, $B^{s_n u}(n+1, N)$ and $B^{s_n d}(n+$

$1, N)$. These values represent a random variable, which will be denoted by $B^{s_n}(n+1, N)$. If $n = N - 1$, then the bond matures at the next step N, when it has just one price independent of the state, namely the face value. We are looking for a probability p_* such that

$$B^{s_n}(n, N) = [p_* B^{s_n u}(n+1, N) + (1 - p_*) B^{s_n d}(n+1, N)]$$
$$\times \exp\{-hr^{s_n}(n)\}. \tag{9.12}$$

This equation can be solved for p_*, which in principle depends on n, N and s_n. (As it will soon turn out, see Proposition 9.31, p_* is in fact independent of maturity N.) Recalling the definition of logarithmic returns, we have

$$B^{s_{n+1}}(n+1, N) = B^{s_n}(n, N)\exp\{k^{s_{n+1}}(n+1, N)\},$$

which gives

$$p_*^{s_n}(n, N) = \frac{\exp\{hr^{s_n}(n)\} - \exp\{k^{s_n d}(n+1, N)\}}{\exp\{k^{s_n u}(n+1, N)\} - \exp\{k^{s_n d}(n+1, N)\}}. \tag{9.13}$$

These numbers are called the *risk-neutral* or *martingale probabilities*. Condition (9.9) for the lack of arbitrage can now be written as

$$0 < p_*^{s_n}(n, N) < 1.$$

Example 9.29

We shall find the tree of risk-neutral probabilities $p_*^{s_n}(n, 3)$ for $n = 0, 1$, using the rectified data in Example 9.23 (the bond prices as shown in Figure 9.10).

First we compute the returns on the money market account. The simplest way is to use the yields (Figure 9.11). With $h = \frac{1}{12}$ we have $hr^{s_n}(n) = y^{s_n}(n, n+1)/12$, $n = 0, 1$, which gives the values as shown in Figure 9.13. Next, we

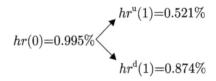

Figure 9.13 Yields in Example 9.29

find the returns $k^{s_1}(1, 3)$ and $k^{s_2}(2, 3)$ on bonds. For example, if $s_2 = ud$, then $k^{ud}(2, 3) = \ln(\frac{0.9875}{0.9848})$. The results are collected in Figure 9.14. We can see that the no-arbitrage conditions are satisfied: $0.84\% < 0.99\% < 1.25\%$, $0.27\% < 0.52\% < 0.58\%$ and $0.84\% < 0.87\% < 1.01\%$. Finally, we can find the risk-neutral probabilities by a direct application of (9.13), see Figure 9.15.

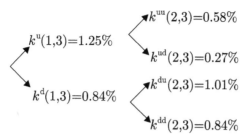

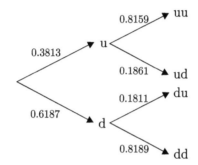

Figure 9.14 Returns in Example 9.29

Figure 9.15 Risk-neutral probabilities in Example 9.29

A crucial observation about the model is this: pricing via replication is equivalent to pricing by means of the risk-neutral probability. This follows from the No-Arbitrage Principle and applies to any cash flow, even a random one, where the amounts depend on the states. This opens a way to pricing any security by means of the expectation with respect to the probabilities $p_*^{s_n}(n, N)$. The expectation is computed step-by-step, starting at the last one and proceeding backwards through the tree.

Example 9.30

Consider a coupon bond maturing at $N = 2$ with face value $F = 100$ and with coupons equal to 5% of the current value of the bond, paid at times 1 and 2. In particular, the coupon at maturity is $C_2 = 5$ in each state. Using the risk-neutral probabilities from Example 9.29, we can find the bond values at time 1.

In the up state we have the discounted value of 105 due at maturity using the short rate $r^u(1) \cong 6.26\%$, which gives 104.4540. In the same way in the down state we obtain 104.0865. Now we add 5% coupons, so the amounts due at time 1 become 109.6767 in the up state and 109.2908 in the down state.

Using the risk-neutral probabilities, we find the present value of the bond:
$108.3545 \cong (0.3813 \times 109.6767 + 0.6187 \times 109.2908)/1.01$.

Exercise 9.29

Use the risk-neutral probabilities in Example 9.29 to find the present value of the following random cash flow: at time 2 we receive \$20 in the state uu, \$10 in the states ud and du, and nothing in the state dd. No payments are due at other times.

Exercise 9.30

Find an arbitrage opportunity for the bond prices in Figure 9.16.

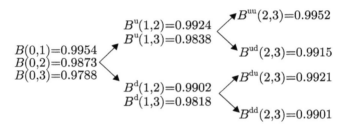

Figure 9.16 Bond prices in Exercise 9.30

Exercise 9.31

Suppose that the risk-neutral probabilities are equal to $\frac{1}{2}$ in every state. Given the short rates shown in Figure 9.17, find the prices of a bond maturing at time 3 (with a one-month time step, $h = \frac{1}{12}$).

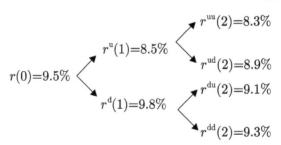

Figure 9.17 Short rates in Exercise 9.31

The next proposition gives an important result, which simplifies the model significantly.

Proposition 9.31

The lack of arbitrage implies that the risk-neutral probabilities are independent of maturity.

Proof

Consider two bonds with maturities $M \leq N$ and fix $n < M$ and s_n. We have to show that $p_*^{s_n}(n, N) = p_*^{s_n}(n, M)$. From (9.13)

$$p_*^{s_n}(n, N) = \frac{e^{hr^{s_n}(n)} - \frac{B^{s_n\mathrm{d}}(n+1,N)}{B^{s_n}(n,N)}}{\frac{B^{s_n\mathrm{u}}(n+1,N)}{B^{s_n}(n,N)} - \frac{B^{s_n\mathrm{d}}(n+1,N)}{B^{s_n}(n,N)}}.$$

Now consider the N-bond as a derivative security with the M-bond as the underlying security. The price of a derivative can be replicated in the binomial model, or, which is equivalent, they can be expressed by means of the risk-neutral probability determined by the underlying, so

$$B^{s_n}(n, N) = e^{-hr^{s_n}(n)}[p_*^{s_n}(n, M)B^{s_n\mathrm{u}}(n+1, N) + (1 - p_*^{s_n}(n, M))B^{s_n\mathrm{d}}(n+1, N)].$$

Solving this for $p_*^{s_n}(n, M)$, we get

$$p_*^{s_n}(n, M) = \frac{e^{hr^{s_n}(n)} - \frac{B^{s_n\mathrm{d}}(n+1,N)}{B^{s_n}(n,N)}}{\frac{B^{s_n\mathrm{u}}(n+1,N)}{B^{s_n}(n,N)} - \frac{B^{s_n\mathrm{d}}(n+1,N)}{B^{s_n}(n,N)}} = p_*^{s_n}(n, N)$$

as claimed.   □

Exercise 9.32

Spot an arbitrage opportunity if the bond prices are as in Figure 9.18.

9.6 Interest Rate Derivative Securities

The tools introduced above make it possible to price any derivative security based on interest rates or, equivalently, on bond prices. Within the binomial tree model the cash flow associated with the derivative security can be replicated using the money market account and a bond with sufficiently long maturity. The bond does not even have to be the underlying security, since the prices of

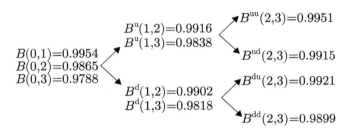

Figure 9.18 Data in Exercise 9.32

the various bonds must be consistent. An alternative is to use the risk-neutral probabilities. The latter approach is often preferable to replication because of its simplicity. The equivalence of both methods should be clear in view of what has been said before.

The pricing of complex securities can essentially be reduced to finding the associated cash flows. Below we present examples of some classical interest rate contingent claims. We begin with the simplest case of options.

9.6.1 Options

The underlying securities for interest rate options are bonds of various kinds.

Example 9.32

With the bond prices as in Example 9.23 (Figure 9.10), consider a call option with exercise time 2 and strike price $X = 0.99$ on a zero-coupon bond maturing at time 3. Starting with the final payoffs shown in the last column in Figure 9.19, we move back step-by-step, computing the risk-neutral expectations of the consecutive values discounted by the appropriate short rates. As a result, the price of the option is 0.00024.

Exercise 9.33

Assume the structure of bond prices as in Example 9.23 (Figure 9.10). Consider a coupon bond maturing at time 2 with face value $F = 100$ and coupons $C = 1$ payable at each step. Find the price of an American call option expiring at time 2 with strike price $X = 101.30$. (Include the coupon in the bond price at each step.)

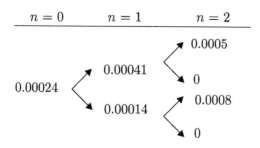

Figure 9.19 Numerical results in Example 9.32

Call options on bonds can be used by institutions issuing bonds to include the possibility of buying the bond back prior to maturity for a prescribed price. A bond that carries such a provision is called a *callable bond*. Its price should be reduced by the price of the attached option.

9.6.2 Swaps

Writing and selling a bond is a method of borrowing money. In the case of a coupon bond trading at par the principal represents the amount borrowed and the coupons represent the interest. This interest may be fixed or floating (variable). The interest is fixed if all coupons are the same. Floating interest can be realised in many ways. Here we assume that it is determined by the short rates as in (9.11). The basis for our discussion is laid by Proposition 9.28, according to which the market value of such a floating-coupon bond must be equal to its face value, the bond trading at par. For a fixed-coupon bond trading at par the size of the coupons can easily be found from (9.10). We could say that the resulting fixed coupon rate is equivalent to the variable short rate over the lifetime of the bond.

Example 9.33

Consider a fixed-coupon bond and a floating-coupon bond, both with annual coupons, trading at par and maturing after two years with face value $F = 100$. Given that the tree of one- and two-year zero-coupon bond prices is as shown in Figure 9.20, where a time step is taken to be one year, $h = 1$, we can evaluate the coupons of the fixed- and floating-coupon bonds. The size of the floating

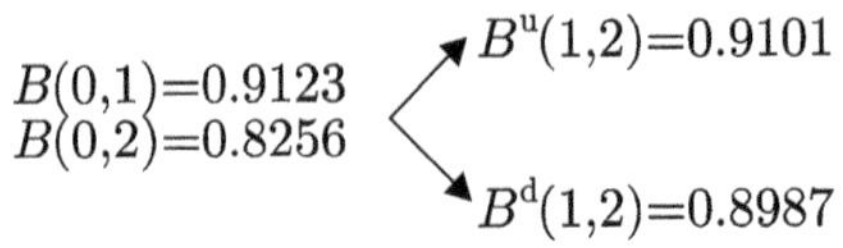

$$B^{\mathrm{u}}(1,2)=0.9101$$

$$B(0,1)=0.9123$$
$$B(0,2)=0.8256$$

$$B^{\mathrm{d}}(1,2)=0.8987$$

Figure 9.20 Bond prices in Example 9.33

coupons C_n can be found from (9.11),

$$C_1 = (B(0,1)^{-1} - 1)F \cong 9.6131,$$
$$C_2^{\mathrm{u}} = (B^{\mathrm{u}}(1,2)^{-1} - 1)F \cong 9.8780,$$
$$C_2^{\mathrm{d}} = (B^{\mathrm{d}}(1,2)^{-1} - 1)F \cong 11.2718.$$

The fixed coupons C can be found by solving equation (9.10), which takes the form

$$F = CB(0,1) + (C+F)B(0,2).$$

This gives

$$C = F\frac{(1 - B(0,2))}{B(0,1) + B(0,2)} \cong 10.0351.$$

The number $\frac{(1-B(0,2))}{B(0,1)+B(0,2)}$ determines the size of the fixed coupon as a percentage of the face value.

Definition 9.34

Consider a fixed-coupon bond with coupons C payable at time intervals τ, maturity $N\tau$ and face value F. The *swap rate* R_{s} is the simple rate which determines the fixed coupon size $C = \tau R_{\mathrm{s}} F$ so that the bond is trading at par.

Proposition 9.35

$$R_{\mathrm{s}} = \frac{1 - B(0,N)}{\sum_{k=1}^{N} \tau B(0,k)}. \tag{9.14}$$

Proof

The proof requires nothing more than a straightforward extension of the argument used in the preceding example. For a bond with fixed coupons $C = R_{\mathrm{s}} F$

to be trading at par, its present value must be equal to the face value,

$$\sum_{k=1}^{N} \tau R_s FB(0,k) + FB(0,N) = F,$$

which proves (9.14). □

By buying a fixed-coupon bond trading at par and selling a floating-coupon bond (or the other way round, selling a fixed-coupon bond trading at par and buying a floating-coupon bond) an investor can create a random cash flow with present value zero, since the two kinds of bond have the same initial price.

A company who has sold fixed-coupon bonds and is paying fixed interest may sometimes wish to switch into paying the floating rate instead. This can be realised by writing a floating-coupon bond and buying a fixed-coupon bond with the same present value. In practice, a financial intermediary will provide this service by offering a contract called a *swap*. Clearly, a swap of this kind will cost nothing to enter. The swap creates a random cash flow

$$\text{Swap}_n = C - C_n,$$

where C_n is the random floating coupon at time n and C is the fixed coupon determined by the swap rate. When added to the floating payments C_n, this cash flow creates the fixed payments C. When subtracted from the fixed payments C, it creates the floating payments C_n.

Example 9.36

When the tree of bond prices is as in Example 9.33, the swap cash flow will be

$$\text{Swap}_1 \cong 10.0351 - 9.6131 = 0.422,$$
$$\text{Swap}_2^u \cong 10.0351 - 9.8780 = 0.1571,$$
$$\text{Swap}_2^d \cong 10.0351 - 11.2718 = -1.2367.$$

The value of the swap may vary with time and state, departing from the initial value of zero. If a company wishes to enter into a swap agreement at a later time, it may purchase a *swaption*, which is a call option on the value of the swap (with prescribed strike price and expiry time).

9.6.3 Caps and Floors

A *cap* is a provision attached to a variable-rate bond which specifies the maximum coupon rate paid in each period over the lifetime of a loan. A *caplet* is a similar provision applying to a particular single period only. In other words, a caplet is a European option on the level of interest paid or received over that period. A cap can be thought of as a series of caplets.

Example 9.37

We take a loan by selling a par floating-coupon bond maturing at time 3. (That is, a bond which always has the par value, the coupons being implied by the short rates as in (9.11).) We use the bond prices and yields in Example 9.23, see Figures 9.10 and 9.11. The cash flow shown in Figure 9.21 includes the initial amount received for selling the bond together with the coupons and face value to be paid. Consider a caplet that applies at time 1 (one month) with

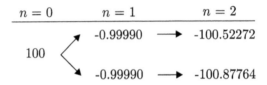

Figure 9.21 Cash flow of a floating-coupon bond issued at par

strike interest rate of 8% (corresponding to 0.67% for a one-month period). The coupon determined by the caplet rate is 0.66889 and we modify the cash flow accordingly. At time 0 we find the bond price by discounting its time 1 value, 100.66889 in each state, that is, 100 plus the coupon. This gives the cash flow shown in Figure 9.22. The price of the bond is reduced by the value of the

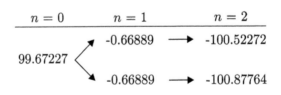

Figure 9.22 Cash flow modified by a caplet exercised at time 1

caplet, that is, by 0.32773.

For a caplet at time 2 with the same strike rate the maximum size of the coupon is 0.66889, as before. In the up state we pay the original interest, exercising the caplet in the down state. The value of the bond at time 1 is not 100, since the final coupons are no longer the same as for the par bond. The time 1 prices are obtained by discounting the time 2 values. At time 0 we find the bond price by evaluating the risk-neutral expectation of the discounted values of the bond at time 1. The resulting cash flow is shown in Figure 9.23.

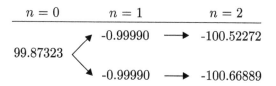

Figure 9.23 Cash flow modified by a caplet exercised at time 2

This fixes the price of this caplet at 0.12677.

Finally, consider a cap for both times 1 and 2 with the same strike rate as above. The cash flow can be obtained in a similar manner, see Figure 9.24.

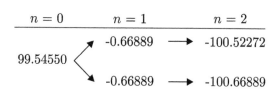

Figure 9.24 Cash flow for a bond with a cap

We can see that the value of the cap, 0.45450, is the sum of the values of the caplets.

Analogously, a *floor* is a provision limiting the coupon from below. This will be of value for a bond holder. It is composed from a series of *floorlets*, each referring to a single period.

Exercise 9.34

In the framework of the above example, value a floor expiring at time 2 with strike rate 8%, based on the bond prices in Example 9.23.

9.7 Final Remarks

We conclude this chapter with some informal remarks on possible ways in which models of the structure of bond prices can be built. This is a complex area and all we can do here is to make some general comments.

As we have seen, the theory of interest rates is more complicated than the theory of stock prices. In order to be able to price interest rate derivatives, we need a model of possible movements of bond prices for each maturity. The bond prices with different maturities have to be consistent with each other. As we have seen above, the specification of

a) a model of possible short rates,

b) a model of possible values of a bond with the longest maturity (consistent with the initial term structure)

determines the structure of possible prices of all bonds maturing earlier. An alternative approach is to specify

a) a model of possible short rates,

b) the probabilities at each state,

and, taking these probabilities as risk-neutral ones, to compute the prices of all bonds for all maturities. The latter method is conceptually simpler, especially if we take the same probability in each state. The flexibility of short-rate modelling allows us to obtain sufficiently many models consistent with the initial term structure.

The simplest choice of probability $1/2$ at each state of the binomial tree appears to be as good as any, and we can focus on constructing short-rate models so that their parameters are consistent with historical data. A general short-rate model in discrete time can be described as follows. Let $t_n = nh$. Then the following relations specify a tree of rate movements:

$$r(t_{n+1}) - r(t_n) = \mu(t_n, r(t_n))h + \sigma(t_n, r(t_n))\xi_n$$

where $\xi_n = \pm 1$ with probability $1/2$ each, and $\mu(t, r)$, $\sigma(t, r)$ are suitably chosen functions. In the continuous time limit (in the spirit of Chapter 8) these relations lead to a stochastic differential equation of the form

$$dr(t) = \mu(t, r(t))dt + \sigma(t, r(t))dW(t).$$

There are many ways in which the functions μ and σ can be specified, but none of them is universally accepted. Here are just a few examples: $\mu(t, r) = b - ar$, $\sigma(t, r) = \sigma$ (Vasiček model), $\mu(t, r) = a(b - r)$, $\sigma(t, r) = \sigma\sqrt{r}$ (Cox–Ingersoll–Ross model), or $\mu(t, r) = \theta(t)r$, $\sigma(t, r) = \sigma r$ (Black–Derman–Toy model).

Given the short rates, the next step is to compute the bond prices. These will depend on the functions μ and σ. Two problems may be encountered:

1. The model is too crude, for example these functions are just constants. Then we may not be able to adjust them so that the resulting bond prices agree with the initial term structure.

2. The model is too complicated, for example we take absolutely general functions μ, σ. Fitting the initial term structure imposes some constraints on the parameters, but many are left free and the result is too general to be of any practical use.

These problems can be avoided if some middle-of-the-road solution is adopted.

Yet another alternative is to specify the dynamics of the entire curve of forward rates. This determines the time evolution of the term structure, with the initial term structure playing the role of initial data. This sounds conceptually simple, but the model (the Heath–Jarrow–Morton model in a continuous time setting) is mathematically complex.

Case 9: Discussion

The annuity offered by the bank is equivalent to an investment at 3.8374% compounded monthly. At that rate the annuity factor is 11.7542, which indeed provides 12 monthly payments of $85.08 per each $1000 invested at the beginning of the year.

To find out if it might be possible to manufacture a better cash flow you look at the term structure at the beginning of the year. The prices $B(0, n)$ of bonds maturing in $n = 1, \ldots, 12$ months turn out to be

$$
\begin{aligned}
B(0,1) &= 0.9943, & B(0,2) &= 0.9891, & B(0,3) &= 0.9869, \\
B(0,4) &= 0.9833, & B(0,5) &= 0.9788, & B(0,6) &= 0.9767, \\
B(0,7) &= 0.9752, & B(0,8) &= 0.9721, & B(0,9) &= 0.9709, \\
B(0,10) &= 0.9683, & B(0,11) &= 0.9661, & B(0,12) &= 0.9651.
\end{aligned}
$$

For a cash flow consisting of 12 equal monthly payments of C to be worth $1000 at the beginning of the year (in effect a fixed-coupon bond with face value 0), we have

$$
1000 = \sum_{n=1}^{12} CB(0,n),
$$

which gives $C = 85.27$ dollars. The corresponding annuity factor is 11.7268, which gives a rate of 4.2736% compounded monthly, that is, more than that offered by the bank.

An alternative way to generate an annuity would be to invest at the floating rate and enter a swap agreement. For a stream of monthly floating rate payments on a principal F to be worth $1000 at the beginning of the year, we have

$$
1000 = F - FB(0,12),
$$

so $F = 28653.30$ dollars. The swap rate can be computed using (9.14), which gives $R_s = 3.5713\%$. At this rate the fixed leg of the swap on the $28653.30 principal corresponds to 12 monthly payments of $85.27, exactly the same as for the last method. The fact that in this case the swap rate happens to be lower than the annuity rate is caused by the decreasing term structure of interest rates implied by the bonds.

10
Appendix

We gather here a few selected facts from undergraduate mathematics courses which are used in the text, formulated in a way needed for the applications.

10.1 Calculus

10.1.1 Taylor formula

We present the Taylor formula in the form employed in Chapter 8.

Let $f(t, x)$ be a function of two real variables twice continuously differentiable. Then

$$f(t + \Delta t, x + \Delta x) = f(t, x) + \frac{\partial f(t, x)}{\partial t} \Delta t + \frac{\partial f(t, x)}{\partial x} \Delta x$$
$$+ \frac{1}{2} \frac{\partial^2 f(t, x)}{\partial t^2} (\Delta t)^2 + \frac{\partial^2 f(t, x)}{\partial t \partial x} \Delta t \Delta x + \frac{1}{2} \frac{\partial^2 f(t, x)}{\partial x^2} (\Delta x)^2$$
$$+ R(t, x, \Delta t, \Delta x),$$

where

$$|R(t, x, \Delta t, \Delta x)| \le M[(\Delta t)^3 + (\Delta t)^2 \Delta x + \Delta t (\Delta x)^2 + (\Delta x)^3]$$

for some $M > 0$. The last term is called the remainder. It converges to zero as $\Delta t \to 0$ and $\Delta x \to 0$.

M. Capiński, T. Zastawniak, *Mathematics for Finance*,
Springer Undergraduate Mathematics Series,
© Springer-Verlag London Limited 2011

Denote Δt as h. If we also have $(\Delta x)^2 = h$, then $R(t, x, h) \leq L h^{\frac{3}{2}}$ for some $L > 0$. In this case we say that the remainder is of order $O(h^{\frac{3}{2}})$. The symbol $O(g(h))$ in general denotes a function converging to zero as $h \to 0$ at the same rate as $g(h)$. Specifically, we assume that $O(g(h)) \leq Cg(h)$ for some constant C.

In addition,

$$\frac{\partial^2 f(t, x)}{\partial t^2}(\Delta t)^2 = \frac{\partial^2 f(t, x)}{\partial t^2}h^2 = O(h^2),$$

$$\frac{\partial^2 f(t, x)}{\partial t \partial x}\Delta t \Delta x = \frac{\partial^2 f(t, x)}{\partial t \partial x}h^{\frac{3}{2}} = O(h^{\frac{3}{2}}).$$

Using the fact that for any small enough $h > 0$, we have $h^2 \leq h^{\frac{3}{2}}$ and so $O(h^2)$ is $O(h^{\frac{3}{2}})$, the Taylor formula can be written as

$$f(t + \Delta t, x + \Delta x) = f(t, x) + \frac{\partial f(t, x)}{\partial t}\Delta t + \frac{\partial f(t, x)}{\partial x}\Delta x$$
$$+ \frac{1}{2}\frac{\partial^2 f(t, x)}{\partial x^2}(\Delta x)^2 + O(h^{\frac{3}{2}})$$

with $h = \Delta t = (\Delta x)^2$.

10.1.2 Lagrange Multipliers

Suppose that f and $g_1, \ldots, g_m$ are continuously differentiable functions of n variables, $m < n$, the set $G = \{\mathbf{x} \in \mathbb{R}^n : g_1(\mathbf{x}) = c_1, \ldots, g_m(\mathbf{x}) = c_m\}$ is contained in the domain of f, and the gradient vectors $\nabla g_1(\mathbf{x}), \ldots, \nabla g_m(\mathbf{x})$ are linearly independent at each $\mathbf{x} \in G$. If f has a constrained extremum at $\mathbf{x}_0$ on the set G, then there exist numbers $\lambda_1, \ldots, \lambda_m$ (called Lagrange multipliers) such that $\nabla f(\mathbf{x}_0) = \lambda_1 \nabla g_1(\mathbf{x}_0) + \cdots + \lambda_m \nabla g_m(\mathbf{x}_0)$.

In order to find $\mathbf{x}_0$ we can solve the system of equations

$$\nabla f(\mathbf{x}_0) = \lambda_1 \nabla g_1(\mathbf{x}_0) + \cdots + \lambda_m \nabla g_m(\mathbf{x}_0),$$
$$g_1(\mathbf{x}_0) = c_1,$$
$$\vdots$$
$$g_m(\mathbf{x}_0) = c_m.$$

This is equivalent to forming the so-called Lagrange function

$$L(\mathbf{x}, \lambda_1, \ldots, \lambda_m) = f(\mathbf{x}) + \lambda_1(g_1(\mathbf{x}) - c_1) + \cdots + \lambda_m(g_m(\mathbf{x}) - c_m)$$

and requiring that

$$\nabla_{\mathbf{x}} L(\mathbf{x}_0, \lambda_1, \ldots, \lambda_m) = 0,$$
$$\frac{\partial L(\mathbf{x}_0, \lambda_1, \ldots, \lambda_m)}{\partial \lambda_1} = 0,$$
$$\vdots$$
$$\frac{\partial L(\mathbf{x}_0, \lambda_1, \ldots, \lambda_m)}{\partial \lambda_m} = 0.$$

10.2 Measure and Integral

For a function $f : [a, b] \to \mathbb{R}$ we define the lower and upper approximating sums

$$m_n = \sum_{i=1}^{n-1} \inf_{x \in [t_i, t_{i+1}]} f(x), \qquad M_n = \sum_{i=1}^{n-1} \sup_{x \in [t_i, t_{i+1}]} f(x),$$

where $a = t_1 < t_2 < \cdots < t_n = b$. If the sequences m_n, M_n converge to a common limit as the grid diameter $\max_i |t_{i+1} - t_i|$ tends to zero, than we say that f is Riemann integrable and denote this limit by $\int_a^b f(x) dx$. A continuous function is always Riemann integrable.

For more general functions we need Lebesgue integration. This is based on Lebesgue measure, which is a real-valued mapping defined on a certain family $\mathcal{L}$ of subsets of $\mathbb{R}$ satisfying two conditions: 1) $m([a, b]) = b - a$, and 2) m is countably additive, that is, for any sequence of pairwise disjoint sets $A_n \in \mathcal{L}$ ($A_i \cap A_j = \emptyset$ if $i \neq j$)

$$m(\bigcup_{n=1}^{\infty} A_n) = \sum_{n=1}^{\infty} m(A_n).$$

For a function of the form $s(x) = \sum_{i=1}^n c_i \mathbb{I}_{A_i}$ (a so-called simple function), where $c_i \geq 0$, $A_i \in \mathcal{L}$, A_i are pairwise disjoint with $\bigcup_{i=1}^n A_i = [a, b]$, and where $\mathbb{I}_{A_i}$ denotes the indicator function of A_i ($\mathbb{I}_{A_i}(\omega) = 1$ if $\omega \in A_i$ or 0 otherwise), we put $\int_{[a,b]} s dm = \sum_{i=1}^n c_i m(A_i)$, and for any $f \geq 0$ we write

$$\int_{[a,b]} f dm = \sup \left\{ \int_{[a,b]} s dm : s \leq f, \ s \text{ is a simple function} \right\}.$$

If this number is finite, we say that f is Lebesgue integrable. For any function f (not necessarily non-negative), we say that f is Lebesgue integrable if both the positive part $f^+ = \max\{f, 0\}$ and the negative part $f^- = \max\{-f, 0\}$ of f are Lebesgue integrable, and then we write

$$\int_{[a,b]} f \mathrm{d}m = \int_{[a,b]} f^+ \mathrm{d}m - \int_{[a,b]} f^- \mathrm{d}m.$$

If f is continuous, then it is Lebesgue integrable, and the Riemann and Lebesgue integrals coincide. The construction can be modified by replacing m by any countably additive non-negative mapping.

The integral can be extended to unbounded intervals in the following way:

$$\int_{-\infty}^{\infty} f(x)\mathrm{d}x = \lim_{n \to \infty} \int_{a_n}^{b_n} f(x)\mathrm{d}x,$$

where $a_n \to -\infty$, $b_n \to \infty$, provided that the limit exists independently of the choice of a_n, b_n.

10.3 Probability

The basic notion is probability space, a triple $(\Omega, \mathcal{F}, P)$ where Ω is an non-empty set, $\mathcal{F}$ is a sigma-field of events of Ω and $P : \mathcal{F} \to [0,1]$ is a probability measure assigning numbers to events.

A sigma-field is a family of subsets of Ω containing $\emptyset$ and Ω, closed with respect to complements and countable unions. Replacing 'countable' by 'finite' gives the notion of a field. The probability measure, or briefly, probability, is required to satisfy $P(\emptyset) = 0$ and $P(\Omega) = 1$, and be countably additive: $P(\bigcup_{n=1}^{\infty} A_n) = \sum_{n=1}^{\infty} P(A_n)$ for pairwise disjoint events A_n. If Ω is finite, a finite summation range is sufficient, in which case we refer to the property as finite additivity.

If Ω is finite, then $\mathcal{F}$ is typically taken to be the family of all subsets of Ω, and P can be defined by prescribing the values on single-element sets, $P(\{\omega\}) = p_\omega \geq 0$ with $\sum_{\omega \in \Omega} p_\omega = 1$, and writing $P(A) = \sum_{\omega \in A} p_\omega$ for any $A \in \mathcal{F}$.

A random variable is a mapping $X : \Omega \to \mathbb{R}$ with $\{\omega : X^\omega \in [a,b]\} \in \mathcal{F}$ for all real $a \leq b$. It generates a probability P_X on $\mathbb{R}$ such that

$$P_X([a,b]) = P(\{\omega : X^\omega \in [a,b]\}) = P(X \in [a,b]),$$

called the probability distribution, defined on intervals, with an extension to a countably additive function defined on the smallest sigma-field $\mathcal{B}$ of subsets of $\mathbb{R}$ containing the intervals (existence of this extension can be proved). The elements of $\mathcal{B}$ are called *Borel sets*.

The cumulative distribution function is defined by

$$F_X(a) = P(X \le a) = P_X((-\infty, a]).$$

A random variable X has a density f_X if for any real $a \le b$

$$P_X([a, b]) = \int_a^b f_X(x)\mathrm{d}x,$$

with $F_X(a) = \int_{-\infty}^a f_X(x)\mathrm{d}x$. An important example is

$$f_X(x) = \frac{1}{\sigma\sqrt{2\pi}} e^{-\frac{(x-\mu)^2}{2\sigma^2}}. \tag{10.1}$$

In this case we say that X has normal distribution $N(\mu, \sigma^2)$ with mean μ and variance σ^2. It has standard normal distribution if $\mu = 0$, $\sigma = 1$, and then we write $N(a) = F_X(a)$, observing that $1 - N(a) = N(-a)$ due to the symmetry of the density.

For a finite Ω we define the expectation of X by

$$\mathbb{E}(X) = \sum_{\omega \in \Omega} X^\omega p_\omega$$

and the variance by $\mathrm{Var}(X) = \mathbb{E}(X - \mathbb{E}(X))^2$. The standard deviation is defined to be the square root of the variance. In general,

$$\mathbb{E}(X) = \int_{-\infty}^{\infty} x \mathrm{d}P_X(x),$$

with the integral understood in the sense of Lebesgue (this integral may not exist, and the expectation may be undefined for some random variables X).

If X has density f_X, then

$$\mathbb{E}(X) = \int_{-\infty}^{\infty} x f_X(x)\mathrm{d}x.$$

If X has normal distribution $N(\mu, \sigma^2)$, then $\mathbb{E}(X) = \mu$ and $\mathrm{Var}(X) = \sigma^2$.

Random variables $X_1, \ldots, X_N$ are said to be independent if for all $2 \le k \le N$, all $\{i_1, \ldots, i_k\} \subset \{1, \ldots, N\}$ and all $a_j \le b_j$, $j = 1, \ldots, k$, we have

$$P\left(\bigcap_{j=1}^k \{\omega : X_{i_j}^\omega \in [a_j, b_j]\}\right) = P(X_{i_1} \in [a_1, b_1]) \times \cdots \times P(X_{i_k} \in [a_k, b_k]).$$

Random variables forming an infinite sequence $X_1, X_2, \ldots$ are independent if any finite number of elements of the sequence are independent.

The conditional probability of an event A given an event B such that $P(B) > 0$ is defined by

$$P(A|B) = \frac{P(A \cap B)}{P(B)}.$$

For a fixed B the conditional probability $P(A|B)$ is a probability on Ω defined for any $A \in \mathcal{F}$. Since $P(B|B) = 1$, the event B can be regarded as a new probability space (events outside B do not matter, $P(A|B) = 0$ if $A \cap B = \emptyset$).

The expectation of a random variable X under the conditional probability $P(\cdot|B)$ is denoted by $\mathbb{E}(X|B)$ and called the conditional expectation of X given B.

If X_n is a sequence of independent random variables with $P(X_n = 1) = P(X_n = -1) = \frac{1}{2}$, then we have the following convergence, which is a special case of the Central Limit Theorem: For all $a \leq b$

$$P\left(a \leq \frac{X_1 + \cdots + X_n}{\sqrt{n}} \leq b\right) \to \int_a^b \frac{1}{\sqrt{2\pi}} e^{-\frac{x^2}{2}} \, dx \quad \text{as } n \to \infty.$$

10.4 Covariance Matrix

Here we collect some properties of the covariance matrix $\mathbf{C}$ used in Chapter 3. The entries of $\mathbf{C}$ are $c_{ij} = \mathrm{Cov}(K_i, K_j)$, where K_i, K_j are the returns on securities $i, j = 1, \ldots, n$.

The covariance matrix is symmetric, $\mathbf{C}^{\mathrm{T}} = \mathbf{C}$, because $c_{ij} = \mathrm{Cov}(K_i, K_j) = \mathrm{Cov}(K_j, K_i) = c_{ji}$. It is also non-negative definite, $\mathbf{x}\mathbf{C}\mathbf{x}^{\mathrm{T}} \geq 0$ for all $\mathbf{x} \in \mathbb{R}^n$, because $\mathbf{x}\mathbf{C}\mathbf{x}^{\mathrm{T}} = \mathrm{Var}(K) \geq 0$ for $K = x_1 K_1 + \cdots + x_n K_n$.

If, in addition, we assume that $\mathbf{C}$ is non-singular, $\det \mathbf{C} \neq 0$, then the inverse matrix $\mathbf{C}^{-1}$ exists, and $\mathbf{C}, \mathbf{C}^{-1}$ have the following properties, which are needed in Chapter 3.

Proposition 10.1

If the covariance matrix $\mathbf{C}$ is non-singular, $\det \mathbf{C} \neq 0$, then:

1) $\mathbf{C}$ is positive definite, that is, $\mathbf{x}\mathbf{C}\mathbf{x}^{\mathrm{T}} > 0$ for each $\mathbf{x} \in \mathbb{R}^n$ such that $\mathbf{x} \neq \mathbf{0}$;

2) $\mathbf{C}^{-1}$ is symmetric and positive definite;

3) $\mathbf{u}\mathbf{C}^{-1}\mathbf{u}^{\mathrm{T}} > 0$, where $\mathbf{u} \in \mathbb{R}^n$ is the vector with all entries equal to 1;

4) If $\mathbf{m}, \mathbf{u} \in \mathbb{R}^n$ are linearly independent vectors, then

$$\begin{vmatrix} \mathbf{m}\mathbf{C}^{-1}\mathbf{m}^{\mathrm{T}} & \mathbf{u}\mathbf{C}^{-1}\mathbf{m}^{\mathrm{T}} \\ \mathbf{m}\mathbf{C}^{-1}\mathbf{u}^{\mathrm{T}} & \mathbf{u}\mathbf{C}^{-1}\mathbf{u}^{\mathrm{T}} \end{vmatrix} > 0.$$

Proof

1) Because $\mathbf{C}$ is a symmetric matrix, it can be diagonalised. Since $\mathbf{C}$ is non-negative definite, all of its eigenvalues are non-negative. If, in addition, $\det \mathbf{C} \neq 0$, it follows that all the eigenvalues of $\mathbf{C}$ are strictly positive. This, in turn, implies that $\mathbf{C}$ is positive definite.

2) Because $\mathbf{C}\mathbf{C}^{-1} = \mathbf{C}^{-1}\mathbf{C} = \mathbf{I}$, it follows that

$$(\mathbf{C}^{-1})^{\mathrm{T}}\mathbf{C}^{\mathrm{T}} = \mathbf{C}^{\mathrm{T}}(\mathbf{C}^{-1})^{\mathrm{T}} = \mathbf{I}^{\mathrm{T}} = \mathbf{I},$$

so that $(\mathbf{C}^{-1})^{\mathrm{T}} = \mathbf{C}^{-1}$, that is, $\mathbf{C}^{-1}$ is a symmetric matrix.

Next, for any $\mathbf{x} \in \mathbb{R}^n$ such that $\mathbf{x} \neq \mathbf{0}$ take $\mathbf{y} = \mathbf{x}\mathbf{C}^{-1} \neq \mathbf{0}$. Then

$$\mathbf{x}\mathbf{C}^{-1}\mathbf{x}^{\mathrm{T}} = \mathbf{y}\,(\mathbf{y}\mathbf{C})^{\mathrm{T}} = \mathbf{y}\mathbf{C}^{\mathrm{T}}\mathbf{y}^{\mathrm{T}} = \mathbf{y}\mathbf{C}\mathbf{y}^{\mathrm{T}} > 0,$$

which shows that $\mathbf{C}^{-1}$ is positive definite.

3) Since $\mathbf{u} \neq \mathbf{0}$ and $\mathbf{C}^{-1}$ is positive definite, we have $\mathbf{u}\mathbf{C}^{-1}\mathbf{u}^{\mathrm{T}} > 0$.

4) If $\mathbf{m}, \mathbf{u}$ are linearly independent, then $\mathbf{u} + x\mathbf{m} \neq \mathbf{0}$ for each $x \in \mathbb{R}$. Because $\mathbf{C}^{-1}$ is positive definite

$$0 < (\mathbf{u} + x\mathbf{m})\,\mathbf{C}^{-1}\,(\mathbf{u} + x\mathbf{m})^{\mathrm{T}}$$
$$= \left(\mathbf{m}\mathbf{C}^{-1}\mathbf{m}^{\mathrm{T}}\right) x^2 + \left(\mathbf{u}\mathbf{C}^{-1}\mathbf{m}^{\mathrm{T}} + \mathbf{m}\mathbf{C}^{-1}\mathbf{u}^{\mathrm{T}}\right) x + \mathbf{u}\mathbf{C}^{-1}\mathbf{u}^{\mathrm{T}}$$

for each $x \in \mathbb{R}$. Using the fact that $\mathbf{C}^{-1}$ is a symmetric matrix, so that

$$\mathbf{u}\mathbf{C}^{-1}\mathbf{m}^{\mathrm{T}} = \mathbf{m}\mathbf{C}^{-1}\mathbf{u}^{\mathrm{T}},$$

we can write the above inequality as

$$0 < \left(\mathbf{m}\mathbf{C}^{-1}\mathbf{m}^{\mathrm{T}}\right) x^2 + 2 \left(\mathbf{u}\mathbf{C}^{-1}\mathbf{m}^{\mathrm{T}}\right) x + \mathbf{u}\mathbf{C}^{-1}\mathbf{u}^{\mathrm{T}}$$

for each $x \in \mathbb{R}$. The discriminant of this quadratic function in x must therefore be negative,

$$0 > 4 \left(\mathbf{u}\mathbf{C}^{-1}\mathbf{m}^{\mathrm{T}}\right)^2 - 4 \left(\mathbf{m}\mathbf{C}^{-1}\mathbf{m}^{\mathrm{T}}\right) \left(\mathbf{u}\mathbf{C}^{-1}\mathbf{u}^{\mathrm{T}}\right)$$
$$= 4 \left(\mathbf{u}\mathbf{C}^{-1}\mathbf{m}^{\mathrm{T}}\right) \left(\mathbf{m}\mathbf{C}^{-1}\mathbf{u}^{\mathrm{T}}\right) - 4 \left(\mathbf{m}\mathbf{C}^{-1}\mathbf{m}^{\mathrm{T}}\right) \left(\mathbf{u}\mathbf{C}^{-1}\mathbf{u}^{\mathrm{T}}\right)$$
$$= -4 \begin{vmatrix} \mathbf{m}\mathbf{C}^{-1}\mathbf{m}^{\mathrm{T}} & \mathbf{u}\mathbf{C}^{-1}\mathbf{m}^{\mathrm{T}} \\ \mathbf{m}\mathbf{C}^{-1}\mathbf{u}^{\mathrm{T}} & \mathbf{u}\mathbf{C}^{-1}\mathbf{u}^{\mathrm{T}} \end{vmatrix}.$$

$\square$

10.5 Regression Between Returns

Consider two assets A and B with returns K_A and K_B. The returns are random variables, each taking possible values $K_A(n) = K_A^{\omega_n}$ and $K_B(n) = K_B^{\omega_n}$ with probability p_n in scenario ω_n, where $n = 1, \ldots, N$. (We assume an N-element probability space Ω.)

The regression line (the line of best fit) between the returns K_A, K_B can be computed by the least squares method. It is a straight line as in Fig. 10.1

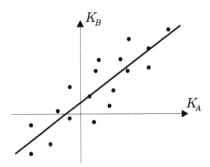

Figure 10.1   Regression line

described by the equation

$$y = \beta x + \alpha,$$

where α and β (the intercept and gradient of the line) are the coefficients that minimise the variance of the residual error $\varepsilon = K_B - (\beta K_A + \alpha)$, which can be written as

$$f(\alpha, \beta) = \mathrm{Var}(\varepsilon) = \sum_{n=1}^{N} \left(K_B(n) - (\beta K_A(n) + \alpha) \right)^2 p_n$$

$$= \mathbb{E}(K_B^2) + \beta^2 \mathbb{E}(K_A^2) + \alpha^2 - 2\beta \mathbb{E}(K_A K_B) - 2\alpha \mathbb{E}(K_B) + 2\alpha\beta \mathbb{E}(K_A).$$

The minimum will be achieved at a point where

$$\frac{\partial f(\alpha, \beta)}{\partial \beta} = 2\beta \mathbb{E}(K_A^2) - 2\mathbb{E}(K_A K_B) + 2\alpha \mathbb{E}(K_A) = 0,$$

$$\frac{\partial f(\alpha, \beta)}{\partial \alpha} = 2\alpha - 2\mathbb{E}(K_B) + 2\beta \mathbb{E}(K_A) = 0.$$

We can solve this system of equations for α and β, to obtain

$$\beta = \frac{\mathbb{E}(K_A K_B) - \mathbb{E}(K_A)\mathbb{E}(K_B)}{\mathbb{E}(K_A^2) - \mathbb{E}(K_A)^2}$$
$$= \frac{c_{AB}}{\sigma_A^2} = \frac{\rho_{AB}\sigma_B}{\sigma_A},$$

$$\alpha = \mathbb{E}(K_B) - \frac{\mathbb{E}(K_A K_B) - \mathbb{E}(K_A)\mathbb{E}(K_B)}{\mathbb{E}(K_A^2) - \mathbb{E}(K_A)^2}\mathbb{E}(K_A)$$
$$= \mu_B - \frac{c_{AB}}{\sigma_A^2}\mu_A = \mu_B - \frac{\rho_{AB}\sigma_B}{\sigma_A}\mu_A.$$

Here

$$\mu_A = \mathbb{E}(K_A), \quad \mu_B = \mathbb{E}(K_B)$$

are the expected returns on assets A and B,

$$\sigma_A^2 = \text{Var}(K_A) = \mathbb{E}[(K_A - \mu_A)^2], \quad \sigma_B^2 = \text{Var}(K_B) = \mathbb{E}[(K_B - \mu_B)^2]$$

are the variances of the returns, and

$$c_{AB} = \text{Cov}(K_A, K_B) = \mathbb{E}[(K_A - \mu_A)(K_B - \mu_B)], \quad \rho_{AB} = \frac{c_{AB}}{\sigma_A\sigma_B}$$

are, respectively, the covariance and correlation between the two returns.

11
Solutions

Chapter 1

1.1 The value of the portfolio at time 0 is
$$V(0) = xS(0) + yA(0) = 1,600$$
dollars. The value of the portfolio at time T will be
$$V(T) = xS(T) + yA(T) = \begin{cases} 1,800 & \text{if stock goes up,} \\ 1,700 & \text{if stock goes down.} \end{cases}$$
Hence, the return on the portfolio will be
$$K_V = \frac{V(T) - V(0)}{V(0)} = \begin{cases} 0.1250 & \text{if stock goes up,} \\ 0.0625 & \text{if stock goes down,} \end{cases}$$
that is, either 12.5% or 6.25%.

1.2 Given the same bond and stock prices as in Exercise 1.1, the value of a portfolio (x, y) at time T will be
$$V(T) = \begin{cases} x30 + y100 & \text{if stock goes up,} \\ x20 + y100 & \text{if stock goes down.} \end{cases}$$
Thus, we obtain a system of equations
$$\begin{cases} x30 + y100 = 1,160, \\ x20 + y100 = 1,040. \end{cases}$$
The solution is $x = 12$ and $y = 8$. A portfolio with 12 shares of stock and 8 bonds will produce the desired value at time T. The time 0 value of this portfolio is
$$V(0) = 12 \times 25 + 8 \times 90 = 1,020$$
dollars.

1.3 An arbitrage opportunity can be realised as follows:

M. Capiński, T. Zastawniak, *Mathematics for Finance*,
Springer Undergraduate Mathematics Series,
© Springer-Verlag London Limited 2011

- use dealer A to change 1 dollar into $\frac{1}{1.5844} \cong 0.6312$ pounds;
- use dealer B to change 0.6312 pounds into $\frac{0.6312}{0.6401} \cong 0.9861$ euros;
- use dealer A to change 0.9861 euros into $0.9861 \times 1.0202 \cong 1.0060$ dollars.

The arbitrage gain will be about 0.0060 dollars.

1.4 We want both $xS(0)$ and $yA(0)$ to be equal to a half of the initial wealth. This gives $x80 = 5,000$ and $y100 = 5,000$, so $x = 62.5$ and $y = 50$. The value of this portfolio at time T will be

$$V(T) = 62.5S(T) + 50A(T) = \begin{cases} 11,750 & \text{if stock goes up,} \\ 9,250 & \text{if stock goes down,} \end{cases}$$

and hence the return on this portfolio will be

$$K_V = \begin{cases} 0.175 & \text{if stock goes up,} \\ -0.075 & \text{if stock goes down.} \end{cases}$$

Now we can compute the expected return

$$\mathbb{E}(K_V) = 0.175 \times 0.8 - 0.075 \times 0.2 = 0.125,$$

which is 12.5%, and the risk

$$\sigma_V = \sqrt{(0.175 - 0.125)^2 \times 0.8 + (-0.075 - 0.125)^2 \times 0.2} = 0.1,$$

that is, 10%.

1.5 The following strategy will realise an arbitrage opportunity. At time 0:

- borrow \$34;
- buy a share of stock for \$34;
- enter into a short forward contract with forward price \$38.60 and delivery date 1.

At time 1

- sell the stock for \$38.60, closing the short forward position;
- pay $34 \times 1.12 = 38.08$ dollars to clear the loan with interest.

The balance of $38.60 - 38.08 = 0.52$ dollars will be your arbitrage profit.

1.6 a) The payoff of a call option with strike price \$90 will be

$$C(T) = \begin{cases} 30 & \text{if stock goes up,} \\ 0 & \text{if stock goes down.} \end{cases}$$

The replicating investment into x shares and y bonds satisfies the system of equations

$$\begin{cases} x120 + y110 = 30, \\ x80 + y110 = 0. \end{cases}$$

The solution is $x = \frac{3}{4}$ and $y = -\frac{6}{11}$. Hence the price of the option must be

$$C(0) = \frac{3}{4} \times 100 - \frac{6}{11} \times 100 \cong 20.45$$

dollars.

b) The payoff of a call option with strike price $110 will be

$$C(T) = \begin{cases} 10 & \text{if stock goes up,} \\ 0 & \text{if stock goes down.} \end{cases}$$

The replicating investment into x shares and y bonds satisfies

$$\begin{cases} x120 + y110 = 10, \\ x80 + y110 = 0. \end{cases}$$

Solving this system of equations, we find that $x = \frac{1}{4}$ and $y = -\frac{2}{11}$. Hence the price of the option is

$$C(0) = \frac{1}{4} \times 100 - \frac{2}{11} \times 100 \cong 6.82$$

dollars.

1.7 a) The replicating investment into x shares and y bonds satisfies the system of equations

$$\begin{cases} x120 + y105 = 20, \\ x80 + y105 = 0, \end{cases}$$

which gives $x = \frac{1}{2}$ and $y = -\frac{8}{21}$. Hence

$$C(0) = \frac{1}{2} \times 100 - \frac{8}{21} \times 100 \cong 11.91$$

dollars.

b) In this case the replicating investment into x shares and y bonds satisfies

$$\begin{cases} x120 + y115 = 20, \\ x80 + y115 = 0, \end{cases}$$

so $x = \frac{1}{2}$ and $y = -\frac{8}{23}$. It follows that

$$C(0) = \frac{1}{2} \times 100 - \frac{8}{23} \times 100 \cong 15.22$$

dollars.

1.8 We need to find an investment into x shares and y bonds replicating the put option, that is, such that $xS(T) + yA(T) = P(T)$, no matter whether the stock price goes up or down. This leads to the system of equations

$$\begin{cases} x120 + y110 = 0, \\ x80 + y110 = 20. \end{cases}$$

The solution is $x = -\frac{1}{2}$ and $y = \frac{6}{11}$. To replicate the put option we need to take a short position of $\frac{1}{2}$ a share in stock and to buy $\frac{6}{11}$ of a bond. The value of this investment at time 0 is

$$xS(0) + yA(0) = -\frac{1}{2} \times 100 + \frac{6}{11} \times 100 \cong 4.55$$

dollars. By a similar argument as in Proposition 1.14, it follows that $xS(0) + yA(0) = P(0)$, or else an arbitrage opportunity would arise. Therefore, the price of the put must be $P(0) \cong 4.55$ dollars.

1.9 If $F > S(0)\frac{1+K_h}{1+K_f}$, then $(\frac{1}{1+K_f}, -\frac{S(0)}{1+K_f}, -1)$ is an arbitrage portfolio (assuming $A_f(0) = A_h(0) = 1$ for simplicity) since it costs nothing to initiate and at time T the interest on the foreign currency increases our holding to one unit, sold at the forward price F, which is more than the home currency loan of $\frac{S(0)}{1+K_f}$ repaid with interest. If $F < S(0)\frac{1+K_h}{1+K_f}$, then we take the opposite positions.

1.10 Suppose that a sterling bond promising to pay £100 at time 1 is selling for x pounds at time 0. To find x consider the following strategy. At time 0:

- borrow $1.6x$ dollars and change the sum into x pounds;
- purchase a sterling bond for x pounds;
- take a short forward position to sell £100 for $1.50 to a pound with delivery date 1.

Then, at time 1:

- cash the bond, collecting £100;
- close the short forward position by selling £100 for $150;
- repay the cash loan with interest, that is, $1.68x$ dollars in total.

The balance of all these transactions will be $150 - 1.68x$ dollars, which must be equal to zero or else an arbitrage opportunity would arise. It follows that a sterling bond promising to pay £100 at time 1 must sell for $x = \frac{150}{1.68} \cong 89.29$ pounds at time 0.

1.11 To replicate the put payoff we solve

$$x(1 + K_£)S^u(T) + y(1 + K_\$) = 0,$$
$$x(1 + K_£)S^d(T) + y(1 + K_\$) = X - S^d(T)$$

with $X = 1.71$, $K_£ = 3\%$, $K_\$ = 5\%$, $S^u(T) = 1.84$, $S^d(T) = 1.46$ to get $x \cong -0.638734$ and $y \cong 1.1528815$ so that

$$P(0) = xS(0) + y \cong 0.118132$$

dollars.

1.12 The investor will buy $\frac{500}{100} = 5$ shares and $\frac{500}{13.6364} \cong 36.6667$ options. Her final wealth will be $5 \times S(T) + 36.6667 \times C(T)$, that is, $5 \times 120 + 36.6667 \times 20 \cong 1,333.33$ dollars if the price of stock goes up to $120, or $5 \times 80 + 36.6667 \times 0 \cong 400.00$ dollars if it drops to $80.

1.13 a) If $p = 0.25$, then the standard deviation of the cost of buying one share is about 51.96 dollars when no option is purchased and about 25.98 dollars with the option.
  b) If $p = 0.5$, then the standard deviation is 60 dollars and 30 dollars, respectively.
  c) For $p = 0.75$ the standard deviation is about 51.96 dollars and 25.98 dollars, respectively.

1.14 The standard deviation of a random variable taking values a and b with probabilities p and $1 - p$, respectively, is $|a - b|\sqrt{p(1 - p)}$. If no option is involved, then the cost of buying one share will be 160 or 40 dollars, depending on whether stock goes up or down. In this case $|a - b| = |160 - 40| = 120$ dollars. If one option is used, then the cost will be 135 or 75 dollars, and $|a - b| = |135 - 75| = 60$ dollars. Clearly, the standard deviation $|a - b|\sqrt{p(1 - p)}$ will be reduced by a half, no matter what the value of p might be.

Chapter 2

2.1 The rate r satisfies

$$\left(1 + \frac{61}{365} \times r\right) \times 9,000 = 9,020.$$

This gives $r \cong 0.0133$, that is, about 1.33%. The return on this investment will be

$$K\left(0, \frac{61}{365}\right) = \frac{9,020 - 9,000}{9,000} \cong 0.0022,$$

that is, about 0.22%.

2.2 Denote the amount to be paid today by P. Then the return will be

$$\frac{1,000 - P}{P} = 0.02.$$

The solution is $P \cong 980.39$ dollars.

2.3 The time t satisfies

$$(1 + t \times 0.09) \times 800 = 830,$$

which gives $t \cong 0.4167$ years, that is, $0.4167 \times 365 \cong 152.08$ days. The return will be

$$K(0, t) = \frac{830 - 800}{800} = 0.0375,$$

that is, 3.75%.

2.4 The principal P to be invested satisfies

$$\left(1 + \frac{91}{365} \times 0.08\right) \times P = 1,000,$$

which gives $P \cong 980.44$ dollars.

2.5 The time t when the future value will be double the initial principal satisfies the equation

$$\left(1 + \frac{0.06}{365}\right)^{365t} = 2.$$

The solution is $t \cong 11.5534$ years. Because no interest will be paid for a fraction of the last day, this needs to be rounded up to a whole number of days, which gives 11 years and 202 days. (We disregard leap years and assume for simplicity that each year has 365 days.)

2.6 The interest rate r satisfies the equation

$$(1 + r)^{10} = 2,$$

which gives $r \cong 0.0718$, that is, about 7.18%.

2.7 a) In the case of annual compounding the value after two years will be

$$V(2) = \left(1 + \frac{0.1}{1}\right)^{2 \times 1} \times 100 = 121.00$$

dollars.

b) Under semi-annual compounding the value will be

$$V(2) = \left(1 + \frac{0.1}{2}\right)^{2\times 2} \times 100 \cong 121.55$$

dollars, which is clearly greater than in case a).

2.8 At 15% compounded daily the deposit will grow to

$$\left(1 + \frac{0.15}{365}\right)^{1\times 365} \times 1,000 \cong 1,161.80$$

dollars after one year. If interest is compounded semi-annually at 15.5%, the value after one year will be

$$\left(1 + \frac{0.155}{2}\right)^{1\times 2} \times 1,000 \cong 1,161.01$$

dollars, which is less than in the former case.

2.9 The initial principal P satisfies the equation

$$(1 + 0.12)^2 P = 1,000.$$

It follows that $P \cong 797.19$ dollars.

2.10 a) Under daily compounding the present value is

$$100,000 \left(1 + \frac{0.05}{365}\right)^{-100\times 365} \cong 674.03$$

dollars.

  b) If annual compounding applies, then the present value is

$$100,000 \left(1 + 0.05\right)^{-100} \cong 760.45$$

dollars.

2.11 The return will be

$$K(0,1) = \left(1 + \frac{0.1}{12}\right)^{12} - 1 \cong 0.1047,$$

that is, about 10.47%.

2.12 Using the binomial formula to expand the mth power, we obtain

$$K(0,1) = \left(1 + \frac{r}{m}\right)^m - 1$$

$$= 1 + r + \frac{(1 - \frac{1}{m})}{2!}r^2 + \cdots + \frac{(1 - \frac{m-1}{m})}{m!}r^m - 1 > r$$

if m is an integer greater than 1.

2.13 Denote the interest rate by r, the amount borrowed by P and the amount of each instalment by C,

$$C = \frac{P}{\text{PA}(r,5)} = \frac{Pr}{1 - (1+r)^{-5}},$$

see Example 2.9. Let $n = 1, 2, 3, 4$ or 5. The present value of the outstanding balance after $n - 1$ instalments are paid is equal to the amount borrowed reduced by the present value of the first $n - 1$ instalments:

$$P - \frac{C}{1+r} - \cdots - \frac{C}{(1+r)^{n-1}} = P\frac{(1+r)^{6-n} - 1}{(1+r)^5 - 1}.$$

The actual outstanding balance remaining after $n-1$ instalments are paid can be found by dividing the above by the discount factor $(1+r)^{-(n-1)}$, which gives

$$P\frac{(1+r)^5 - (1+r)^{n-1}}{(1+r)^5 - 1}. \tag{S.1}$$

The interest included in the nth instalment is, therefore,

$$P\frac{(1+r)^5 - (1+r)^{n-1}}{(1+r)^5 - 1}r. \tag{S.2}$$

The capital repaid as part of the nth instalment is the difference between the outstanding balance of the loan after the $(n - 1)$st and after the nth instalment. By (S.1) this difference is equal to

$$P\frac{(1+r)^n - (1+r)^{n-1}}{(1+r)^5 - 1} = P\frac{r(1+r)^{n-1}}{(1+r)^5 - 1}. \tag{S.3}$$

Putting P to be $\$1,000$ and r to be 15% in (S.1), (S.2) and (S.3), we can compute the figures collected in the table:

t (years)	interest paid ($)	capital repaid ($)	outstanding balance ($)
0	—	—	1,000.00
1	150.00	148.32	851.68
2	127.75	170.56	681.12
3	102.17	196.15	484.97
4	72.75	225.57	259.40
5	38.91	259.40	0.00

2.14 The amount you can afford to borrow is

$$\text{PA}(18\%, 10) \times 10,000 = \frac{1 - (1 + 0.18)^{-10}}{0.18} \times 10,000 \cong 44,941$$

dollars.

2.15 The present value of the balance after 40 years is

$$\text{PA}(5\%, 40) \times 1,200 = \frac{1 - (1 + 0.05)^{-40}}{0.05} \times 1,200 \cong 20,591$$

dollars. Dividing by the discount factor $(1 + 0.05)^{-40}$, we find that the actual balance after 40 years will be

$$\frac{20,591}{(1 + 0.05)^{-40}} \cong 144,960$$

dollars.

2.16 The annual payments will amount to

$$C = \frac{100,000}{PA(6\%,10)} \cong 13,586.80$$

dollars each. The outstanding balance to be paid to clear the mortgage after 8 years (once the 8th annual payment is made) will be

$$PA(6\%,2) \times C \cong 24,909.93$$

dollars.

2.17 Suppose that payments $C, C(1+g), C(1+g)^2, \ldots$ are made after 1 year, 2 years, 3 years, and so on. If the interest rate is constant and equal to r, then the present values of these payments are $C\,(1+r)^{-1}, C(1+g)\,(1+r)^{-2}, C(1+g)^2\,(1+r)^{-3}, \ldots$. The present value of the infinite stream of payments is, therefore,

$$\frac{C}{1+r} + \frac{C(1+g)}{(1+r)^2} + \frac{C(1+g)^2}{(1+r)^3} + \ldots = \frac{C}{r-g}.$$

The condition $g < r$ must be satisfied or otherwise the series will be divergent. Using this formula and the tail-cutting procedure, we can find that for a stream of n payments

$$\frac{C}{r-g} - \frac{C(1+g)^n}{r-g}\frac{1}{(1+r)^n} = C\frac{1 - \left(\frac{1+g}{1+r}\right)^n}{r-g}.$$

2.18 The time t it will take to earn \$1 in interest satisfies

$$e^{0.1t} \times 1,000,000 \cong 1,000,001.$$

This gives $t \cong 0.00001$ years, that is, 315.36 seconds.

2.19 a) The value in the year 2000 of the sum of \$24 for which Manhattan was bought in 1626 would be

$$24e^{(2000-1626)\times 0.05} \cong 3,173,350,575$$

dollars, assuming continuous compounding at 5%.

  b) The same amount compounded at 5% annually would be worth

$$24(1+0.05)^{2000-1626} \cong 2,018,408,628$$

dollars in the year 2000.

2.20 \$100 deposited at 10% compounded continuously will become

$$100e^{0.1} \cong 110.52$$

dollars after one year. The same amount deposited at 10% compounded monthly will become

$$100\left(1 + \frac{0.1}{12}\right)^{12} \cong 110.47$$

dollars. The difference is about \$0.05.

  If the difference is to be less than \$0.01, the compounding frequency m should satisfy

$$100\left(1 + \frac{0.1}{m}\right)^m > 110.51.$$

This means that m should be greater than about 55.19.

2.21 The present value is

$$1,000,000e^{-20\times0.06} \cong 301,194$$

dollars.

2.22 The rate r satisfies

$$950e^{0.5r} = 1,000.$$

It follows that

$$r = \frac{1}{0.5}\ln\frac{1,000}{950} \cong 0.1026,$$

that is, about 10.26%.

2.23 The interest rate is

$$r = \frac{0.03}{2/12} = 0.18,$$

that is, 18%.

2.24 The rate r satisfies

$$e^r = \left(1 + \frac{0.12}{12}\right)^{12}.$$

The solution is $r \cong 0.1194$, about 11.94%.

2.25 The frequency m satisfies

$$\left(1 + \frac{0.2}{m}\right)^m = 1 + 0.21.$$

Whence $m = 2.0$.

2.26 If monthly compounding at a rate r applies, then $\left(1 + \frac{r}{12}\right)^{12} = 1 + r_e$ and the present value of the annuity is

$$\begin{aligned}
V(0) &= \frac{C}{1 + \frac{r}{12}} + \frac{C}{\left(1 + \frac{r}{12}\right)^2} + \cdots + \frac{C}{\left(1 + \frac{r}{12}\right)^{12n}} \\
&= \frac{C}{(1 + r_e)^{1/12}} + \frac{C}{(1 + r_e)^{2/12}} + \cdots + \frac{C}{(1 + r_e)^n} \\
&= C\frac{1 - (1 + r_e)^{-n}}{(1 + r_e)^{1/12} - 1}.
\end{aligned}$$

2.27 If bimonthly compounding at a rate r applies, then $\left(1 + \frac{r}{6}\right)^6 = 1 + r_e$ and the present value of the perpetuity is

$$\begin{aligned}
V(0) &= \frac{C}{1 + \frac{r}{6}} + \frac{C}{\left(1 + \frac{r}{6}\right)^2} + \frac{C}{\left(1 + \frac{r}{6}\right)^3} + \cdots \\
&= \frac{C}{(1 + r_e)^{1/6}} + \frac{C}{(1 + r_e)^{2/6}} + \frac{C}{(1 + r_e)^{3/6}} + \cdots \\
&= \frac{C}{(1 + r_e)^{1/6} - 1}.
\end{aligned}$$

2.28 We solve the equation

$$100 = 95\left(1 + r\right)^{\frac{1}{2}}$$

for r to find the implied effective rate to be about 10.80%. If this rate remains constant, then the bond price will reach $99 at a time t such that

$$100 = 99 \left(1 + r\right)^{\left(\frac{1}{2} - t\right)}.$$

The solution is $t \cong 0.402$ years, that is, about $0.402 \times 365 \cong 146.73$ days. The bond price will reach $99 on day 147.

2.29 The interest rate for annual compounding implied by the bond can be found by solving the equation

$$(1 + r)^{-(1-0.5)} = 0.9455$$

for r. The solution is about 11.86%. By solving the equation

$$\left(1 + \frac{r}{2}\right)^{-2(1-0.5)} = 0.9455,$$

we obtain the semi-annual rate of about 11.53%, and solving

$$e^{-r(1-0.5)} = 0.9455,$$

we find the continuous rate to be about 11.21%.

2.30 a) If the continuous compounding rate is 8%, then the price of the bond will be

$$5e^{-0.08} + 5e^{-2 \times 0.08} + 5e^{-3 \times 0.08} + 105e^{-4 \times 0.08} \cong 89.06$$

dollars.

b) If the rate is 5%, then the price of the bond will be

$$5e^{-0.05} + 5e^{-2 \times 0.05} + 5e^{-3 \times 0.05} + 105e^{-4 \times 0.05} \cong 99.55$$

dollars.

2.31 The price of the bond as a function of the continuous compounding rate r can be expressed as $5e^{-r} + 5e^{-2r} + 5e^{-3r} + 105e^{-4r}$. The graph of this function is shown in Figure S.1. When $r \searrow 0$, the price approaches $5 + 5 + 5 + 105 = 120$ dollars. In the limit as $r \nearrow \infty$ the price tends to zero.

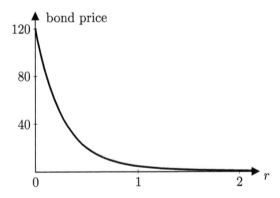

Figure S.1 Bond price versus interest rate in Exercise 2.31

2.32 The time t price of the coupon bond in Examples 2.24 and 2.25 is

$$10e^{r(t-1)} + 10e^{r(t-2)} + 10e^{r(t-3)} + 10e^{r(t-4)} + 110e^{r(t-5)} \quad \text{if} \quad 0 \le t < 1,$$
$$10e^{r(t-2)} + 10e^{r(t-3)} + 10e^{r(t-4)} + 110e^{r(t-5)} \quad \text{if} \quad 1 \le t < 2,$$
$$10e^{r(t-3)} + 10e^{r(t-4)} + 110e^{r(t-5)} \quad \text{if} \quad 2 \le t < 3,$$
$$10e^{r(t-4)} + 110e^{r(t-5)} \quad \text{if} \quad 3 \le t < 4,$$
$$110e^{r(t-5)} \quad \text{if} \quad 4 \le t < 5.$$

The graph is shown in Figure S.2.

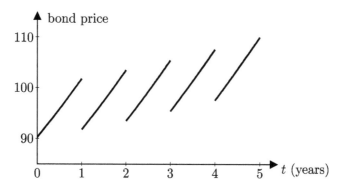

Figure S.2 Coupon bond price versus time in Exercise 2.32

2.33 In Figure S.2 we can see that the bond price will reach \$95 for the first time during year one, when the bond price is given by $10e^{r(t-1)} + 10e^{r(t-2)} + 10e^{r(t-3)} + 10e^{r(t-4)} + 110e^{r(t-5)}$. Putting $r = 0.12$, we can find the desired time t when the price will reach \$95 by solving the equation

$$10e^{r(t-1)} + 10e^{r(t-2)} + 10e^{r(t-3)} + 10e^{r(t-4)} + 110e^{r(t-5)} = 95.$$

This gives $t \cong 0.4257$ years or 155.4 days.

2.34 Since the bond is trading at par, its initial price is the same as the face value $F = 100$. The implied continuous compounding rate r can be found by solving the equation
$$8e^{-r} + 8e^{-2r} + 108e^{-3r} = 100.$$
This gives $r \cong 0.0770$ or 7.70%.

2.35 By solving the equation
$$(1+r)^{-1} = 0.89$$
we find that $r \cong 0.1236$, that is, the effective rate implied by the bond is about 12.36%. The price of the bond after 75 days will be

$$B(75/365, 1) = B(0,1)(1+r)^{\frac{75}{365}} = 0.89(1+0.1236)^{\frac{75}{365}} \cong 0.9115$$

and the return will be

$$K(0, 75/365) = \frac{B(75/365, 1) - B(0,1)}{B(0,1)} \cong \frac{0.9115 - 0.89}{0.89} \cong 0.0242,$$

about 2.42%.

2.36 The initial price of a 6-month unit bond is $e^{-0.5r}$, where r denotes the implied continuous rate. If the bond is to produce a 7% return over six months, then

$$\frac{1 - e^{-0.5r}}{e^{-0.5r}} = 0.07,$$

which gives $r \cong 0.1353$, or 13.53%.

2.37 The continuous rate implied by the bond satisfies

$$e^{-r} = 0.92.$$

The solution is $r \cong 0.0834$. At time t the bond will be worth $0.92e^{rt}$. It will produce a 5% return at a time t such that

$$\frac{0.92e^{rt} - 0.92}{0.92} = 0.05,$$

which gives $t \cong 0.5851$ years or 213.6 days.

2.38 At time 0 we buy $1/B(0,1) = e^r$ bonds. At time 1 we increase our holdings to $e^r/B(1,2) = e^{2r}$ bonds. Generally, at time n we purchase $e^{(n+1)r}$ one-year bonds.

2.39 Because the bond is trading at par and the interest rate remains constant, the price of the bond at the beginning of each year will be $100. The sum of $1,000 will buy 10 bonds at the beginning of year one. At the end of the year the coupons will pay $10 \times 8 = 80$ dollars, enough to buy 0.8 bonds at $100 each. As a result, the investor will be holding $10 + 0.8 = 10.8$ bonds in year two. At the end of that year the coupons will pay $10.8 \times 8 = 86.4$ dollars and 0.864 additional bonds will be purchased at $100 each. The number of bonds held in year three will be $10.8 + 0.864 = 11.664$, and so on.

In general, the number of bonds held in year n will be

$$10\left(1 + \frac{8}{100}\right)^{n-1},$$

which gives 10.0000, 10.8000, 11.6640, 12.5971, 13.6049 bonds held in years one to five.

Chapter 3

3.1 a) Our initial investment is $wS(0)$, the final value is of the form $S(T) - (1 + r_l)(1 - w)S(0)$, so the return is

$$K_{mp} = \frac{S(T) - (1 + r_l)(1 - w)S(0) - wS(0)}{wS(0)}$$

$$= K_S + \frac{1 - w}{w}\frac{S(T) - (1 + r_l)S(0)}{S(0)},$$

$$\mathbb{E}(K_{mp}) = \mathbb{E}(K_S) + \frac{1 - w}{w}\frac{\mathbb{E}(S(T)) - (1 + r_l)S(0)}{S(0)}.$$

We have additional expected return if $\mathbb{E}(S(T)) > (1 + r_l)S(0)$. The coefficient $\frac{1-w}{w}$ measures the degree of leverage, being large if our financial input is small.

b) We invest $cS(0)$ and receive $S(0)(1+r) - S(T) + cS(0)(1+r_c)$ with return

$$K_{ss} = \frac{S(0)(1+r) - S(T) + cS(0)(1+r_c) - cS(0)}{cS(0)}$$
$$= \frac{1}{c}(-K_S) + \frac{S(0)r + cS(0)r_c}{cS(0)}.$$

In particular, if $r_c = 0$ and $c = 1$, then

$$\mathbb{E}(K_{ss}) = -\mathbb{E}(K_S) + r,$$

which has clear meaning: we gain if the stock goes down.

3.2 In the first investment project

$$\mu_{K_1} = 0.12 \times 0.25 + 0.12 \times 0.75 = 0.12,$$
$$\sigma_{K_1} = \sqrt{(0.12 - 0.12)^2 \times 0.25 + (0.12 - 0.12)^2 \times 0.75} = 0.$$

In the second project

$$\mu_{K_2} = 0.11 \times 0.25 + 0.13 \times 0.75 = 0.125,$$
$$\sigma_{K_2} = \sqrt{(0.11 - 0.125)^2 \times 0.25 + (0.13 - 0.125)^2 \times 0.75} \cong 0.008660.$$

Finally, in the third project

$$\mu_{K_3} = 0.02 \times 0.25 + 0.22 \times 0.75 = 0.17,$$
$$\sigma_{K_3} = \sqrt{(0.02 - 0.17)^2 \times 0.25 + (0.22 - 0.17)^2 \times 0.75} = 0.086603.$$

The first project is the least risky one, in fact, it is risk free. The third project involves the highest risk.

3.3 First we put $K_2^{\omega_2} = x$ and compute

$$\mathrm{Var}(K_1) = 0.001875,$$
$$\mathrm{Var}(K_2) = 0.1875x^2 + 0.015x + 0.0003.$$

The two securities will have the same risk if $\mathrm{Var}(K_1) = \mathrm{Var}(K_2)$. This gives the following equation

$$0.0003 + 0.187\,5x^2 + 0.015x = 0.001875$$

with two solutions $x = -0.14$ or 0.06. This means that $K_2^{\omega_2} = -14\%$ or 6%.

3.4 Let x_1 and x_2 be the number of shares of type 1 and 2 in the portfolio. Then

$$V(T) = x_1 S_1(T) + x_2 S_2(T) = V(0) \left(w_1 \frac{S_1(T)}{S_1(0)} + w_2 \frac{S_2(T)}{S_2(0)} \right)$$
$$= 100 \left(0.25 \times \frac{48}{45} + 0.75 \times \frac{32}{33} \right) = 99.394.$$

3.5 The return on the portfolio is $K_V = w_1 K_1 + w_2 K_2$. This gives

$$K_V^{\omega_1} = 0.30 \times 12\% - 0.7 \times 4\% = 0.8\% \quad \text{in scenario } \omega_1,$$
$$K_V^{\omega_2} = -0.30 \times 10\% + 0.7 \times 7\% = 1.9\% \quad \text{in scenario } \omega_2.$$

3.6 First we find $\mathbb{E}(K_1) = 8\%$ and $\mathbb{E}(K_2) = 11.5\%$. If the expected return on the portfolio is to be $\mathbb{E}(K_V) = 10\%$, then by (3.3) and (3.1) the weights must satisfy the system of equations

$$8w_1 + 11.5w_2 = 10,$$
$$w_1 + w_2 = 1.$$

The solution is $w_1 \cong 42.86\%$ and $w_2 \cong 57.14\%$.

3.7 First, we compute $\mu_1 = 6\%$ and $\mu_2 = 11\%$ from the data in Example 3.13. Next, (3.5) and (3.1) give the system of equations

$$6w_1 + 11w_2 = 50,$$
$$w_1 + w_2 = 1,$$

for the weights w_1 and w_2. The solution is $w_1 = -7.8$ and $w_2 = 8.8$. Finally, we use (3.6) with the values $\sigma_1^2 \cong 0.0144$, $\sigma_2^2 \cong 0.0254$ and $\rho_{12} \cong -0.6065$ computed in Example 3.13 to find the risk of the portfolio:

$$\sigma_V^2 \cong (-7.8)^2 \times 0.0144 + (8.8)^2 \times 0.0254$$
$$+2 \times (-7.8) \times 8.8 \times (-0.6065) \times \sqrt{0.0144} \times \sqrt{0.0254}$$
$$\cong 4.4354.$$

3.8 The returns on risky securities are non-constant random variables, that is, $K_1^{\omega_1} \neq K_1^{\omega_2}$ and $K_2^{\omega_1} \neq K_2^{\omega_2}$. Because of this, the system of equations

$$K_1^{\omega_1} = aK_2^{\omega_1} + b,$$
$$K_1^{\omega_2} = aK_2^{\omega_2} + b,$$

must have a solution $a \neq 0$ and b. It follows that $K_1 = aK_2 + b$.
  Now, use the properties of covariance and variance to compute

$$\text{Cov}(K_1, K_2) = \text{Cov}(aK_2 + b, K_2) = a\text{Cov}(K_2, K_2) = a\text{Var}(K_2) = a\sigma_2^2,$$
$$\sigma_1^2 = \text{Var}(K_1) = \text{Var}(aK_2 + b) = a^2\text{Var}(K_2) = a^2\sigma_2^2.$$

It follows that $\sigma_1 = |a|\,\sigma_2$ and

$$\rho_{12} = \frac{\text{Cov}(K_1, K_2)}{\sigma_1\sigma_2} = \frac{a\sigma_2^2}{|a|\,\sigma_2^2} = \pm 1.$$

3.9 Using the values $\sigma_1^2 \cong 0.0144$, $\sigma_2^2 \cong 0.0254$ and $\rho_{12} \cong -0.6065$ computed in Example 3.13, we find s_0 from (3.10):

$$s_0 = \frac{\sigma_2^2 - \rho_{12}\sigma_1\sigma_2}{\sigma_1^2 + \sigma_2^2 - 2\rho_{12}\sigma_1\sigma_2} \cong 0.5873.$$

This means that the weights in the portfolio with minimum risk are $w_1 \cong 0.5873$ and $w_2 \cong 0.4127$, and the portfolio involves no short selling.

3.10 $\mu_V = 0.06$, $\sigma_V \cong 0.1013$.

3.11 The weights of the three securities in the minimum variance portfolio are $\mathbf{w} \cong \begin{bmatrix} 0.228 & 0.235 & 0.537 \end{bmatrix}$, the expected return on the portfolio is $\mu_V \cong 0.167$ and the standard deviation is $\sigma_V \cong 0.152$.

3.12 The weights in the portfolio with the minimum variance among all feasible portfolios with expected return $\mu_V = 20\%$ are $\mathbf{w} \cong [\ 0.722 \quad -0.208 \quad 0.486\ ]$. The standard deviation of this portfolio is $\sigma_V \cong 0.211$.

3.13 The weights and standard deviations of portfolios along the minimum variance line, parameterised by the expected return μ_V, are

$$\mathbf{w} \cong [\ -2.314 + 15.180\mu_V \quad 2.515 - 13.615\mu_V \quad 0.799 - 1.566\mu_V\ ],$$

$$\sigma_V = \sqrt{0.584 - 6.696\mu_V + 19.996\mu_V^2}\ .$$

This minimum variance line is presented in Figure S.3, along with the set of feasible portfolios with short selling (light shading) and without (darker shading).

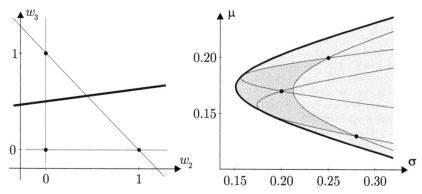

Figure S.3 Minimum variance line and feasible portfolios on the w_2, w_3 and σ, μ planes

3.14 The market portfolio weights are $\mathbf{w} \cong [\ 0.438 \quad 0.012 \quad 0.550\ ]$. The expected return on this portfolio is $\mu_M \cong 0.183$ and the standard deviation is $\sigma_M \cong 0.156$.

3.15 Let $\mathbf{m}$ be the one-row matrix formed by the expected returns of the three securities. By multiplying the equality $\gamma\mathbf{w}\mathbf{C} = \mathbf{m} - \mu\mathbf{u}$ by $\mathbf{C}^{-1}\mathbf{u}^T$ and, respectively, $\mathbf{C}^{-1}\mathbf{m}^T$, we get

$$\mu_V(\mathbf{m} - \mu\mathbf{u})\mathbf{C}^{-1}\mathbf{u}^T = (\mathbf{m} - \mu\mathbf{u})\mathbf{C}^{-1}\mathbf{m}^T,$$

since $\mathbf{w}\mathbf{u}^T = 1$ and $\mathbf{w}\mathbf{m}^T = \mu_V$. This can be solved for μ to get

$$\mu = \frac{\mathbf{m}\mathbf{C}^{-1}(\mathbf{m}^T - \mu_V^T\mathbf{u})}{\mathbf{u}\mathbf{C}^{-1}(\mathbf{m}^T - \mu_V^T\mathbf{u})} \cong 0.142.$$

Then, γ can be computed as follows:

$$\gamma = (\mathbf{m} - \mu\mathbf{u})\mathbf{C}^{-1}\mathbf{u}^T \cong 1.367.$$

3.16 The return on the portfolio can be expressed as $K_V = w_1 K_1 + \cdots + w_n K_n$ in terms of the returns on the individual securities. Because covariance is linear in each of its arguments,

$$\beta_V = \frac{\mathrm{Cov}(K_V, K_M)}{\sigma_M^2} = w_1 \frac{\mathrm{Cov}(K_1, K_M)}{\sigma_M^2} + \cdots + w_n \frac{\mathrm{Cov}(K_n, K_M)}{\sigma_M^2}$$

$$= w_1\beta_1 + \cdots + w_n\beta_n.$$

3.17 Taking expectation on both sides of (3.28) gives $\mathbb{E}(\varepsilon_V) = 0$. For the second claim use (3.28) to get

$$\begin{aligned}
\mathrm{Cov}(K_M, \varepsilon_V) &= \mathrm{Cov}(K_M, (K_V - \mu_V) - \beta_V(K_M - \mu_M)) \\
&= \mathrm{Cov}(K_M, K_V - \beta_V K_M) \\
&= \mathrm{Cov}(K_M, K_V) - \beta_V \mathrm{Cov}(K_M, K_M)
\end{aligned}$$

where we employed linearity of covariance with respect to each variable and the fact that the covariance between a random variable and a constant is zero. Now inserting the definition of β_V and using $\mathrm{Cov}(K_M, K_M) = \sigma_M^2$ gives the result.

Chapter 4

4.1 Yes, there is an arbitrage opportunity. We enter into a long forward contract and sell short one share, investing 70% of the proceeds at 8% and paying the remaining 30% as a security deposit to attract interest at 4%. At the time of delivery the cash investments plus interest will be worth about $18.20, out of which $18 will be paid for one share to close the short position in stock. This leaves an arbitrage profit of $0.20.

The rate d for the security deposit such that there is no arbitrage opportunity should satisfy

$$30\% \times 17 \times e^d + 70\% \times 17 \times e^{8\%} \le 18.$$

The highest such rate is $d \cong 0.1740\%$.

4.2 We take 1 January 2000 to be time 0. By (4.1)

$$F\left(0, \frac{3}{4}\right) = S(0)e^{\frac{3}{4} \times 6\%}, \qquad F\left(\frac{1}{4}, \frac{3}{4}\right) = 0.9 \times S(0)e^{\frac{2}{4} \times 6\%}.$$

It follows that the forward price drops by

$$\frac{F(0, \frac{3}{4}) - F(\frac{1}{4}, \frac{3}{4})}{F(0, \frac{3}{4})} = \frac{e^{\frac{3}{4} \times 6\%} - 0.9 e^{\frac{1}{2} \times 6\%}}{e^{\frac{3}{4} \times 6\%}} \cong 11.34\%.$$

4.3 The present value of the dividends is

$$div_0 = 1 \times e^{-\frac{6}{12} \times 12\%} + 2 \times e^{-\frac{9}{12} \times 12\%} \cong 2.77$$

dollars. Then

$$[S(0) - div_0]e^{rT} \cong (120 - 2.77)e^{\frac{10}{12} \times 12\%} \cong 129.56$$

dollars, which is less than the quoted forward price of $131. As a result, there will be an arbitrage opportunity, which can be realised as follows:

- on 1 January 2000 enter into a short forward position and borrow $120 to buy stock;
- on 1 July 2000 collect the first dividend of $1 and invest it risk free;
- on 1 October 2000 collect the second dividend of $2 and invest it risk free;
- on 1 November 2000 close out all positions.

You will be left with an arbitrage profit of

$$131 - 120e^{\frac{10}{12} \times 12\%} + 1e^{\frac{4}{12} \times 12\%} + 2e^{\frac{1}{12} \times 12\%} \cong 1.44$$

dollars.

4.4 It is not possible to realise arbitrage in these circumstances. Though the theoretical no-arbitrage forward price is about \$87.83, the first strategy in the proof of Theorem 4.4 brings a loss of $89 - 83e^{10\%} + 2e^{0.5 \times 7\%} \cong -0.66$ dollars and the second one results in a loss of $-89 + 83e^{7\%} - 2e^{0.5 \times 10\%} \cong -2.08$ dollars.

4.5 The euro plays the role of the underlying asset with dividend yield 3%. Hence the forward price (the exchange rate) is

$$F(0, 1/2) = 0.9834e^{0.5(4\% - 3\%)} \cong 0.9883$$

euros to a dollar.

4.6 By (4.2) the initial forward price is $F(0, 1) \cong 45.72$ dollars. This takes into account the dividend paid at time $1/2$.
  a) If $S(9/12) = 49$ dollars, then $F(9/12, 1) \cong 49.74$ dollars by (4.1). It follows by (4.6) that $V(9/12) \cong 3.96$ dollars.
  b) If $S(9/12) = 51$ dollars, then $F(9/12) \cong 51.77$ dollars and $V(9/12) \cong 5.96$ dollars.

4.7 Since $f(t, T) = S(t)e^{r(T-t)}$, the random variables $S(t)$ and $f(t, T)$ are perfectly correlated with $\rho_{S(t)f(t,T)} = 1$ and $\sigma_{f(t,T)} = e^{r(T-t)}\sigma_{S(t)}$. It follows that $N = e^{-r(T-t)}$.

4.8 Observe that Theorem 4.8 on the equality of futures and forward prices applies also in the case of an asset with dividends paid continuously. We can, therefore, use (4.4) to obtain

$$r_{\text{div}} = 8\% - \frac{1}{0.75} \ln \frac{14,100}{13,500} \cong 2.20\%.$$

4.9 The return on the index will be 3.37%. For $R = 1\%$ this gives the futures prices $f(0, 3t) \cong 916.97$ and $f(t, 3t) \cong 938.49$. If the beta factor for a portfolio is $\beta_V = 1.5$, then the expected return on this portfolio will be $\mu_V \cong 4.56\%$. To construct a portfolio with $\beta_{\tilde{V}} = 0$ and initial value $V(0) = 100$ dollars, supplement the original portfolio with $N \cong 0.1652$ short futures contracts (observe that N is the same as in Example 4.18).

If the actual return on the original portfolio during the first time step happens to be equal to the expected return, then its value after one step will be $V(t) \cong 104.56$ dollars. Marking to market requires the holder of $N \cong 0.1652$ short forward contracts to pay \$3.56. As a result, after one step the value of the portfolio with forward contracts will be $\tilde{V}(t) \cong 104.56 - 3.56 = 101.00$ dollars, once again matching the risk-free growth.

Chapter 5

5.1 The investment will bring a profit of

$$(36 - S(T))^+ - 4.50e^{0.12 \times \frac{3}{12}} = 3,$$

where $S(T)$ is the stock price on the exercise date. This gives $S(T) \cong 28.36$ dollars.

5.2 $\mathbb{E}((S(T) - 90)^+ - 8e^{0.09 \times \frac{6}{12}}) \cong -5.37$ dollars.

5.3 By put-call parity $2.83 - P_E = 15.60 - 15.00e^{-\frac{3}{12} \times 0.0672}$, so $P_E \cong 1.98$ dollars.

5.4 Put-call parity is violated, $5.09 - 7.78 > 20.37 - 24e^{-0.0748 \times \frac{6}{12}}$. Arbitrage can be realised as in the first part of the proof of Theorem 5.2. At time 0:

- buy a share for $20.37;
- buy a put option for $7.78;
- write and sell a call option for $5.09;
- borrow $23.06 at the interest rate of 7.48%.

The balance of these transactions is zero. After six months:

- sell one share for $24 by exercising the put option or settling the short position in calls, depending on whether the share price turns out to be below or above the strike price;
- repay the loan with interest, amounting to $23.06e^{\frac{1}{2} \times 0.0748} \cong 23.94$ dollars in total.

The balance of $24 - 23.96 = 0.06$ dollars will be your arbitrage profit.

5.5 If $C_E - P_E > S(0) - \text{div}_0 - Xe^{-rT}$, then at time 0 buy a share and a put option, write and sell a call option, and invest (or borrow, if negative) the balance on the money market at the interest rate r. As soon as you receive the dividend, invest it at the rate r. At the exercise time T close the money market investment and sell the share for X, either by exercising the put if $S(T) < X$, or by settling the call if $S(T) \geq X$. The final balance $(C_E - P_E - S(0) + \text{div}_0)e^{rT} + X > 0$ will be your arbitrage profit.

On the other hand, if $C_E - P_E < S(0) - \text{div}_0 - Xe^{-rT}$, then at time 0 sell short a share, write and sell a put, and buy a call option, investing the balance on the money market. When a dividend becomes due on the shorted share, borrow the amount and pay it to the owner of the stock. At time T close the money market position, buy a share for X by exercising the call if $S(T) > X$ or settling the put if $S(T) \leq X$, and close the short position in stock. Your arbitrage profit will be $(-C_E + P_E + S(0) - \text{div}_0)e^{rT} - X > 0$.

5.6 If $C_E - P_E < S(0)e^{-r_{\text{div}}T} - Xe^{-rT}$, then at time 0 sell short $e^{-r_{\text{div}}T}$ of a share, write and sell a put, and buy a call option, investing the balance at the rate r. Between time 0 and T pay dividends to the stock owner, raising cash by shorting the stock. This will lead to one shorted share held at time T. If the put option is exercised at time T, you will have to buy a share for X. Use this share to close the short position in stock. You will be left with a call option and a positive amount $(-C_E + P_E + S(0)e^{-r_{\text{div}}T} - Xe^{-rT})e^{rT} > 0$. If the put option is not exercised at all, then you can use the call to buy a share for X at time T, closing the short position in stock. You will also be left with a positive final balance $(-C_E + P_E + S(0)e^{-r_{\text{div}}T} - Xe^{-rT})e^{rT} > 0$.

On the other hand, if $C_E - P_E > S(0)e^{-r_{\text{div}}T} - Xe^{-rT}$, then the opposite strategy will also lead to arbitrage.

5.7 The strike price is equal to the forward price (more precisely, the exchange rate) of 0.9883 euros to a dollar computed in Solution 4.5.

5.8 If $S(0) - Xe^{-rT} < C_A - P_A$, then write and sell a call, buy a put, and buy a share, investing (or borrowing, if negative) the balance at the rate r. Now the same argument as in the first part of the proof of Theorem 5.4 applies, except that the arbitrage profit may also include the dividend if the call is exercised after the dividend becomes due. (Nevertheless, the dividend cannot be included in this inequality because the option may be exercised and the share sold before the dividend is due.)

If $C_A - P_A < S(0) - \text{div}_0 - X$, then at time 0 sell short a share, write and sell a put, and buy a call option, investing the balance at the rate r. If the put is exercised at time $t < T$, you will have to buy a share for X, borrowing the amount at the rate r. As the dividend becomes due, borrow the amount at the rate r and pay it to the owner of the share. At time T return the share to the owner, closing the short sale. You will be left with the call option and a positive amount $(S(0)+P_A-C_A-\text{div}_0)\mathrm{e}^{rT}-X\mathrm{e}^{r(T-t)} > X\mathrm{e}^{rT}-X\mathrm{e}^{r(T-t)} \geq 0$. (If the put is exercised before the dividend becomes due, you can increase your arbitrage profit by closing out the short position in stock immediately, in which case you would not have to pay the dividend.) If the put is not exercised before expiry T, then the second part of Solution 5.5 applies.

5.9 If $S(0) - X\mathrm{e}^{-rT} < C_A - P_A$, then use the same strategy as in the first part of the proof of Theorem 5.4. The resulting arbitrage profit will in fact be increased by the dividends accumulated up to the time when the call option is exercised.

If $C_A - P_A < S(0)\mathrm{e}^{-r_{\text{div}}T} - X$, then at time 0 sell short $\mathrm{e}^{-r_{\text{div}}T}$ of a share, write and sell a put, and buy a call option, investing the balance at the rate r. Between time 0 and T pay dividends to the stock owner, raising cash by shorting the stock. This will lead to one shorted share held at time T. If the put option is exercised at time $t \leq T$, you will have to buy a share for X, borrowing this amount at the rate r. At time T use this share to close the short position in stock. You will be left with a call option and a positive amount $(-C_A+P_A+S(0)\mathrm{e}^{-r_{\text{div}}T})\mathrm{e}^{rT}-X\mathrm{e}^{r(T-t)} \geq (-C_A+P_A+S(0)\mathrm{e}^{-r_{\text{div}}T}-X)\mathrm{e}^{rT} > 0$ plus any dividends accumulated since the share was bought at time t. If the put option is not exercised at all, then you can use the call to buy a share for X at time T, closing the short position in stock. You will also be left with a positive final balance $(-C_A + P_A + S(0)\mathrm{e}^{-r_{\text{div}}T})\mathrm{e}^{rT} - X > (-C_A + P_A + S(0)\mathrm{e}^{-r_{\text{div}}T} - X)\mathrm{e}^{rT} > 0$.

5.10 If $C_E > C_A$, then write and sell a European call and purchase an American call. The difference $C_E - C_A$ will be your arbitrage income. To keep it, do not exercise the American option until expiry, when either both options will turn out worthless, or the loss from settling the European call will be recovered by exercising the American call. The argument for put options is similar.

5.11 Suppose that $C_E \geq S(0) - \text{div}_0$. In this case, write and sell a call option and buy stock, investing the balance on the money market. As soon as you receive the dividends, also invest them on the money market. On the exercise date you can sell the stock for at least $\min(S(T), X)$, settling the call option. Your final wealth will be positive, $(C_E - S(0) + \text{div}_0)\mathrm{e}^{rT} + \min(S(T), X) > 0$. This proves that $C_E < S(0) - \text{div}_0$.

The remaining bounds follow by the put-call parity relation (5.5) for dividend-paying stock: $S(0) - \text{div}_0 - X\mathrm{e}^{-rT} \leq C_E$, since $P_E \geq 0$; $-S(0) + \text{div}_0 + X\mathrm{e}^{-rT} \leq P_E$, since $C_E \geq 0$; and $P_E < X\mathrm{e}^{-rT}$, since $C_E < S(0) - \text{div}_0$.

5.12 For dividend-paying stock the regions determined by the bounds on call and put prices in Proposition 5.5 are shown as shaded areas in Figure S.4.

5.13 If $C_A \geq S(0)$, then buy a share, write and sell an American call and invest the balance $C_A - S(0)$ without risk. If the option is exercised before or at expiry, then settle your obligation by selling the share for X. If the option is not exercised at all, you will end up with the share worth $S(T)$ at expiry. In either case the final cash value of this strategy will be positive. The final

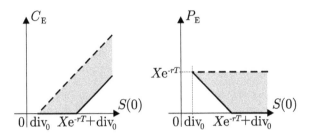

Figure S.4 Bounds on European call and put prices for dividend-paying stock

balance will in fact also include the dividend collected, unless the option is exercised before the dividend becomes due.

5.14 Suppose that $X' < X''$, but $C_E(X') < C_E(X'')$. We can write and sell a call with strike price X'' and buy a call with strike price X', investing the difference $C_E(X'')-C_E(X') > 0$ without risk. If the option with strike price X'' is exercised at expiry, we will need to pay $(S(T) - X'')^+$. This amount can be raised by exercising the option with strike price X' and cashing the payoff $(S(T) - X')^+$. Since $X' < X''$, it follows that $(S(T) - X')^+ \geq (S(T) - X'')^+$, so an arbitrage profit will be realised.

The inequality for puts follows by a similar arbitrage argument.

5.15 Consider four cases:

1) If $S(T) \leq X' < X < X''$, then (5.9) reduces to $0 \leq 0$.

2) If $X' < S(T) \leq X < X''$, then (5.9) becomes $0 \leq \alpha(S(T) - X')$, obviously satisfied for $X' < S(T)$.

3) If $X' < X < S(T) \leq X''$, then (5.9) can be written as $S(T) - X \leq \alpha(S(T) - X')$. This holds because $X = \alpha X' + (1 - \alpha)X''$ and $S(T) \leq X''$.

4) Finally, if $X' < X < X'' < S(T)$, then (5.9) becomes an equality, $S(T) - X = \alpha(S(T) - X') + (1 - \alpha)(S(T) - X'')$ because $X = \alpha X' + (1 - \alpha)X''$.

5.16 Suppose that $P_E(S') < P_E(S'')$ for some $S' < S''$, where $S' = x'S(0)$ and $S'' = x''S(0)$. Write and sell a put option on a portfolio with x'' shares and buy a put option on a portfolio with x' shares, investing the balance $P_E(S'')-P_E(S') > 0$ without risk. If the written option is exercised at time T, then the liability can be met by exercising the other option. Since $x' < x''$, the payoffs satisfy $(X - x'S(T))^+ \geq (X - x''S(T))^+$. It follows that this is an arbitrage strategy.

5.17 Suppose that $X' < X''$, but $C_A(X') < C_A(X'')$. We can write and sell the call option with strike price X'' and purchase the call option with strike price X', investing the balance $C_A(X'')-C_A(X') > 0$ without risk. If the written option is exercised at time $t \leq T$, we will have to pay $(S(t) - X'')^+$. This liability can be met by exercising the other option immediately, receiving the payoff $(S(t) - X')^+$. Since $(S(t) - X'')^+ \leq (S(t) - X')^+$, this strategy leads to arbitrage.

The inequality for put options can be proved in a similar manner.

5.18 We shall prove Proposition 5.29 for American put options. The argument for European puts is similar. By Proposition 5.22 $P_A(S)$ is a decreasing function of S. When $S \geq X$, the intrinsic value of a put option is zero, and so the time value, being equal to $P_A(S)$, is also a decreasing function of S. On the other hand, $P_A(S') - P_A(S'') \leq S'' - S'$ for any $S' < S''$ by Proposition 5.23. This implies that $P_A(S') - (X - S')^+ \leq P_A(S'') - (X - S'')^+$ if $S' < S'' \leq X$, so the time value is an increasing function of S for $S \leq X$. As a result, the time value has a maximum at $S = X$.

Chapter 6

6.1 Let us compute the derivative of the price $C_E(0)$ of a call option with respect to U. The formula for the price, assuming that $S^d < X < S^u$, is

$$C_E(0) = \frac{1}{1+R} \frac{R-D}{U-D} [S(0)(1+U) - X].$$

The derivative with respect to U is equal to

$$\frac{(R-D)[X - S(0)(1+D)]}{(1+R)(U-D)^2} = \frac{(R-D)[X - S^d]}{(1+R)(U-D)^2}.$$

This is positive in our situation, since $R > D$ and $X > S^d$, so $C_E(0)$ increases as U increases.

The derivative of $C_E(0)$ with respect to D is equal to

$$-\frac{(U-R)[S(0)(1+U) - X]}{(1+R)(U-D)^2} = -\frac{(U-R)[S^u - X]}{(1+R)(U-D)^2},$$

which is negative, since $R < U$ and $X < S^u$. The option price decreases as D increases.

6.2 If $R = 0$ and $S(0) = X = 1$, then $C_E(0) = \frac{-UD}{U-D}$. For $U = 0.05$ and $D = -0.05$ we have $U - D = 0.1$ and $C_E(0) = 0.025$ dollars. However, if $U = 0.01$ and $D = -0.19$, then $U - D = 0.2$. The variance of the return on stock is equal to $(U - D)^2 p(1 - p)$ and is higher in the latter case, but the option price is lower: $C_E(0) = 0.0095$ dollars.

6.3 To replicate a call option the writer needs to buy stock initially and sell it when the option is exercised. The following system of equations needs to be solved to find the replicating portfolio:

$$\begin{cases} 110(1 - c)x + 1.05y = 10, \\ 90(1 - c)x + 1.05y = 0. \end{cases}$$

For $c = 2\%$ we obtain $x \cong 0.5102$ and $y \cong -42.8471$, so the initial value of the replicating portfolio will be $100x + y \cong 8.1633$ dollars. If $c = 0$, then the replicating portfolio will be worth 7.1429 dollars. Note that the money market position y is the same for each c.

6.4 The borrowing rate should be used to replicate a call, since the money market position will be negative. This gives $x(1) \cong 0.6667$ and $y(1) \cong -40.1786$, so the replicating portfolio value is 9.8214 dollars. For a put we obtain $x(1) \cong -0.3333$ and $y(1) \cong 27.7778$ using the rate for deposits, so the replicating portfolio will be worth 2.7778 dollars initially.

The results are consistent with the put and call prices obtained from Proposition 6.9 with the appropriate risk-neutral probability (computed using the corresponding interest rate), $p_* \cong 0.7333$ for a call and $p_* \cong 0.6$ for a put.

6.5 The option price at time 0 is 22.92 dollars. In addition to this amount, the option writer should borrow 74.05 dollars and buy 0.8081 of a share. At time 1, if $S(1) = 144$, then the amount of stock held should be increased to 1 share, the purchase being financed by borrowing a further 27.64 dollars, increasing the total amount of money owed to 109.09 dollars. If, on the other hand, $S(1) = 108$ dollars at time 1, then some stock should be sold to reduce the number of shares held to 0.2963, and 55.27 dollars should be repaid, reducing the amount owed to 26.18 dollars. (In either case the amount owed at time 1 includes interest of 7.40 dollars on the amount borrowed at time 0.)

6.6 At time 1 the stock prices $S^{u} = 144$ and $S^{d} = 108$ dollars (the so-called cum-dividend prices) drop by the amount of the dividend to 129 and 93 dollars (the so-called ex-dividend prices). These prices generate the prices at time 2, hence $S^{uu} = 154.80$, $S^{ud} = 116.10$, $S^{du} = 111.60$ and $S^{dd} = 83.70$ dollars. The option will be exercised with payoff 34.80 dollars if the stock goes up twice. In the remaining scenarios the payoff will be zero. The option value at time 1 will be 21.09 dollars in the up state and zero dollars in the down state. The replicating portfolio constructed at time 1 and based on ex-dividend prices consists of 0.8992 shares and a loan of 94.91 dollars in the up state, and no shares and no money market position in the down state. The option price at time 0 is 12.78 dollars. Replication (based on the cum-dividend prices at time 1, since the dividend is available to the owner of the stock purchased at time 0) involves buying 0.5859 shares and borrowing 57.52 dollars.

6.7 From the Cox–Ross–Rubinstein formula $C_E(0) \cong 5.93$ dollars, $P_E(0) \cong 7.76$ dollars.

6.8 The least integer m such that $S(0)(1+u)^{m}(1+d)^{N-m} > X$ is $m = 35$. From the Cox–Ross–Rubinstein formula we obtain $C_E(0) \cong 3.4661$ dollars.

6.9 The stock values are

n	0	1	2	3
				79.86
			72.60 <	
		66.00 <		68.97
$S(n)$	60.00 <		62.70 <	
		57.00 <		59.57
			54.15 <	
				51.44

The American put prices are

n	0	1	2	3
				0.00
			0.00 <	
		0.50 <		0.00
$P_A(n)$	2.52 <		1.10 <	
		5.00 <		2.43
			7.85 <	
				10.56

The option will be exercised early in the d node at time 1 or in the dd node at time 2 (bold figures).

6.10 The values of the European and American calls are the same,

n	0	1	2
			52.80
		34.91 <	
$C_E(n) = C_A(n)$	22.92 <		9.60
		5.82 <	
			0.00

At time 2 both options have, of course, the same payoff. At time 1 the American call will not be exercised in the up state, as this would bring only 24 dollars, which is less than the value of holding the option until expiry. In the down state the American call will be out of the money and will not be exercised either. Similarly, it will not be exercised at time 0. As a result, the American call is equivalent to its European counterpart.

6.11 The ex-dividend stock prices are

n	0	1	2
			12.32
		11.20 <	
			10.64
$S(n)$ ex-div	12.00		
		9.40 <	10.34
			8.93

The corresponding European and American put prices will be

n	0	1	2
			1.68
			1.68
		2.53	
		2.80 <	
			3.36
			3.36
$P_E(n)$	3.42		
$P_A(n)$	3.69		
			3.66
			3.66
		4.33	
		4.60 <	
			5.07
			5.07

At time 1 the payoff of the American put option in both the up and down states will be higher than the value of holding the option to expiry, so the option should be exercised in these states (indicated by bold figures).

6.12 The risk-neutral expected return at each step is

$$\mathbb{E}_*(K(2)) = \mathbb{E}_*(K(3)) = p_*U + (1 - p_*)D = R,$$

equal to the risk-free rate. It follows that

$$\mathbb{E}_*(\widetilde{S}(2)|\mathcal{P}_1) = \frac{S(1)(1 + \mathbb{E}_*(K(2)))}{(1 + R)^2} = \frac{S(1)(1 + R)}{(1 + R)^2} = \widetilde{S}(1),$$

$$\mathbb{E}_*(\widetilde{S}(3)|\mathcal{P}_2) = \frac{S(2)(1 + \mathbb{E}_*(K(3)))}{(1 + R)^3} = \frac{S(2)(1 + R)}{(1 + R)^3} = \widetilde{S}(2).$$

6.13 At time 0 the hedging strategy should start with

$$x(1) = \frac{C_{\mathrm{E}}^{\mathrm{u}}(1) - C_{\mathrm{E}}^{\mathrm{d}}(1)}{S^{\mathrm{u}}(1) - S^{\mathrm{d}}(1)} = 0.8301,$$

$$y(1) = C_{\mathrm{E}}(0) - x(1)S(0) = -59.7045.$$

If the stock price turns out to be $S^{\mathrm{d}}(1)$ at time 1, the portfolio needs to be rebalanced to

$$x^{\mathrm{d}}(2) = \frac{C_{\mathrm{E}}^{\mathrm{du}}(2) - C_{\mathrm{E}}^{\mathrm{dd}}(2)}{S^{\mathrm{du}}(2) - S^{\mathrm{dd}}(2)} = 0.5522,$$

$$y^{\mathrm{d}}(2) = y(1) + \frac{1}{1+R}(x(1) - x^{\mathrm{d}}(2))S^{\mathrm{d}}(1) = -36.9647.$$

Finally, if the stock price turns out to be $S^{\mathrm{du}}(2)$ at time 2, the portfolio needs to be rebalanced to

$$x^{\mathrm{du}}(3) = \frac{C_{\mathrm{E}}^{\mathrm{duu}}(3) - C_{\mathrm{E}}^{\mathrm{dud}}(3)}{S^{\mathrm{duu}}(3) - S^{\mathrm{dud}}(3)} = 0.7593,$$

$$y^{\mathrm{du}}(3) = y^{\mathrm{d}}(2) + \frac{1}{(1+R)^2}(x^{\mathrm{d}}(2) - x^{\mathrm{du}}(3))S^{\mathrm{du}}(2) = -55.4470.$$

6.14 At time 0

$$x(1) = \frac{P_{\mathrm{A}}^{\mathrm{u}}(1) - P_{\mathrm{A}}^{\mathrm{d}}(1)}{S^{\mathrm{u}}(1) - S^{\mathrm{d}}(1)} = -0.3248,$$

$$y(1) = P_{\mathrm{A}}(0) - x(1)S(0) = 35.6624.$$

If the stock price turns out to be $S^{\mathrm{u}}(1)$ at time 1, the portfolio needs to be rebalanced to

$$x^{\mathrm{u}}(2) = \frac{P_{\mathrm{A}}^{\mathrm{uu}}(2) - P_{\mathrm{A}}^{\mathrm{ud}}(2)}{S^{\mathrm{uu}}(2) - S^{\mathrm{ud}}(2)} = -0.0236,$$

$$y^{\mathrm{u}}(2) = y(1) + \frac{1}{1+R}(x(1) - x^{\mathrm{u}}(2))S^{\mathrm{u}}(1) = 2.8049.$$

If the stock price turns out to be $S^{\mathrm{ud}}(2)$ at time 2, the portfolio needs to be rebalanced to

$$x^{\mathrm{ud}}(3) = \frac{P_{\mathrm{A}}^{\mathrm{udu}}(3) - P_{\mathrm{A}}^{\mathrm{udd}}(3)}{S^{\mathrm{udu}}(3) - S^{\mathrm{udd}}(3)} = -0.0864,$$

$$y^{\mathrm{ud}}(3) = y^{\mathrm{u}}(2) + \frac{1}{(1+R)^2}\left(x^{\mathrm{u}}(2) - x^{\mathrm{ud}}(3)\right)S^{\mathrm{ud}}(2) = 8.4147.$$

For scenario udd the holder should wait till expiry before exercising the option.

Chapter 7

7.1 If $R \geq U$ (or $R \leq D$), then investing risk free and shorting stock (or investing in stock the funds borrowed risk free) clearly leads to arbitrage, so no-arbitrage implies $D < R < U$. For the inverse implication assume this inequality, consider a portfolio (x, y) with zero initial value, so that $y = -xS(0)$ (assuming $A(0) = 1$). Now at time 1 we have $xS(0)(1+U) - xS(0)(1+R) > 0$, $xS(0)(1+D) - xS(0)(1+R) < 0$, so arbitrage is excluded.

7.2 Equation (7.1) can be transformed to

$$p_*(U - D) = -q_*(M - D) + R - D$$

and the existence of $p_* \in (0, 1)$ is guaranteed by

$$0 < \frac{-q_*(M - D) + R - D}{U - D} < 1,$$

which leads to

$$\frac{R - U}{M - D} < q_* < \frac{R - D}{M - D}.$$

We can find suitable solutions q_* (meaning $q_* \in (0, 1)$) since $\frac{R-D}{M-D} > 0$ and $\frac{R-U}{M-D} < 0$.

7.3 Let $p_*, q_*, 1 - p_* - q_*$ be the probabilities of upward, middle and downward price movements, respectively. Condition (7.1) implies that $0.2p_* - 0.1(1 - p_* - q_*) = 0$, that is, $q_* = 1 - 3p_*$ and $1 - p_* - q_* = 2p_*$. Observe that $p_*, 1 - 3p_*, 2p_* \in (0, 1)$ if and only if $p_* \in (0, 1/3)$. It follows that $p_*, q_*, 1 - p_* - q_*$ are risk-neutral probabilities if and only if $q_* = 1 - 3p_*$ and $p_* \in (0, 1/3)$.

7.4 First we show that, in general, for any two non-constant random variables X, Y defined on a two-element probability space, $\Omega = \{u, d\}$ say, there are real numbers a, b such that $Y = aX + b$. The postulated equality gives two equations

$$\begin{cases} Y^u = aX^u + b, \\ Y^d = aX^d + b, \end{cases}$$

which can be solved to give $a = \frac{Y^u - Y^d}{X^u - X^d}$, $b = Y^u - aX^u$. Next, we compute the covariance

$$\mathrm{Cov}(X, Y) = \mathrm{Cov}(X, aX + b) = a\mathrm{Var}(X),$$

hence

$$\rho_{XY} = \frac{\mathrm{Cov}(X, Y)}{\sqrt{\mathrm{Var}(X)\mathrm{Var}(Y)}} = \frac{a\mathrm{Var}(X)}{\sqrt{\mathrm{Var}(X)\mathrm{Var}(aX + b)}} = \frac{a\mathrm{Var}(X)}{\sqrt{a^2\mathrm{Var}^2(X)}}$$
$$= \frac{a}{|a|} = \pm 1.$$

7.5 The determinants of both systems of equations (7.2) and (7.3) are the same up to a non-zero constant, namely

$$\det \begin{bmatrix} S_1^u(1) & S_2^u(1) & A(0)(1 + R) \\ S_1^m(1) & S_2^m(1) & A(0)(1 + R) \\ S_1^d(1) & S_2^d(1) & A(0)(1 + R) \end{bmatrix}$$
$$= S_1(0)S_2(0)A(0)(1 + R)[(U_1 - D_1)(M_2 - D_2) - (M_1 - D_1)(U_2 - D_2)]$$

and

$$\det \begin{bmatrix} U_1 - D_1 & M_1 - D_1 \\ U_2 - D_2 & M_2 - D_2 \end{bmatrix} = (U_1 - D_1)(M_2 - D_2) - (M_1 - D_1)(U_2 - D_2),$$

so one system has a unique solution if and only if the other one has this property.

7.6 Solving the system (7.3), we get $p_* = 0.125$, $q_* = 0.525$. The resulting triple represents genuine risk-neutral probabilities.

7.7 We can use the formulae in the proof of Proposition 7.16 to find

$$y(1) = \frac{200 - 35.24 \times 60 - 24.18 \times 20}{100} \cong -23.98,$$

$$V(1) \cong 35.24 \times 65 + 24.18 \times 15 - 23.98 \times 110 \cong 15.50,$$

$$y(2) = \frac{15.50 + 40.50 \times 65 - 10.13 \times 15}{110} \cong 22.69,$$

$$V(2) \cong -40.50 \times 75 + 10.13 \times 25 + 22.69 \times 121 \cong -38.60.$$

7.8 For a one-step strategy admissibility reduces to a couple of inequalities, $V(0) \geq 0$ and $V(1) \geq 0$. The first inequality can be written as

$$10x + 10y \geq 0.$$

The second inequality means that both values of the random variable $V(1)$ should be non-negative, which gives two more inequalities to be satisfied by x and y,

$$13x + 11y \geq 0,$$
$$9x + 11y \geq 0.$$

The set of portfolios (x, y) satisfying all these inequalities is shown in Figure S.5.

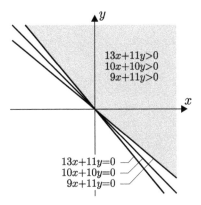

Figure S.5 Admissible portfolios in Exercise 7.8

7.9 Suppose that there is a self-financing predictable strategy with initial value $V(0) = 0$ and final value $V(2) \geq 0$, such that $V(1) < 0$ with positive probability. The last inequality means that this strategy is not admissible, but we shall construct an admissible one that violates the No-Arbitrage Principle. Here is how to proceed to achieve arbitrage:

- do not invest at all at time 0;
- at time 1 check whether the value $V(1)$ of the non-admissible strategy is negative or not. If $V(1) \geq 0$, then refrain from investing at all once again. However, if $V(1) < 0$, then take the same position in stock as in the non-admissible strategy and a risk-free position that is lower by $-V(1)$ than that in the non-admissible strategy.

This defines a predictable self-financing strategy. Its time 0 and time 1 value is 0. The value at time 2 will be

$$\begin{cases} 0 & \text{if } V(1) \geq 0, \\ V(2) - V(1) > 0 & \text{if } V(1) < 0. \end{cases}$$

This is, therefore, an admissible strategy realising an arbitrage opportunity.

7.10 If you happen to know that an increase in stock price will be followed by a fall at the next step, then adopt this strategy:

- at time 0 do not invest in either asset;
- at time 1 check whether the stock has gone up or down. If down, then again do not invest in either asset. But if stock has gone up, then sell short one share for $S(0)U$, investing the proceeds risk free.

Clearly, the time 0 and time 1 value of this strategy is $V(0) = V(1) = 0$. If stock goes down at the first step, then the value at time 2 will be $V(2) = 0$. But if stock goes up at the first step, then it will go down at the second step and $V(2) = S(0)U(R - D)$ will be positive, as required, since $U > R > D$.

Clearly, this is not a predictable strategy, which means that no arbitrage has been achieved.

7.11 a) If there are no short-selling restrictions, then the following strategy will realise an arbitrage opportunity:

- at time 0 do not invest at all;
- at time 1 check the price $S(1)$. If $S(1) = 120$ dollars, then once again do not invest at all. But if $S(1) = 90$ dollars, then sell short one share of the risky asset and invest the proceeds risk free.

The time 0 and time 1 value of this admissible strategy is 0. The value at time 2 will be

$$\begin{cases} 0 & \text{in scenarios } \omega_1 \text{ and } \omega_2, \\ 3 & \text{in scenario } \omega_3, \end{cases}$$

which means that arbitrage can be achieved.

b) In a) above, arbitrage has been achieved by utilising the behaviour of stock prices at the second step in scenario ω_2: the return on the risky asset is lower than the risk-free return. Thus, shorting the risky asset and investing the proceeds risk free creates arbitrage. However, when short selling of risky assets is disallowed, this arbitrage opportunity will be beyond the reach of investors.

7.12 The arbitrage strategy described in Solution 7.11 involves buying a fraction of a bond. If $S(1) = 90$ dollars, then one share of stock should be shorted and $\frac{9}{11}$ of a bond purchased at time 1. To obtain an arbitrage strategy involving an integer number of units of each asset, multiply these quantities by 11, that is, sell short 11 shares of stock and buy 9 bonds.

7.13 Suppose that transaction costs of 5% apply whenever stock is bought or sold. An investor who tried to follow the strategy in Solution 7.11, short selling one share of stock at time 1 if $S(1) = 90$ dollars, would have to pay transaction costs of $90 \times 5\% = 4.50$ dollars. If the remaining amount of $90 - 4.50 = 85.50$ dollars were invested risk free, it would be worth $85.5 \times \frac{121}{110} = 94.05$ dollars at time 2. But closing the short position in stock would cost $96, making the final wealth negative. As a result, there is no arbitrage strategy.

Chapter 8

8.1 If a random variable X takes two values $a < b$ with probabilities p, q, then $\mathbb{E}(X) = pa + qb$ and $\text{Var}(X) = (b-a)^2 pq$, so in our case with $p = q = \frac{1}{2}$ and $X = k_N(h)$ we get

$$\frac{a+b}{2} = \mu h, \qquad \frac{b-a}{2} = \sigma\sqrt{h}.$$

This system can be solved, $a = \mu h + \sigma\sqrt{h}$, $b = \mu h - \sigma\sqrt{h}$, so the prices change according to

$$S_N(h) = \begin{cases} e^{\mu h + \sigma\sqrt{h}} S(0), \\ e^{\mu h - \sigma\sqrt{h}} S(0), \end{cases}$$

which shows (8.2).

8.2 First note that $\mathbb{E}\left(S(0)e^{\mu t + \sigma W(t)}\right) = S(0)e^{\mu t}\mathbb{E}(e^{\sigma W(t)})$. Next, $\sigma W(t)$ has normal distribution with zero expectation and variance equal to $\sigma^2 t$, so it has density $\frac{1}{\sqrt{2\pi t}\sigma}e^{-\frac{1}{2\sigma^2 t}x^2}$, which can be used to compute the expectation

$$\mathbb{E}\left(e^{\sigma W(t)}\right) = \int_{-\infty}^{+\infty} \frac{1}{\sqrt{2\pi t}\sigma} e^x e^{-\frac{1}{2\sigma^2 t}x^2}\, dx$$

$$= e^{\frac{1}{2}\sigma^2 t} \int_{-\infty}^{+\infty} \frac{1}{\sqrt{2\pi t}\sigma} e^{-\frac{1}{2\sigma^2 t}(x - \frac{\sigma}{\sqrt{2t}})^2}\, dx$$

after adding and subtracting $\frac{1}{2}\sigma^2 t$ in the exponential and completing the square. The integral is equal to 1 since the integrand is a density, hence the result.

8.3 We know that $S(t) = e^{\mu t + \sigma W(t)}$ and $W_*(t) = \frac{\mu - r + \frac{1}{2}\sigma^2}{\sigma}t + W(t)$, so inserting the expression for $W(t)$ from the latter equality into the formula for $S(t)$, we can see that μt cancels out and $S(t) = e^{rt - \frac{1}{2}\sigma^2 t + \sigma W_*(t)}$ as claimed.

8.4 Using $S(t) = S(0)e^{rt - \frac{1}{2}\sigma^2 t + \sigma W_*(t)}$, we apply the Itô formula with $f(t,x) = S(0)e^{rt - \frac{1}{2}\sigma^2 t + \sigma x}$ with $W_*(t)$ in place of $W(t)$, which is legitimate since $W_*(t)$ is a Wiener process under P_*, and $S(t) = f(t, W_*(t))$. We have

$$f_t(t,x) = S(0)(r - \frac{1}{2}\sigma^2)e^{rt - \frac{1}{2}\sigma^2 t + \sigma x},$$

$$f_x(t,x) = \sigma S(0)e^{rt - \frac{1}{2}\sigma^2 t + \sigma x},$$

$$f_{xx}(t,x) = \sigma^2 S(0)e^{rt - \frac{1}{2}\sigma^2 t + \sigma x},$$

so that

$$dS(t) = (r - \frac{1}{2}\sigma^2)S(t)dt + \sigma S(t)dW_*(t) + \frac{1}{2}\sigma^2 S(t)dt$$
$$= rS(t)dt + \sigma S(t)dW_*(t).$$

This argument is similar to the derivation of (8.16).

8.5 We begin with the right-hand side:

$$P_*\{S(T) \geq X\} = P_*\{S(0)e^{rT - \frac{1}{2}\sigma^2 T + \sigma W_*(T)} \geq X\}$$

$$= P_*\left\{\frac{1}{\sqrt{T}}W_*(T) \geq \frac{\ln\frac{X}{S(0)} - rT + \frac{1}{2}\sigma^2 T}{\sigma\sqrt{T}} = -d_-\right\}$$

by (8.31). Next, $\frac{1}{\sqrt{T}}W_*(T)$ has standard normal distribution, so $P_*\{S(T) \geq X\} = 1 - N(-d_-) = N(d_-)$ by the symmetry of the density (see Appendix 10.3).

8.6 By put-call parity, for $t = 0$

$$P_E(0) = C_E(0) - S(0) + Xe^{-rT}$$
$$= S(0)(N(d_+) - 1) - Xe^{-rT}(N(d_-) - 1)$$
$$= -S(0)N(-d_+) + Xe^{-rT}N(-d_-).$$

Now, by substituting t for 0 and $T - t$ for T, we obtain the Black–Scholes formula for $P_E(t)$.

8.7 We shall verify the formulae for $t = 0$ for simplicity. We know that $x(0) = u_x(0, S(0))$ and $y(0) = u(0, S(0)) - x(0)S(0)$ (assuming that $A(0) = 1$), where $u(0, S(0)) = C_E(0)$. From the Black–Scholes formula for $C_E(0)$ we can see that

$$u(0, x) = xN\left(\frac{\ln\frac{x}{X} + \left(r + \frac{1}{2}\sigma^2\right)T}{\sigma\sqrt{T - t}}\right) - Xe^{-rT}N\left(\frac{\ln\frac{x}{X} + \left(r - \frac{1}{2}\sigma^2\right)T}{\sigma\sqrt{T - t}}\right).$$

Simple but tedious differentiation with respect to x gives

$$u_x(0, x) = N\left(\frac{\ln\frac{x}{X} + \left(r + \frac{1}{2}\sigma^2\right)T}{\sigma\sqrt{T - t}}\right).$$

It follows that

$$x(0) = u_x(0, S(0)) = N(d_+),$$
$$y(0) = u(0, S(0)) - x(0)S(0) = -Xe^{-rT}N(d_-),$$

where d_+, d_- are given by (8.31). Finally, substituting t for 0 and $T - t$ for T gives the result for $x(t), y(t)$.

8.8 By put-call parity (5.1)

$$\frac{d}{dS}P_E(S) = \frac{d}{dS}C_E(S) - 1 = N(d_+) - 1 = -(1 - N(d_+)) = -N(-d_+),$$

where d_+ is given by (8.31). The delta of a put option is negative, consistently with the fact that the value of a put option decreases as the price of the underlying asset increases.

8.9 We maximise $581.96 \times S - 30,779.62 - 1,000 \times C_E(S, \frac{1}{365})$, where S stands for the stock price after one day, and $C_E(S, t)$ is the price of a call at time t, one day in our case, with 89 days to maturity, and where $\sigma = 30\%$ and $r = 8\%$, as before. Equating the derivative with respect to S to zero, we infer that the delta of the option after one day should be the same as the delta on day zero, $\frac{d}{dS}C_E(S, \frac{1}{365}) = 0.58196$. This gives the following condition for the stock price (after inverting the normal distribution function):

$$\frac{\ln\frac{S}{60} + \left(r + \frac{1}{2}\sigma^2\right) \times \frac{89}{365}}{\sigma\sqrt{\frac{89}{365}}} = \frac{\ln\frac{60}{60} + \left(r + \frac{1}{2}\sigma^2\right) \times \frac{90}{365}}{\sigma\sqrt{\frac{90}{365}}}.$$

The result is $S \cong 60.0104$ dollars.

8.10 The premium for a single put is 0.031648 dollars (from the Black–Scholes formula), so the bank will receive 1,582.40 dollars by writing and selling 50,000 puts. The delta of a single put is -0.355300, so the delta-hedging portfolio requires shorting 17,765.00 shares, which will raise 32,332.29 dollars. This gives a total of 33,914.69 dollars received to be invested at 5%. The value of the delta neutral portfolio consisting of the shorted stock, invested cash and sold options will be $-32,332.29 + 33,914.69 - 1,582.40 = 0.00$ dollars.

One day later the shorted shares will be worth $17,765 \times 1.81 = 32,154.64$ dollars, while the cash investment will grow to $33,914.69e^{0.05/365} \cong 33,919.34$ dollars. The put price will increase to 0.035182 dollars, so the price of 50,000 puts will be 1,759.11 dollars. The value of the delta neutral portfolio will be $-32,154.64 + 33,919.34 - 1,759.11 \cong 5.59$ dollars.

8.11 The price of a single put after one day will now be 0.038885 dollars, the 50,000 options sold will therefore be worth 1944.26 dollars, the stock and cash deposit positions remaining as in Solution 8.10. The delta neutral portfolio will bring a loss of 179.56 dollars.

8.12 If the stock price does not change, $S(t) = S(0) = S$, then the value of the portfolio after time t will be given by

$$V(t) = SN(d_+) - Xe^{rt}e^{-rT}N(d_-) - C_E(S,t),$$

where $C_E(S,t)$ is given by the Black–Scholes formula and d_+, d_- by (8.31). Then

$$\frac{\mathrm{d}}{\mathrm{d}t}V(t)\bigg|_{t=0} = -rXe^{-rT}N(d_-) - \frac{\mathrm{d}}{\mathrm{d}t}C_E(S,t)\bigg|_{t=0}$$

$$= -rXe^{-rT}N(d_-) - \mathrm{theta}_{C_E}$$

$$= \frac{\sigma S}{2\sqrt{2\pi T}}e^{-d_+/2},$$

which is positive.

8.13 Using put-call parity and the Greek parameters for a call, we can find those for a put:

$$\mathrm{delta}_{P_E} = N(d_+) - 1 = \mathrm{delta}_{C_E} - 1 = -N(-d_+),$$

$$\mathrm{gamma}_{P_E} = \mathrm{gamma}_{C_E},$$

$$\mathrm{theta}_{P_E} = -\frac{S\sigma}{2\sqrt{2\pi T}}e^{-\frac{d_+}{2}} + rXe^{-rT}N(-d_-),$$

$$\mathrm{vega}_{P_E} = \mathrm{vega}_{C_E},$$

$$\mathrm{rho}_{P_E} = -TXe^{-rT}N(-d_-).$$

(The Greek parameters are computed at time $t = 0$.) These equalities can also be verified directly by differentiating the Black–Scholes formula for the put price.

8.14 The rho of the original option is 7.5878, the delta of the additional option is 0.4104 and the rho is 7.1844. The delta-rho neutral portfolio requires buying approximately 148.48 shares of stock and 1,056.14 additional options, while borrowing $7,693.22. The position after one day is presented in the following table, in which we also recall the results of the delta hedge:

$S(\frac{1}{365})$	delta-rho			delta
	$r = 8\%$	$r = 9\%$	$r = 15\%$	$r = 9\%$
58.00	-7.30	-9.65	-26.14	-133.72
58.50	-2.71	-4.63	-17.95	-97.22
59.00	0.18	-1.23	-10.93	-72.19
59.50	1.59	0.77	-4.85	-58.50
60.00	1.76	1.60	0.52	-55.96
60.50	0.92	1.50	5.45	-64.38
61.00	-0.68	0.72	10.16	-83.51
61.50	-2.78	-0.47	14.90	-113.07
62.00	-5.13	-1.84	19.91	-152.78

8.15 With 95% probability the logarithmic return on the exchange rate satisfies $k > m + x\sigma \cong -23.68\%$, where $x \cong -1.645$, so that $N(x) \cong 5\%$. The 1,000 dollars converted into euros, invested without risk at the rate r_{EUR}, and converted back into dollars after one year, will give $1,000e^{r_{EUR}}e^k$ dollars. With probability 95% this amount will satisfy

$$1,000e^{r_{EUR}}e^k > 1,000e^{r_{EUR}}e^{m+x\sigma} \cong 821.40 \text{ dollars.}$$

On the other hand, $1,000$ dollars invested at the rate r_{USD} would have grown to $1,000e^{r_{USD}} \cong 1,051.27$ dollars. As a result,

$$\text{VaR} = 1,000e^{r_{USD}} - 1,000e^{r_{EUR}}e^{m+x\sigma} \cong 229.88 \text{ dollars.}$$

8.16 A single call costs $21.634. We purchase approximately 46.22 options. With probability 5% the stock price will be less than $49.74. We shall still be able to exercise the options, cashing $450.18 in the borderline case. The alternative risk-free investment of $1,000 at 8% would grow to $1,040.81. Hence VaR $\cong$ 590.63 dollars. If the stock grows at the expected rate, reaching $63.71, then we shall obtain $1,095.88 when the options are exercised. With 5% probability the stock price will be above $81.6 and then our options will be worth at least $1,922.75.

Chapter 9

9.1 The yields are $y(0) \cong 14.08\%$ and $y(3) \cong 13.63\%$. Thus $B(0,3) = e^{-3ry(0)} \cong 0.9654$ dollars. Arbitrage can be achieved as follows:

- at time 0 buy a 6-month bond for $B(0,6) = 0.9320$ dollars, raising the money by issuing 0.9654 of a 3-month bond, which sells at $B(0,3) \cong 0.9654$ dollars;
- at time 3 (after 3 months) issue 0.9989 of a 3-month bond, which sells at $B(3,6) = 0.9665$ dollars, and use the proceeds of $0.9654 to settle the fraction of a 3-month bond issued at time 0;
- at time 6 (after half a year) the 6-month bond bought at time 0 will pay $1, out of which $0.9989 will settle the fraction of a 3-month bond issued at time 3.

The balance of $0.0011 will be the arbitrage profit.

9.2 The implied rates are $y(0) \cong 12.38\%$ and $y(6) \cong 13.06\%$. Investing $100, we can buy 106.38 bonds now and 113.56 after six months. The logarithmic return over one year is $\ln(113.56/100) \cong 12.72\%$, the arithmetic mean of the semi-annual returns.

9.3 To achieve a return of 14%, we would have to sell the bond for $0.8700e^{14\%} \cong$ 1.0007 dollars, which is impossible. (A zero-coupon bond can never fetch a price higher than its face value.)

In general, we have to solve the equation $B(0,12)e^k = e^{-\tau y(6)}$ to find $y(6)$, where k is the prescribed logarithmic return. The left-hand side must be smaller than 1.

9.4 During the first six months the rate is $y(n) \cong 8.34\%$, for $n = 0, \ldots, 179$, and during the rest of the year $y(n) \cong 10.34\%$, for $n = 180, \ldots, 360$. The bond should be sold for $0.92e^{4.88\%} \cong 0.9660$ dollars or more. This cannot be achieved during the first six months, since the highest price before the rate changes is $B(179, 360) \cong 0.9589$ dollars. On the day of the rate change $B(180, 360) \cong 0.9496$ dollars, and we have to wait until day $n = 240$, on which the bond price will exceed the required \$0.9660 for the first time.

9.5 The rate can be found by using a spreadsheet with goal seek facility to solve the equation

$$10.896 \times \left(10 + 10e^{-y(1)} + 10e^{-2y(1)} + 110e^{-3y(1)}\right) = 1,000e^k.$$

This gives $y(1) \cong 12.00\%$ for $k = 12\%$ in case a), $y(1) \cong 12.81\%$ for $k = 10\%$ in case b) and $y(1) \cong 11.19\%$ for $k = 14\%$ in case c).

9.6 The numbers were found using an Excel spreadsheet with accuracy higher than the displayed 2 decimal points.

9.7 Scenario 1: \$1,427.10; Scenario 2: \$1,439.69.

9.8 Formula (9.2) can be applied directly to find $D \cong 1.6846$.

9.9 The duration is equal to 4 if the face value is \$73.97. The smallest possible duration, which corresponds to face value $F = \$0$, is about 2.80 years. For very high face values F, the duration is close to 5, approaching this number as F goes to infinity.

When $F = 100$, the coupon value $C \cong 13.52$ gives duration of 4 years. If the coupon value is zero, then the duration is 5 years. For very high coupon values C tending to infinity, the duration approaches about 2.80 years.

9.10 Since the second derivative of $P(y)$ is positive,

$$\frac{d^2}{dy^2}P(y) = (\tau n_1)^2 C_1 e^{-\tau n_1 y} + (\tau n_2)^2 C_2 e^{-\tau n_2 y} + \cdots$$

$$+ (\tau n_N)^2 (C_N + F)e^{-\tau n_N y} > 0,$$

P is a convex function of y.

9.11 Solving the system $6 = 2w_A + 3.4w_B$, $w_A + w_B = 1$, we find $w_A \cong -1.8571$ and $w_B \cong 2.8571$. As a result, we invest \$14,285.71 to buy 14,005.60 bonds B, raising the shortfall of \$9,285.71 by issuing 9,475.22 bonds A.

9.12 The yield on the coupon bond A is about 13.37%, so the price of the zero-coupon bond B is \$87.48. The coupon bond has duration 3.29, and we find the weights to be $w_A \cong 0.4366$ and $w_B \cong 0.5634$. This means that we invest \$436.59 to buy 4.2802 bonds A and \$563.41 to buy 6.4403 bonds B.

9.13 Directly from the definition of duation (9.2) we compute the duration D_t at time t (note that the bond price grows by a factor of e^{yt}),

$$D_t = \frac{1}{e^{yt}P(y)}\left((\tau n_1 - t)C_1 e^{-y(\tau n_1 - t)} + \cdots + (\tau n_N - t)(C_N + F)e^{-y(\tau n_N - t)}\right)$$

$$= \frac{1}{P(y)}\left((\tau n_1 - t)C_1 e^{-\tau n_1 y} + \cdots + (\tau n_N - t)(C_N + F)e^{-\tau n_N y}\right)$$

$$= D - t,$$

since the weights $C_1 e^{-\tau n_1 y}/P(y), C_2 e^{-\tau n_2 y}/P(y), \ldots, (C_N + F)e^{-\tau n_N y}/P(y)$ add up to one.

9.14 Denote the annual payments by C_1, C_2 and the face value by F, so that

$$P(y) = C_1 e^{-y} + (C_2 + F)e^{-2y},$$

$$D(y) = \frac{C_1 e^{-y} + 2(C_2 + F)e^{-2y}}{P(y)}.$$

Compute the derivative of $D(y)$ to see that it is negative:

$$\frac{d}{dy}D(y) = \frac{-C_1(C_2 + F)e^{-3y}}{P(y)^2} < 0.$$

9.15 We first find the prices and durations of the bonds: $P_A(y) \cong 120.72$, $P_B(y) \cong 434.95$, $D_A(y) \cong 1.8471$, $D_B(y) \cong 1.9894$. The weights $w_A \cong -7.46\%$, $w_B \cong 107.46\%$ give duration 2, which means that we have to buy 49.41 bonds B and issue 12.35 bonds A. After one year we shall receive \$247.05 from the coupons of B and will have to pay the same amount for the coupons of A. Our final amount will be \$23,470.22, exactly equal to the future value of \$20,000 at 8%, independently of any rate changes.

9.16 If the term structure is to be flat, then the yield $y(0,6) = 8.16\%$ applies to any other maturity, which gives $B(0,3) = 0.9798$ dollars and $B(0,9) = 0.9406$ dollars.

9.17 Issue and sell 500 bonds maturing in 6 months with \$100 face value, obtaining \$48,522.28. Use this sum to buy 520.4054 one-year bonds. After 6 months settle the bonds issued by paying \$50,000. After one year cash the face value of the bonds purchased. The resulting rate is 8%.

9.18 You need to deposit $100,000\frac{1}{1+3.46\%/12} \cong 99,712.50$ dollars for one month, which will grow to the desired level of \$100,000, and borrow the same amount for 6 months at 4.54%. Your customer will receive \$100,000 after 1 month and will have to pay $99,304.04(1 + 4.54\%/2) \cong 101,975.97$ dollars after 6 months, which implies a forward rate of 5.45% (simple interest).

The simple interest rate for a 4-month loan starting in 2 months is approximately 6.81%, so a deposit at 6.91% would give an arbitrage opportunity.

9.19 To see that the forward rates $f(n, N)$ may be negative, let us analyse the case with $n = 0$ for simplicity. Then

$$f(0, N) = (N + 1)y(0, N + 1) - Ny(0, N)$$

and $f(0, N) < 0$ requires that $(N + 1)y(0, N + 1) < Ny(0, N)$. The borderline case is when $y(0, N + 1) = \frac{N}{N+1}y(0, N)$, which enables us to find a numerical example. For instance, for $N = 8$ and $y(0, 8) = 9\%$ a negative value $f(0, 8)$ will be obtained if $y(0, 9) < \frac{8}{9} \times 9\% = 8\%$.

9.20 Suppose that $f(n, N)$ increases with N. We want to show that the same is true for
$$y(n, N) = \frac{f(n, n) + f(n, n + 1) + \cdots + f(n, N - 1)}{N - n}.$$
This follows from the fact that if a sequence a_n increases, then so does the sequence of averages $S_n = \frac{a_1 + \cdots + a_n}{n}$. To see this multiply the target inequality $S_{n+1} > S_n$ by $n(n+1)$ to get (after cancellations) $na_{n+1} > a_1 + \cdots + a_n$. The latter is true, since $a_{n+1} > a_i$ for all $i = 1, \ldots, n$.

9.21 The values of $B(0, 2)$, $B(0, 3)$, $B(1, 3)$ have no effect on the values of the money market account.

9.22 a) For an investment of \$100 in zero-coupon bonds, divide the initial cash by the price of the bond $B(0, 3)$ to get the number of bonds held, 102.82, which gives final wealth of \$102.82. The logarithmic return is 2.78%.

b) For an investment of \$100 in single-period zero-coupon bonds, compute the number of bonds maturing at time 1 as $100/B(0, 1) \cong 100.99$. Then, at time 1 find the number of bonds maturing at time 2 in a similar way, $100.99/B(1, 2) \cong 101.54$. Finally, we arrive at $101.54/B(2, 3) \cong 102.51$ bonds, each giving a dollar at time 3. The logarithmic return is 2.48%.

c) An investment of \$100 in the money market account, for which we receive $100A(3) \cong 102.51$ at time 3, produces the same logarithmic return of 2.48% as in b).

9.23 We begin from the right, that is, from the face values of the bonds, first computing the values of $B(2, 3)$ in all states. These numbers together with the known returns give $B^u(1, 3)$ and $B^d(1, 3)$. These, in turn, determine the missing returns $k^{ud}(2, 3) = 0.20\%$ and $k^{dd}(2, 3) = 0.16\%$. The same is done for the first step, resulting in $k^d(1, 3) = 0.23\%$. The bond prices are given in Figure S.6.

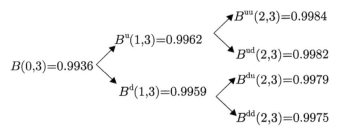

Figure S.6 Bond prices in Solution 9.23

9.24 Because of the additivity of the logarithmic returns, $k^u(1, 3) + k^{uu}(2, 3) + k^{uuu}(3, 3) = 0.64\%$ gives the return in the period of three weeks. To obtain the yield we have to rescale it to the whole year by multiplying by 52/3, hence $y(0, 3) = 11.09\%$. Note that we must have $k^u(1, 3) + k^{ud}(2, 3) + k^{udu}(3, 3) = 0.64\%$ which allows us to find $k^{ud}(2, 3) = 0.20\%$. The other missing returns can be computed in a similar manner, first $k^d(1, 3)$, then $k^{dd}(2, 3)$.

9.25 The bond prices are given in Figure S.7.

9.26 The money market account is given in Figure S.8. Note that the values for the 'up' movements are lower than for the 'down' movements. This is related to the fact that the yield decreases as the bond price increases, and our trees are based on bond price movements.

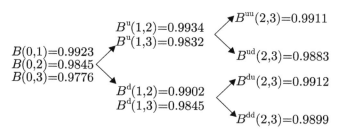

Figure S.7 Bond prices in Solution 9.25

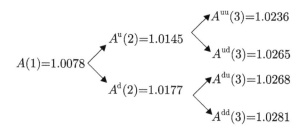

Figure S.8 Money market account in Solution 9.26

9.27 The prices $B^u(1,2) = 0.9980$ and $B^d(1,2) = 0.9975$ are found by discounting the face value 1 to be received at time 2, using the short rates $r^u(1)$ and $r^d(1)$. The price $B(0,2) = 0.9944$ can be found by the replication procedure.

9.28 At time 2 the coupons are 0.5227 or 0.8776, depending on whether we are in the up or down state at time 1. At time 1 the coupon is 0.9999.

9.29 At time 1 we find $18.0647 = (0.8159 \times 20 + 0.1841 \times 10)/1.0052$ in the up state and $1.7951 = (0.1811 \times 10 + 0.8189 \times 0)/1.0088$ in the down state. Next, applying the same formula again, we obtain $7.9188 = (0.3813 \times 18.3928 + 0.6187 \times 1.7951)/1.01$.

9.30 There is an arbitrage opportunity in the up state at time 1. The price $B^u(1,2) = 0.9924$ implies that the growth factor in the money market is 1.00766, whereas the prices of the bond maturing at time 3 imply growth factors 1.01159 and 1.00783. To realise arbitrage, bonds with maturity 3 should be bought, the purchase financed by a loan in the money market.

9.31 Using formula (9.12) and the short rates given, we find the structure of bond prices as shown in Figure S.9.

9.32 It is best to compute the risk-neutral probabilities. The probability at time 1 of the up movement based on the bond maturing at time 3 is 0.76, whereas the probability based on the bond maturing at time 2 is 0.61. The present price of the bond maturing at time 2 computed using the prices of the bond maturing at time 3 and the risk-neutral probabilities computed from these prices is 0.9867. So, shorting at time 0 the bond maturing at time 3 and buying the bond maturing at time 2 will give an arbitrage profit.

9.33 At time 2 the option is worthless. At time 1 we evaluate the bond prices by adding the coupon to the discounted final payment of 101.00 at the appropriate (monthly) money market rate: 0.521% in the up state and 0.874% in the down

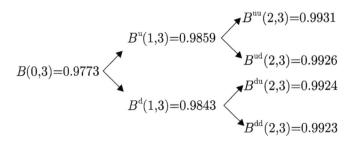

Figure S.9 Bond prices in Solution 9.31

state. The results are 101.4748 and 101.1213, respectively. The option can be exercised at that time in the up state, so the cash flow is 0.1748 and 0, respectively. Expectation with respect to the risk-neutral probabilities of the discounted cash flow gives the initial value 0.06598 of the option.

9.34 The coupons of the bond with the floor provision differ from the par bond at time 2 in the up state: 0.66889 instead of 0.52272. This results in the following bond prices at time 1: 101.14531 in the up state and 100.9999 in the down state. (The latter is the same as for the par bond.) Expectation with respect to the risk-neutral probability gives the initial bond price 100.05489, so the floor is worth 0.05489.

Bibliography

Background Reading: Probability and Stochastic Processes

Ash, R. B. (1970), *Basic Probability Theory*, John Wiley & Sons, New York.

Brzeźniak, Z. and Zastawniak, T. (1999), *Basic Stochastic Processes*, Springer Undergraduate Mathematics Series, Springer-Verlag, London.

Capiński, M. and Kopp, P. E. (2007), *Measure, Integral and Probability*, 2nd edition, Springer Undergraduate Mathematics Series, Springer-Verlag, London.

Capiński, M. and Zastawniak, T. (2001), *Probability Through Problems*, Springer-Verlag, New York.

Chung, K. L. (2000), *A Course in Probability Theory*, 2nd edition, Academic Press, New York.

Stirzaker, D. (1999), *Probability and Random Variables: A Beginner's Guide*, Cambridge University Press, Cambridge.

Background Reading: Calculus and Linear Algebra

Blyth, T. S. and Robertson, E. F. (1998), *Basic Linear Algebra*, Springer Undergraduate Mathematics Series, Springer-Verlag, London.

Jordan, D. W. and Smith, P. (2002), *Mathematical Techniques*, Oxford University Press, Oxford.

Stewart, J. (1999), *Calculus*, Brooks Cole, Pacific Grove, California.

Further Reading: Mathematical Finance

Back, K. (2005), *A Course in Derivative Securities: Introduction to Theory and Computation*, Springer-Verlag, Berlin, London, New York.

M. Capiński, T. Zastawniak, *Mathematics for Finance*,
Springer Undergraduate Mathematics Series,
© Springer-Verlag London Limited 2011

Baxter, M. W. and Rennie, A. J. O. (1996), *Financial Calculus: An Introduction to Derivative Pricing*, Cambridge University Press, Cambridge.

Bingham, N. H. and Kiesel, R. (1998), *Risk-Neutral Valuation: Pricing and Hedging of Financial Derivatives*, Springer-Verlag, Berlin.

Björk, T. (1998), *Arbitrage Theory in Continuous Time*, Oxford University Press, Oxford.

Dana, R.-A. and Jeanblanc, M. (2007), *Financial Markets in Continuous Time*, Springer-Verlag, Berlin, London, New York.

Delbaen, F. and Schachermayer, W. (2009), *The Mathematics of Arbitrage*, Springer-Verlag, Berlin, London, New York.

Elliott, R. J. and Kopp, P. E. (2009), *Mathematics of Financial Markets,* 2nd edition, Springer-Verlag, New York.

Elton, E. J. and Gruber, M. J. (1995), *Modern Portfolio Theory and Investment Analysis*, John Wiley & Sons, New York.

Etheridge, A. (2002), *A Course in Financial Calculus*, Cambridge University Press, Cambridge.

Higham, D. J. (2004), *An Introduction to Financial Option Valuation: Mathematics, Stochastics and Computation*, Cambridge University Press, Cambridge.

Hoek, J. van der, and Elliott, R. J. (2009), *Binomial Models in Finance*, Springer-Verlag, Berlin, London, New York.

Hull, J. (2000), *Options, Futures and Other Derivatives*, Prentice Hall, Upper Saddle River, New Jersey.

Jarrow, R. A. (1995), *Modelling Fixed Income Securities and Interest Rate Options*, McGraw-Hill, New York.

Jarrow, R. A. and Turnbull, S. M. (1996), *Derivative Securities*, South-Western College, Cincinnati, Ohio.

Karatzas, I. and Shreve, S. (1998), *Methods of Mathematical Finance*, Springer-Verlag, Berlin.

Lamberton, D. and Lapeyre, B. (2007), *Introduction to Stochastic Calculus Applied to Finance*, 2nd edition, Chapman and Hall, London.

Musiela, M. and Rutkowski, M. (2008), *Martingale Methods in Financial Modelling*, 2nd edition, Springer-Verlag, Berlin.

Øksendal, B. (2007), *Stochastic Differential Equations: An Introduction with Applications*, 6th edition, Springer-Verlag, Berlin, London, New York.

Pliska, S. R. (1997), *Introduction to Mathematical Finance: Discrete Time Models*, Blackwell, Maldon, Massachusetts.

Roman, S. (2004), *Introduction to the Mathematics of Finance: From Risk Management to Options Pricing*, Springer-Verlag, Berlin, London, New York.

Steele, J. M. (2003), *Stochastic Calculus and Financial Applications*, Springer-Verlag, Berlin, London, New York.

Glossary of Symbols

A	fixed income security (bond) price; money market account
B	bond price
β	beta factor
c	covariance
C	call price; coupon value
$\mathbf{C}$	covariance matrix
C_A	American call price
C_E	European call price
Cov	covariance
d	'down' node
D	return for 'down' node; duration
delta	Greek parameter delta
div	dividend
div_0	present value of dividends
$\mathbb{E}$	expectation
$\mathbb{E}_*$	risk-neutral (martingale) expectation
f	futures price; forward rate
F	forward price; future value; face value
gamma	Greek parameter gamma
Φ	cumulative binomial distribution
h	time step
H	derivative security price
H_A	price of an American type derivative security
k	logarithmic return
K	return
i	coupon rate

M. Capiński, T. Zastawniak, *Mathematics for Finance*,
Springer Undergraduate Mathematics Series,
© Springer-Verlag London Limited 2011

M	market portfolio
$\mathbf{m}$	expected returns as a row matrix
μ	expected return
N	cumulative normal distribution
$\binom{N}{k}$	the number of k-element combinations out of N elements
ω	scenario
Ω	probability space
p	branching probability in binomial tree
p_*	risk-neutral probability in binomial tree
P	put price; principal; probability
P_*	risk-neutral (martingale) probability
P_A	American put price
P_E	European put price
PA	present value factor of an annuity
r	interest rate
r_div	dividend yield
r_e	effective rate
R	risk-free return
rho	Greek parameter rho
ρ	correlation
S	risky security (stock) price
σ	standard deviation; risk; volatility
t	current time
T	maturity time; expiry time; exercise time; delivery time
theta	Greek parameter theta
u	'up' node
$\mathbf{u}$	row matrix with all entries 1
U	return for 'up' node
V	portfolio value
$\widetilde{V}$	discounted portfolio value
Var	variance
VaR	value-at-risk
vega	Greek parameter vega
$\mathbf{w}$	weights in a portfolio as a row matrix
W	Wiener process; Brownian motion
W_N	symmetric random walk
x	position in a risky security
X	strike price
y	position in a fixed income (risk-free) security; yield of a bond
z	position in a derivative security (contingent claim)

Index

M. Capiński, T. Zastawniak, *Mathematics for Finance*,
Springer Undergraduate Mathematics Series,
© Springer-Verlag London Limited 2011

Cox–Ingersoll–Ross model, 278
Cox–Ross–Rubinstein formula, 157
cum-dividend price, 312

delta, 148, 217, 218, 223
delta hedging, 217
delta neutral portfolio, 217
delta-gamma hedging, 224
delta-gamma neutral portfolio, 223
delta-vega hedging, 226
delta-vega neutral portfolio, 223
density, 285
derivative security, 271
– American, 159
discount factor, 28, 32, 38
discounted value, 28, 32, 150
discrete compounding, 29
distribution
– binomial, 157
– log normal, 202
– normal, 201
distribution function, 285
dividend yield, 97
divisibility, 4, 179, 181, 192
dominance relation, 74
duration, 241
dynamic hedging, 245

effective rate, 41
efficient
– frontier, 74
– portfolio, 74
equation, Black–Scholes, 213
equivalent compounding methods, 40
European
– call option, 114, 157, 215
– put option, 114, 157, 216
European type option, 152
ex-coupon price, 266
ex-dividend price, 312
exercise
– price, 114
– time, 14, 114
exotic option, 151
expectation, 285
– conditional, 162
expected return, 10, 60, 72
expiry time, 114

face value, 43
feasible
– portfolio, 64, 71
– set, 64, 71
field, 284

filtration, 146
fixed interest, 273
fixed-coupon bond, 273
flat term structure, 248
floating interest, 273
floating-coupon bond, 273
floor, 277
floorlet, 277
formula
– Black–Scholes, 215
– Cox–Ross–Rubinstein, 157
– Itô, 205
forward
– contract, 11
– exchange rate, 20
– price, 11
– rate, 251
future value, 26, 29
futures
– contract, 100
– price, 100

gamma, 223
Girsanov theorem, 209
Greek parameters, 222
growth factor, 26, 29, 36

Heath–Jarrow–Morton model, 279
hedging
– delta, 217
– delta-gamma, 224
– delta-vega, 226
– dynamic, 245

in the money, 136
independent random variables, 285
indicator function, 283
initial
– forward rate, 250
– margin, 102
– term structure, 247
instantaneous forward rate, 251
integral
– Itô, 205
– stochastic, 205
interest
– compounded, 29, 36
– fixed, 273
– floating, 273
– simple, 26
– variable, 273
interest rate, 26
– option, 272
– swap, 273

Lightning Source UK Ltd.
Milton Keynes UK
UKHW020844090821
388022UK00012B/66